PWS CIVIL ENGINEERING SERIES LIST

Geotechnology of Waste Management

SECOND EDITION

ISSA S. OWEIS

Converse Consultants

RAJ P. KHERA

New Jersey Institute of Technology

PWS Publishing Company

I(T)P *An International Thomson Publishing Company*

Boston • Albany • Bonn • Cincinnati • London • Madrid • Melbourne • Mexico City • New York • Paris
San Francisco • Singapore • Tokyo • Toronto • Washington

PWS Publishing Company
20 Park Plaza, Boston, MA 02116-4324

Sponsoring Editor: *Bill Barter*
Assistant Editor: *Suzanne Jeans*
Editorial Assistant: *Tricia Kelley*
Market Development Manager: *Nathan Wilbur*
Production Editor/Interior Designer: *Pamela Rockwell*
Cover Designer: *Eileen Hoff*
Manufacturing Buyer: *Andrew Christensen*
Compositor: *Santype International Ltd.*
Text Printer: *Quebecor/Fairfield*
Cover Printer: *Phoenix Color Corp.*

I(T)P® International Thomson Publishing
The trademark ITP is used under license.

For more information, contact:
PWS Publishing Company
20 Park Plaza
Boston, MA 02116-4324

International Thomson Publishing Europe
Berkshire House 168-173
High Holborn
London WC1V 7AA
England

Thomas Nelson Australia
102 Dodds Street
South Melbourne, 3205
Victoria, Australia

Nelson Canada
1120 Birchmont Road
Scarborough, Ontario
Canada M1K 5G4

International Thomson Editores
Campos Eliseos 385, Piso 7
Col. Polanco
11560 Mexico D.F., Mexico

International Thomson Publishing GmbH
Königswinterer Strasse 418
53227 Bonn, Germany

International Thomson Publishing Asia
221 Henderson Road
#05-10 Henderson Building
Singapore, 0315

International Thomson Publishing Japan
Hirakawacho Kyowa Building, 31
2-2-1 Hirakawacho
Chiyoda-ku, Tokyo 102
Japan

Printed and bound in the United States of America.
98 99 00 — 10 9 8 7 6 5 4 3 2 1

Library of Congress Cataloging-in-Publication Data

Oweis, Issa S.
    Geotechnology of waste management / Issa S. Oweis, Raj P. Khera. – – 2nd ed.
        p. cm.
    Includes bibliographical references and index.
    ISBN 0-534-94524-4
    1. Waste disposal in the ground.   2. Environmental geotechnology.
I. Khera, Raj P. II. Title.
TD795.7.093 1998
628.4′456—dc21

97-31813
CIP

Contents

CHAPTER **5**   **SHEAR STRENGTH   75**

CHAPTER **6**   **HYDRAULIC PROPERTIES   100**

CHAPTER **7**   **SITE INVESTIGATION   139**

CHAPTER 11

LEACHATE GENERATION AND COLLECTION   295

CHAPTER 12

CAPS   341

CHAPTER **13**   **FOUNDATION AND SLOPE STABILITY**   **379**

CHAPTER **14**   **GAS MANAGEMENT**   **409**

Preface

Since the first publication of this book eight years ago, a considerable number of research studies and cases histories on various aspects of waste management have been published. In addition some regulations have been changed or rescinded, and new regulations have appeared. The aim of the book, however, has not changed since its first appearance: to equip the student and practicing engineer with the basic knowledge needed for the geotechnical design of waste facilities, the closure and improvement of waste facilities, and construction on waste. Although we expect the student to have a basic training in geotechnical engineering, essential geotechnical principles are covered to the extent that reference to a basic geotechnical text is unnecessary.

In the second edition, we have rearranged several chapters and added new material to each of the original chapters as well as several new chapters. Throughout, we have emphasized the limitations of design methods, particularly with respect to liners and covers. Expanded chapters include Chapter 1, "Forms of Waste"; Chapter 2, "Index Properties"; Chapter 4, "Compressibility and Settlement"; Chapter 7, "Site Investigation"; and Chapter 9, "Ground Modification and Compaction." Old Chapter 7 has been expanded into two chapters: Chapter 11, "Leachate Generation and Collection," and Chapter 14, "Gas Generation and Management." Similarly, old Chapter 11 has been developed into three chapters: Chapter 10, "Liners"; Chapter 12, "Caps"; and Chapter 13, "Foundation and Slope Stability." Solved examples and exercise problems have been included to address a need identified by users and reviewers of the first edition.

The expanded version of the text may make using it in two graduate courses, rather than one, advisable. Chapters 1 through 9 could be covered in one course, while Chapters 10 through 14 could be reserved for a second course. Geotechnical/environmental instructors will find that the book does not conflict with other courses, such as solid waste planning or groundwater management because we have attempted to restrict the book to geotechnical issues. Storm water management and erosion control, typically taught in civil design, are, however, briefly covered because they greatly impact cap performance. But the book is intended to be a living document: an instructor can readily introduce new material in teaching any of the chapters.

Issa S. Oweis
Raj P. Khera

List of Reviewers

Wayne R. Bergstrom
Engineering Geo Techniques, Inc.

Robert B. Gilbert
The University of Texas at Austin

C. W. Lovell
C. W. Lovell and Associates

Sibel Pamukcu
Lehigh University

Vijay K. Puri
Southern Illinois University/Carbondale

Charles D. Shackelford
Colorado State University

1   Forms of Waste

1.1   GENERATION OF WASTE

With increasing industrialization, the quantity of waste has increased immensely. Waste may take the form of solids, sludges, liquids, gases, or any combination thereof. Depending on the source of generation, some wastes may degrade into harmless products whereas others may be nondegradable and hazardous. Approximately 10 to 15% of the waste generated in the United States is considered hazardous. Hazardous wastes pose potential risks to human health and that of other living organisms. Such wastes are not degradable and may have cumulative detrimental effects.

Nonhazardous waste quantities and their primary disposal techniques, as described by the Resource Conservation and Recovery Act (RCRA), are given in Table 1.1. Land disposal, in the form of landfills, surface impoundment, land application, deep well injection, etc., is the most common form of waste management for both nonhazardous as well as hazardous waste. About 90% of off-site disposal of hazardous waste in the United States is on land (U.S. Congress, 1986). Because of the immensely large volume of waste being generated, many federal and state agencies are actively encouraging the recycling of recoverable materials and finding new ways to use harmless waste products as construction materials, such as aggregate for cement concrete and asphalt concrete, stabilizing road bases, and placement in highway embankments and structural fills. In 1987 the Federal Highway Administration implemented the Beneficial Use of Coal Ash Act, which provides the incentive of increased federal funding to states for projects where large amounts of coal ash are used. Some of the by-products, such as fly ash, kiln dust, and air pollution control waste from dry and wet scrubbers, may be utilized to stabilize and/or solidify liquid wastes before disposal in landfills. They can also be used as stabilizing agents for soils, which, in turn, can be used as pavement subgrade, embankments, and backfills for retaining walls.

The federal Pollution Prevention Act (1990) promotes prevention of pollution and encourages further source reduction. It defines *source reduction* as "any practice which reduces the amount of any hazardous substance, pollutant, or contaminant entering any waste stream or otherwise released into the environment prior to recycling, treatment, or disposal; and reduces the hazards to public health and the environment associated with the release of such substances, pollutants, or contaminants."

1.2   MUNICIPAL SOLID WASTE

Total municipal solid waste (MSW) generated in the United States has increased from 87.8×10^6 tons in 1960 to 195.7×10^6 tons in 1990 (EPA, 1992). In 1990 the amount discarded after material recovery and combustion

Table 1.1   Annual nonhazardous waste quantities

Waste type	Quantity × 10^6 tons	1	2	3	4	5	6	7	8	9
Municipal[b]	185	x	x	x	x					
Compost	2.5								x	
Scrap tires	2.5	x			x					x
Used oil	2				x					
Household hazardous	0.1–0.001	x				x	x			
Sewage sludge, dry wt basis	8.4	x	x	x				x	x	
Water treatment sludge	0.21–0.83	x								x
Industrial nonhazardous	430	x						x	x	
Coal ash (fly ash and bottom ash)	72	x						x		x
Ferrous and nonferrous slags	34	x								x
Reclaimed paving materials	103									x
Construction and demolition	31.5	x								
Cement and limekiln dusts	24	x								x
Sulfate	18				x			x		
Lime	2				x					x
Roofing shingle	8	x								x
Foundry	10	x								x
Ceramic	3									
Silica fume	1	x								x
Small quantity generator hazardous (<1000 kg/month)[e]	0.66									
Dredged material	500 yd^3	x	x	x				x	x	x
Mineral										
Waste rock	1020	x								x
Mill tailings	520	x								x
Coal refuse	120	x								x
Washery rejects	105									
Phosphogypsum	35	x								
Oil and gas — production, processing, etc., gal/day	6250							x		
Agricultural										
Animal manure	1600								x	
Crop	400								x	
Logging and wood	70								x	
Miscellaneous organic	30								x	
Liquid waste, gal/day	1000							x		

[a] 1—land disposal; 2—ocean disposal; 3—incineration; 4—resource recovery; 5—sewer system; 6—septic tanks; 7—lagoon/surface impoundment; 8—land application; 9—construction application.
[b] Includes municipal solid waste combustion residue, reclaimed plastic, yard waste, and waste glass.
[e] Onsite disposal (sewer, recycling, treatment), off-site disposal (recycling, solid waste facility, or subtitle C).
Source: Subtitle D, 1986; Collins and Ciesielski, 1992

was 162.3×10^6 tons. The amount of production of MSW per capita per day, which was 2.6 lb in 1960, is expected to reach 4.5 lb. by the year 2000 (EPA, 1992), and an increase in waste generation of 13% is projected between 1990 and 2000. Because of material recovery, yard trimming composting, and incineration, the amount of material to be disposed of in landfills is projected to decrease from 130×10^6 tons to 109×10^6 tons within this period, as shown in Figure 1.1.

In the past, most municipal waste was landfilled, so municipal landfills are heterogeneous mixtures of wastes that are primarily of residential and commercial origin. Typically, a municipal landfill consists of food and garden wastes, paper products, plastics and rubber, textiles, wood, ashes, and the soils used as cover material. More than 50% of municipal solid waste, by weight, is paper and yard waste. Larger objects, such as tree stumps, refrigerators, automobile bodies, and demolition waste may also be present. The proportions of these materials may vary from one region to the other and depend on the type of local commerce and industry. Household hazardous waste, which may include cleaners, automotive products, paints, and garden products, is either landfilled or discharged into sewer or septic systems.

Refuse composition is determined, when necessary, by manually sorting large quantities of refuse (1/2 to 1 ton) over several days or weeks. Four or five experienced persons could separate and weigh 1 ton of refuse during a day's work, but to be meaningful, several samplings are needed over time from various locations in an existing landfill or waste stream. For samples above 200 lb, no significant statistical differences exist in the reliability of the composition determinations (Klee and Carruth, 1970). Another technique for determining composition is to observe the refuse in test pits or the waste stream and adjust percentages of national or local averages.

The constituents of the landfill may vary considerably from country to country. A comparison of municipal waste components from different countries is given in Table 1.2. Developing countries like India show a lower proportion of salvageable material because there it is more economical to recycle these materials. It is interesting to note that as a country becomes more prosperous, the proportion of salvageable materials in the refuse increases. However, as land available for disposal of waste has become more scarce near highly populated areas and mandatory recycling legislation increases, this

Figure 1.1
Projected disposition of MSW (from EPA, 1992)

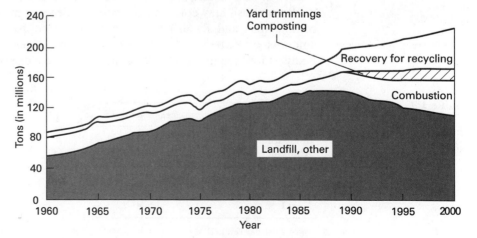

Table 1.2   Typical municipal waste composition percentage by weight

Component	COUNTRY OR CITY[a]															
	1	2	3	4	5	6	7	8	9	10	11	12	13	14	15	16
Metals		6	1	3	6	3	3		3		5	7		2	8	5
Paper and paper board	37[b]	32	5	22	39	33	40	38	10	14[b]	30	32	14[b]	8	30	31
Plastics		13	1	7	5		9	8			6	3				
Rubber, leather, wood		4	1		1	7	1	12	4			7		3	1	4
Textile		3			4	10	3				4	4		4	2	2
Food and yard waste	45[e]	33	45	42	34	15	30	18	74	56[e]	25	36	50[e]	25	16	16
Glass		9	1	7	6	10	7		7		12	4		3	8	13
Nonfood inorganic	18[d]		46	2	22		24	2		30[d]		7	21[d]	55	35	29
Other			19		3		7				18					

[a] 1—Australia; 2—Switzerland; 3—Beijing; 4—Italy; 5—Canada; 6—Hong Kong; 7—Sweden; 8—Japan; 9—Korea; 10—Madras India; 11—France; 12—Singapore; 13—Spain; 14—Taiwan; 15—UK; 16—West Germany.
[b] Metal is included under paper and paper board.
[e] Includes wood, bones, etc.
[d] Includes glass, coconut, shells, fibers, etc.
Sources: Aziz, 1986; Carra and Cossu, 1990; Hillenbrand, 1986; Sargunan et al., 1986; Subtitle D, 1986; *Waste Age*, 1986

trend has begun to reverse itself. Japan, one of the most industrialized countries in the world, has placed major emphasis on waste reduction at each source and on recycling. The estimated recycling rate in Japan is about 50% of the total waste (Cossu, 1990).

1.2.1 *Incinerated Municipal Waste*

Thermal destruction and resource recovery is gaining favor as a means of reducing the volume of municipal waste. In Japan, the preferred means of disposal after recycling is through incineration; more than two-thirds of municipal waste in Japan is incinerated (Cossu, 1990). Incineration reduces the weight of the waste to roughly 20 to 25% of its original weight and produces a residue of bottom ash, slag, and fly ash (see Table 1.3).

Bottom ash may contain glass, stones, pieces of metal, etc., which are noncombustible, and fly ash from air pollution control devices contains trace amounts of PCB, metals, dioxins, PVC, and other toxic compounds. Thus many states require that incinerator waste be disposed of like hazardous

Table 1.3
Composition of municipal incinerator waste

Constituent	Incinerator residue %
Silica (SiO_2)	32–43
Alumina (Al_2O_3)	9–19
Iron oxide (Fe_2O_3)	23–24
Calcium (CaO)	4–7
Magnesium (MgO)	0–1
Na_2O	1–8

Source: Lum and Tay, 1992

Table 1.4
Total metals in
municipal
incinerator waste

Compound	Bottom ash mg/kg	Fly ash mg/kg
Pb	31–36,600	2.0–26,000
Cd	0.81–100	5–2210
As	0.8–50	4.8–750
Cr	1.3–1500	21–1900
Ba	47–2000	88–9000
Ni	ND (1.5)–12,910	ND (1.5)–3600
Cu	40–10,700	187–2300

ND = Not Detectable; ( ) = Detection Limit.
Source: Wiles et al., 1991

waste. The concentrations of these metals and compounds are higher in fly ash than in bottom ash and slag. Table 1.4 shows the range of amounts of metals in municipal waste incineration residue in the United States. Initial data from solidification/stabilization of fly ash indicate that in some instances up to 60% of the materials it contains could dissolve or erode with time. Salt release through dissolution significantly affects specimen integrity. In tests where pH alkalinity increased, lead and cadmium showed greater mobility than other metals, exhibiting leachate concentrations in excess of those considered hazardous (Kosson et al., 1991). Contaminants leaching from landfills containing such fly ashes may reach groundwater (Clapp et al., 1987).

1.2.2 *Discarded Tires*

About 250 million tires are scrapped each year in the United States, the majority of them being stockpiled or landfilled and about 25% recycled. In municipal waste, whole tires do not lend themselves to compaction, so they reduce landfill capacity as well as rise to the surface with time and damage the final cover. Stockpiled tires are susceptible to hard-to-control fires, which release polyaromatic hydrocarbons into the atmosphere and produce residue that may cause soil and water pollution. Because of these special disposal problems, many states require that tires be processed by cutting or shredding before permitting their disposal in landfills. Increasing numbers of scrap tires are being recycled through retreading or as soil reinforcement, erosion control, breakwaters, acoustic barriers, lightweight aggregate, bituminous additives, drainage material, and energy recovery.

1.3 MINERAL WASTE

Large amounts of mineral wastes are generated during the extraction of materials such as metals, fuels, fertilizers, chemicals, and clays and are stored near mining or extraction operations where, although they are not generally considered hazardous, they can become a source of soil, water, and air pollution if not properly managed. Increasing demand on natural resources has led to ores of poorer quality being mined and processed, generating ever-larger amounts of waste.

In mining operations, chunks of rock, or overburden, are removed to reach ore. This material is transported and placed in loose dumps called overburden piles, spoil heaps, or tips. Both for stability and to avoid leachate generation, moisture in these dumps must be controlled through resloping and erosion control. Groundwater and precipitation infiltrate through mines during extraction. During refining, the country or the host rock containing the ore is crushed and ground in the bar mills or rod mills to release the concentrated mineral. As the ore is processed, the remaining solid/water slurries consisting of mine waste are transported by pipelines to disposal areas. The slurries contain both sharp, angular sand-size particles (coarse-grained materials called *tailings*) and smaller clay and colloidal-size particles (fine-grained materials called *slimes*).

In the past, before disposal the fine-grained materials were separated from the coarse-grained fraction. The coarse-grained fraction could be backfilled into active mines, but the fine-grained fraction could not, as it might cause flooding. Any nonhazardous coarse-grained fraction of the waste could also be used as construction material. The suspended fine-grained materials could be disposed of by impoundment behind tailing dams, injection into deep wells, or backfilling into abandoned mines. The major difficulty in disposing the fine-grained material is that it is suspended in water. The enormous volume of these slimes must be reduced and the shear strength increased. Another challenge is the processing chemicals, dissolved salts, detergents, heavy metals, and residuals from smeltering and electrolytic refining the tailings contain, that may pose short- and long-term threat to the quality of surface and groundwater and may not receive regulatory approval or public acceptance. The recent trend in disposal has been to keep the coarse-grained fraction suspended and dispose of both fractions together. When the water content of the slurry is reduced sufficiently before depositing, there is no need for ponding and thus the environmental difficulties are minimized (Morgenstern and Scott, 1995).

1.3.1 *Coal Waste*

Increasing demand for coal and improved mining technology have made possible the mining of less productive seams resulting in larger amounts of coal and the related wastes of sandstone, siltstone, shale, mudstone, clay, etc. in various proportions. Coarse coal waste is generated during mining and separation and consists of particles larger than 0.02 in. (0.5 mm) (rock particles larger than 3 in. (76 mm) are known as *mine rock*). The elongated shapes of the particles break under pressure and form smaller-size particles, and the larger surface area of such particles makes them more susceptible to weathering, which is due to oxidation of pyrite in shale and carbonaceous rocks.

The fine components of coal refuse are in the form of slurries produced during the washing of coal; these are disposed of behind impoundment dikes constructed out of coarse refuse. The specific gravity of fine waste is lower than that of the coarse waste. If coarse and fine refuse are mixed, they form an unstable material and require disposal as slurries. In strip mining, additional waste is generated from the removal of overburden prior to the recovery of coal. This waste may be disposed of separate from the coal waste itself.

1.3.2 Phosphatic Waste

Phosphate rocks are mined to produce chemical fertilizers. About two-thirds of the phosphate ore, or *matrix* as it is commonly known, consists of about equal amounts of sand and clay. The sand-size tailings are separated and readily stabilized, but a very large volume of clay suspension is still left for disposal. These clays primarily contain montmorillonite, attapulgite, illite, and kaolinite, and their large surface areas keep them in suspension, creating a difficult problem of dewatering and stabilizing. Dikes 3 to 15 m (10 to 50 ft) high are built to store the clay suspensions, which initially have solid contents of 2 to 6%. Sedimentation, or free falling, of clay particles takes a few days to a few weeks, after which the solid content is about 10%. Substantial volume reduction of the clay through subsequent settlement under self-weight requires 10 to 100 years (McVay et al., 1986).

The use of sulfuric acid in the wet processing of phosphoric acid from phosphate rock yields phosphogypsum (PG) as a by-product. PG has been deposited in mined-out areas or is slurried and pumped to storage areas known as *stacks*. Although PG has been used for agricultural purposes, it has been outlawed in construction applications because it exhibits residual radio-activity. The estimated stock pile of PG in Florida alone by the year 2000 is expected to be more than a billion tons (Taha and Seals, 1992).

1.3.3 Bauxite Processing Waste

Waste from the processing of bauxite results in so-called red mud, which has high alkalinity and 12 to 15% solid contents. Its sand fraction may range up to 30%; the remaining particles are silt size. The fines settle very slowly. Although the volumes of such waste are not enormous, they occupy large areas (Somogyi and Gray, 1977).

1.4 INDUSTRIAL WASTE

About 15% of industrial waste is classified as hazardous, and our increasing understanding of the impact of these wastes on the environment will likely lead to more types of wastes being classified as such. The largest amount of industrial waste is produced in the generation of electricity through the burning of coal. Several other industries, such as chemical manufacturing and processing, food processing, iron and steel manufacturing, petroleum refining, and the manufacturing of plastics and resins, paper and pulp, lumber and wood, and pharmaceuticals, generate large quantities of waste. Agricultural waste can be considered industrial waste as well. It may include animal waste, irrigation waste, collected field runoff, crop production waste, and fertilizers. Some of these wastes may contain various levels of organic and inorganic chemicals and heavy metals.

Until 1978 about 80% of industrial waste was disposed of in unlined landfills or in surface impoundments as aqueous solution of organics and inorganics, as organic liquids with organic solutes, and as sludges (Brown and Anderson, 1983). Since November, 1981, RCRA prohibits the land disposal of drums containing waste liquids, and even with an adequate liner and waste collection system, bulk liquids may not be disposed of in landfills.

1.4.1 *Pulp and Paper Mill Waste*

Paper mills produce a particular kind of industrial waste with unique disposal requirements. Paper mill sludge is composed of wood bark particles, fibers and fiber particles, talc, silt, and clay. It has a consistency of fibrous organic material, with a high initial water content, very unstable structure, and the potential for large settlements under applied loads. The processing of printed paper for recycling also generates some hazardous waste.

1.5   COAL BURNING WASTE

Coal is increasingly used to generate electricity. About 12 to 15% of coal in power plants ends up as ash consisting of boiler slag, bottom ash, cinders, and fly ash. Boiler slag is produced as residue from wet-bottom or cyclone boilers, flows out from the furnace into cold water, and solidifies to form crystals and angular, black, glasslike particles. The typical compositions of bottom ash and fly ash are shown in Table 1.5.

Fly ash produced by burning subbituminous coal, or lignite, is called *Class C* fly ash. This type of fly ash contains lime and other compounds that impart self-cementing properties. According to ASTM standard C 618, the chemical composition of Class C fly ash must be such that $SiO_2 + Al_2O_3 + Fe_2O_3 \geq$ 50% and sulfur trioxide may not exceed 5%. This class of fly ash is more common in the western and midwestern United States. Fly ash formed by burning bituminous coal, or anthracite, is known as *Class F* fly ash. Its chemi-

Table 1.5
Composition of
combustion waste

Constituent	Bottom ash %	Fly ash %
Silica (SiO_2)	21–60	34–58
Alumina (Al_2O_3)	10–37	20–40
Iron oxide (Fe_2O_3)	5–37	4–24
Calcium (CaO)	0–22	2–30
Magnesium (MgO)	0–4	1–6
Sulfates (SO_3)	—	0.5–4
Water soluble alkalis	0–7	2–6

Source: Di Gioia et al., 1977; Bowders et al., 1987; Edil et al., 1987

cal composition must be such that $SiO_2 + Al_2O_3 + Fe_2O_3 \geq 70\%$ and sulfur trioxide may not exceed 5% (ASTM C 618). This category of fly ash is primarily generated in the eastern United States. Although it is pozzolanic, it shows few self-cementing properties. The engineering properties of fly ash are influenced most by free lime and unburned carbons. Low-volume coal combustion waste, such as boiler cleaning waste, exhibits toxicity levels that would characterize them as hazardous based on EPA tests.

In the past, fly ash was generally disposed of by the wet method—it was allowed to settle in ponds. Environmental concerns have caused a shifting to dry disposal: the moisture content of the fly ash is controlled (known as *conditioning*), and it is compacted like a soil.

Where certain extraction processes require a considerable amount of energy and coal is used to produce the needed energy, coal, ash, and other coal-related wastes are mixed with processing waste at the extraction sites.

1.5.1 *Flue Gas Waste*

Current regulations require that particulate matter and sulfur dioxide be removed from flue gases. Granular particle sizes greater than 100 microns settle out due to gravity. Dust particles, ranging in size from 1 to 100 microns and consisting primarily of metal oxides, ash with the specific gravity (G_s) between 2 and 3, smaller proportions of organics ($G_s \approx 1$), and heavy metals and their oxides ($G_s > 5$) must be removed by other means. Particles smaller than 1 micron are contained in the dispersion of solid or liquid particles and are called *fumes*. Table 1.6 shows the range of sizes for various industrial particles (Mantell, 1975).

Some sources of particulate matter are the processing of iron ore, the manufacturing of cement, lead smeltering, electric furnaces, and the combustion of coal, etc. The removal of particulate matter may be done by mechanical separators, fabric filter collectors, electrostatic precipitators, or wet scrubbers. The type of equipment used is determined by the nature of the manufacturing process. Mechanical separators are more efficient for high gas flow, fabric filters are effective in removing zinc oxide fumes from lead smeltering, electrostatic precipitators perform well in collecting fly ash, and wet

Table 1.6
Sizes of particles in industrial waste

Type of materials	Particle size (μm)
Carbon black	0.01–0.2
Zinc oxide smoke	0.01–0.3
Metallurgical fumes	0.01–2.5
Oil smoke	0.03–1.0
Metallurgical dust	0.7–100
Pulverized coal fly ash	1–100
Cement dust	6–200
Pulverized coal	10–200
Ground limestone	20–800

Source: Mantell, 1975

scrubbers are commonly selected when both gas and particulate matter are present. Some of these materials can be recycled; others require proper disposal.

The major by-product at power plants is flue gas desulfurization (FGD) sludge waste generated from scrubbers. Sulfur oxides may be removed from flue gases through contact with lime or limestone slurries, which absorb sulfur oxides. This process of wet scrubbing yields waste sludges consisting of calcium sulfite, calcium sulfate (gypsum), and often unreacted lime or limestone and trace elements from fly ash. Sulfite sludges contain huge amounts of water requiring large areas for land disposal, and their thixotropic nature makes this difficult. Double-alkali methods using a water-soluble absorber such as sodium hydroxide, sodium carbonate, or sodium sulfite are alternatives. Double-alkali waste products are easier to dewater than the FGD wastes produced with lime or limestone. With either method, chemicals that such waste may contain include arsenic, selenium, mercury, and heavy metals, etc., making its disposal on land a potential source of water pollution.

The Clean Air Act Amendment of 1990 is likely to result in changes in the FGD process so that more gypsum will be produced from it. Gypsum is coarser in grain size and can be more readily disposed of by ponding or landfilling as gypsum cake or after blending with fly ash and lime. When blended with fly ash and lime, it is strong enough to be used in structural fills, as base coarse for roadways, and as aggregate (Smith, 1992). In Europe, the most common by-product of flue gas desulfurization is gypsum. The purity of FGD gypsum is usually higher than that of natural gypsum (Gera et al., 1991), and it is used extensively in the manufacture of building products. One German firm, the Veba Kraftwerke Ruhr (AG), recovers gypsum from its flue gases and sells it to manufacturers of plasterboard. However, gypsum has not found much use as building materials in the United States.

1.6   DREDGED WASTE

Dredged materials are bottom sediments or materials that have been dredged or excavated from navigable waters and include natural sediments such as rocks, gravel, sands, silts, and clays, which may contain contaminants from urban, industrial, and/or agricultural runoff from land. Nevertheless, a high proportion of the large amount of waste produced by dredging has been disposed of in water. (Table 1.7 shows heavy metal levels in dredged material from two sources in the United States and in Germany.)

The land disposal of dredged waste is a recent phenomenon and is the result of public environmental concerns and government regulations. Polluted components or those that cannot be released economically in open waters without causing unacceptable turbidity require land disposal. About 20% of the waste is now placed in landfills, which are filled hydraulically in the form of slurry in confined disposal facilities. The water content of these slurries ranges from 50 to 400%, and they consist primarily of organic silts and clays that might be contaminated with the heavy metals listed in Table 1.7 as well as

Table 1.7
Some heavy metals in dredged waste

Contaminant	Two sources in United States (mg/kg)	Germany (mg/kg)
Cadmium	0.4–20	7–25
Copper	ND–43	250–600
Lead	43–879	150–250
Mercury	<0.1–0.5	10–20
Zinc	98–4125	1500–2500

Source: After Rollings, 1994; Rizkallah, 1987

other pollutants. As these materials consolidate, the water expelled will have a high concentration of pollutants and must be treated appropriately.

The materials within a dredged landfill may vary from fairly well distributed to rather inhomogeneous and segregated (Bromwell, 1978). The grain-size characteristic of a dredged deposit can be controlled by the location of inlets and outlets, discharge velocity, water level in the deposition pond dividing large sites into smaller sections, and the ratio of clay solid to water. Furthermore, by proper drainage control and/or drying (depending on the desired results), the deposit can be made to be more stable, to have a larger storage volume, or to serve as a source of good backfill material, provided it is free of contaminant. Frequently the main focus in placing dredged materials is on reducing its volume, not controlling the quality so as to produce a fill with desirable engineering properties. Thus, the engineering properties of dredged materials vary with site location, type of dredging used, method of placement, and over time as consolidation and drying occur.

The soil improvement techniques described in Chapters 4 and 9 can be applied to dredged material to enhance their strength properties and reduce their volume. When aqueous disposal is considered, site selection, subaqueous disposal points, displacement layer thickness, slopes, capping materials, fate of dredgings, effect on habitat, and long-term monitoring should all be part of the design. The beneficial use of dredged material can include engineering uses, agricultural applications, forestry, strip mine reclamation, parks, beach replenishment, shoreline stabilization, and erosion control (U.S. Congress, 1986).

1.7 MUNICIPAL SLUDGES

Municipal sludges are generated from the treatment of both potable water and wastewaters. These sludges have very low solid contents and require some form of conditioning to reduce their volume before disposal. Chemical coagulation is used in the treatment of potable water, resulting in so-called water treatment sludges that have very high water content, plasticity, and organic contents and thus are difficult to handle. Although landfilling, lagooning, and ocean dumping have been the usual means of water treatment sludge disposal, other alternatives, such as use in embankments, land application, and landfill liners, are being investigated for these nontoxic sludges.

The type of solids produced in wastewater treatment are a function of the treatment and the subsequent thickening, digestion, and conditioning. They may be disposed of in sludge landfills or as sludge refuse mix or sludge soil mix. In some areas, these sludges are being composted by mixing them with wood chips.

1.8 HAZARDOUS WASTE

Industrial wastes are the major source of hazardous waste (EPA, 1980). Table 1.8 shows various hazardous substances produced by various industries.

About 90% of the industrial hazardous wastes in the United States are produced as liquids—60% organic liquids and 40% inorganic liquids. Under the RCRA, unless specifically excluded, a solid waste that meets any of the following four criteria is considered hazardous.

1. The waste has been specifically listed as hazardous.

2. The waste is mixed with a hazardous waste.

3. The waste is the result of treatment, storage, or disposal of hazardous waste.

4. The waste is ignitable, corrosive, reactive, or has toxic characteristics.

The Environmental Protection Agency has classified four types of hazardous waste (EPA, 1980):

Type 1. Aqueous–inorganic

Type 2. Aqueous–organic

Type 3. Organic

Type 4. Hazardous sludges, slurries, and solids

Water is the solvent in Type 1. Solutes are mostly inorganics, such as those produced by the electroplating industry (cadmium, cyanides, and metals). Organics (carbon compounds) are the solutes for Type 2 and are produced by

Table 1.8 Representative hazardous substances within the industrial waste stream

Industry	As	Cd	CH[a]	Cr	Cu	CN	Pb	Hg	MO[b]	Se	Zn
Battery		X		X	X						X
Chemical manufacturing			X	X	X			X	X		
Electrical and electronic			X		X	X	X	X		X	
Electroplating and metal finishing		X		X	X	X					X
Explosives	X				X		X	X	X		
Leather				X					X		
Mining and metallurgy	X	X		X	X	X	X	X		X	X
Paint and dye		X		X	X	X	X	X	X	X	
Pesticide	X		X			X	X	X	X		X
Petroleum and coal	X	X					X				
Pharmaceutical	X							X	X		
Printing and duplicating	X			X	X		X		X	X	
Pulp and paper								X	X		
Textile				X	X				X		

[a] Chlorinated hydrocarbons and polychlorinated biphenyls.
[b] Miscellaneous organics such as acrolein, chloropicrin, dimethyl sulfate, dinitrobenzine, dinitrophenol, nitroaniline, and pentachlorophenol.
Source: Matrecon Inc, 1980

the pesticide industry. In Type 3, both the solutes and solvents are organics, such as oil-based paints and motor oil. Type 4 includes sludges generated by various dewatering, filtration, or treatment processes, such as sludges from the petroleum refining industry and treatment plants.

1.9 SUMMARY

Waste is generated in any of the solid, sludge, liquid, or gaseous forms. The amount of waste generated and the types of hazardous materials present in the waste stream increase with increasing industrialization of a country. Large amounts of wastes have been stockpiled throughout the United States. As in other industrialized countries, there may be greater use of combustion/energy conversion of municipal waste in the future. However, the problem of disposing of municipal combustion waste needs resolution. As new regulations governing waste disposal go into effect, suitable sites for its disposal in densely populated areas are becoming more difficult to find. Thus emphasis has shifted from waste disposal to waste reduction, reuse, and recycling.

PROBLEMS

1.1 Determine which Class (C or F) each of the two fly ashes conforms to.

Constituent	Fly ash A	Fly ash B
Silica (SiO_2)	31	47
Alumina (Al_2O_3)	17	28
Iron oxide (Fe_2O_3)	5	16
Calcium (CaO)	30	2
Magnesium (MgO)	7	1
Sulfur trioxide (SO_3)	3	1
Sodium oxide (Na_2O)	3	1
Others	4	4

1.2 What are the residues from municipal solid waste combustion? Do they present any problem when placed in landfill?

1.3 For FGD, what type of process would you select — one that produces (a) sulfite rich waste or (b) sulfate rich waste? Explain.

1.4 What are the components of dredged waste? Which components are land disposed and why?

1.5 List the four types of hazardous waste. To which type do the following wastes belong? (a) petroleum sludge, (b) oil-based paint, (c) cadmium, (d) pesticide.

2 Index Properties

2.1 SOILS

For engineering the disposal of waste, much of which is in the form of particulate matter such as mining waste and coal waste, we have relied on our knowledge of the behavior of soils. Many of the principles of soil mechanics are presented in this book and then extended to the management of solid waste.

Soils are three-phase systems consisting of mineral particles and voids filled with air and/or water. A typical soil element, its phase diagram, and the corresponding notations are shown in Figure 2.1. The basic volume relationships of soil are

V = total volume of soil
V_s = volume of solids
V_v = volume of voids
V_w = volume of water
V_a = volume of air

$$\text{Porosity} \qquad n = \frac{V_v}{V} \tag{2.1}$$

$$\text{Void ratio} \qquad e = \frac{V_v}{V_s} \tag{2.2}$$

$$\text{Degree of saturation} \qquad S = \frac{V_w}{V_v} \tag{2.3}$$

$$\text{Volumetric moisture content} \qquad \theta = \frac{V_w}{V} \tag{2.4}$$

The basic weight relationships are

W = total weight
W_s = weight of solids
W_w = weight of water
W_a = weight of air (negligible)

$$\text{Water content} \qquad w = \frac{W_w}{W_s} \times 100(\%) \tag{2.5}$$

$$\text{Unit weight of water} \qquad \gamma_w = \frac{W_w}{V_w} \tag{2.6}$$

Figure 2.1
Phase diagram of a
soil showing its various
components

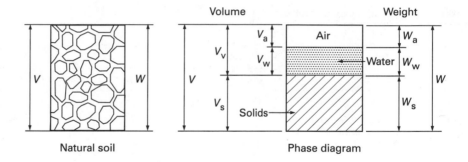

$$\text{Unit weight of solids} \quad \gamma_s = \frac{W_s}{V_s} \quad \text{(2.7)}$$

$$\text{Specific gravity of solids} \quad G_s = \frac{\gamma_s}{\gamma_w} \quad \text{(2.8)}$$

$$\text{Total unit weight} \quad \gamma = \frac{W}{V} = \frac{\gamma_w(G_s + Se)}{(1 + e)} \quad \text{(2.9)}$$

$$\text{Dry unit weight} \quad \gamma_d = \frac{W_s}{V} = \frac{\gamma_w G_s}{(1 + e)} = \frac{\gamma}{(1 + w)} \quad \text{(2.10)}$$

$$\text{Submerged unit weight} \quad \gamma' = \gamma - \gamma_w \quad \text{(2.11)}$$

$$\text{Volumetric moisture content} \quad \theta = \frac{W_w \times \gamma_d}{\gamma_w \times W_s} \quad \text{(2.12)}$$

$$\text{A useful relationship} \quad G_s w = Se \quad \text{(2.13)}$$

The specific gravity for most inorganic soils varies over a narrow range (2.6–2.9) and is commonly assumed to be 2.65 for sands and 2.7 for clays. Some typical values of index properties are provided in Table 2.1.

Useful indices for certain engineering properties are provided by soil texture and particle size distribution for coarse-grained soils and by plastic properties and natural water content for fine-grained soils. Coarse-grained soils consist of particles ranging from boulders to fine sands. A well-graded soil will yield lower permeability and higher strength than will soil with uniform particle sizes. The nominal particle size corresponding to 10% passing by weight is known as the *effective particle diameter* and is represented by D_{10}.

Fine-grained soils may exist in a liquid, plastic, semisolid, or solid state. The limiting water contents for these states, known as *Atterberg limits*, are liquid limit (LL), plastic limit (PL), and shrinkage limit (SL). These limits are indications of the type and amount of clay minerals present in a soil. At liquid limit the strength for most fine-grained soils is about 2 kPa and hydraulic

Table 2.1   Typical index properties

Soil description	VOID RATIO		DRY UNIT WEIGHT lb/ft³		SATURATED UNIT WEIGHT lb/ft³ (kN/m³)	
	Max.	Min.	Min.	Max.	Min.	Max.
Uniform sand	1.0	0.4	83	118	84 (13.2)	136 (21.4)
Silty sand	0.9	0.3	87	127	88 (13.8)	142 (23.3)
Clean, well-graded sand	0.95	0.2	85	138	86 (13.5)	148 (23.3)
Silty sand and gravel	0.85	0.14	89	146	90 (14.2)	155 (24.4)
Sandy or silty clay	1.8	0.25	60	135	100 (15.7)	147 (23.1)
Well-graded gravel, sand, silt, and clay mixture	0.7	0.13	100	148	125 (19.7)	156 (24.5)
Inorganic clay	2.4	0.5	50	112	94 (14.8)	133 (20.9)
Colloidal clay (50% <2 μ)	12	0.6	13	106	71 (11.2)	128 (20.1)

Source: NAVFAC DM 7.1, 1982

conductivity is about 2×10^{-9} m/s, even though the void ratios may be quite different (Mitchell, 1993).

The plasticity index (PI) is used in classification of fine-grained soils and is defined as

$$PI = LL - PL \qquad (2.14)$$

In a soil mix, a greater proportion of soils with high plasticity can make proper compaction more difficult. The ratio of plasticity index to the percentage of clay fraction by weight less than 2 μm is called *activity* (*A*):

$$A = \frac{PI}{\% < 2 \ \mu m} \qquad (2.15)$$

With higher activity numbers, the properties of soils are likely to be affected much more by the changes in pore fluid characteristics. For example, clays with a large proportion of bentonite or sodium montmorillonite may have activity as high as 7, whereas for soils with much kaolin the activity is about one-half. Soils with high activity are not recommended for use in landfill liners or containment structures as they are more readily affected by pollutants.

For organic soils, the specific gravity is less than or equal to 2.6. The moisture content is usually high (Table 2.2) and the unit weight is lower than inorganic soils. The index properties are usually assessed as part of a site investigation process.

2.2   SOIL CLASSIFICATION

The most commonly used classification method is the Classification of Soils for Engineering Purposes (ASTM D 2487), which is based on the Unified Soil Classification System (USCS). This may be used in conjunction with the Description and Identification of Soils (ASTM D 2488). Another classification

Table 2.2 Index properties of organic materials

Soil type	Natural water content (%)	Total unit weight lb/ft^3 (kN/m^3)	Specific gravity, G_s	Liquid limit, LL	Plasticity index, PI
Fibrous peat	500–1200	60–70 (9.4–11.0)	1.2–1.8	—	—
Fine-grained peat	400–800	60–70 (9.4–11.0)	1.2–1.8	400–900	200–500
Silty peat	250–500	65–90 (10.2–14.1)	1.8–2.3	250–500	150–350
Sandy peat	100–400	70–100 (9.4–15.7)	1.8–2.4	100–400	50–150
Organic clay	50–200	70–100 (9.4–15.7)	2.3–2.6	65–150	50–150
Organic sand or silt	30–125	90–110 (14.1–17.3)	2.4–2.6	30–100	NP–40

Source: NAVFAC DM 7.1, 1982

system described in this section is from the United States Department of Agriculture (USDA). An important method of estimating leachate generation and its movement in landfill uses the USDA classification system.

2.2.1 *Classification of Soils for Engineering Purposes* (USCS)

Textural classification of the USCS system is given in Table 2.3. In this system, the soils are placed in three major categories:

1. *Fine-grained soils.* If 50% or more by dry weight of the test specimen passes a No. 200 sieve, the soil is classified as a fine-grained soil.

2. *Coarse-grained soils.* If more than 50% by dry weight of the test specimen is retained on a No. 200 sieve, the soil is classified as a coarse-grained soil.

3. *Highly organic soils.* A soil is classified as Peat (Pt) if it is composed of decomposing vegetable tissues, has fibrous to amorphous texture, has a spongy consistency, is dark brown to black in color, and has organic odor.

Table 2.3 Textural division of soils

Soil name	Particle sizes, mm (in.)	Sieve size, in. or U.S. sieve no.
Boulders	> 300 (12)	
Cobbles	300–75 (12–3)	
Gravel		
Coarse	75–19 (3–3/4)	3 to 3/4
Fine	19–4.75 (3/4–0.187)	3/4 to No. 4
Sand		
Coarse	4.75–2.00	No. 4 to No. 10
Medium	2.00–0.425	No. 10 to No. 40
Fine	0.425–0.075	No. 40 to No. 200
Clays and silts	Less than 0.075	

The criteria for assigning group symbols and group names are given in Table 2.4

2.2.1.1 Fine-grained Soils   In the classification of fine-grained soils, Atterberg limits are determined for the materials passing a No. 40 (0.425 mm) sieve; these are plotted on the plasticity chart shown in Figure 2.2. Clays plot on or above the A line and have PI > 7. For silty clay, $4 \leq PI \leq 7$. Silts fall below the A line or have PI < 4.

2.2.1.2 Coarse-grained Soils   As shown in Table 2.4, coarse-grained soils are those in which more than 50% of the sample is retained on a No. 200 (75 μm) sieve. If the coarse fraction plus No. 4 sieve material is greater than 50%, the soil is classified as gravel; otherwise, it is sand. If fines are present (materials pass a No. 200 sieve), their proportion by weight is determined. A particle size distribution curve is plotted for soils with fines $\leq 12\%$. On a particle size distribution curve, D_{10}, D_{30}, and D_{60} are the nominal particle diameters for 10%, 30%, and 60% finer by weight.

Example 2.1

Index properties and grain size distribution curves for three soils are shown in Table 2.5 and Figure 2.3, respectively (p. 20). Classify the soils in accordance with the ASTM system of soil classification.

Solution: Use Table 2.4, Figure 2.2, and Figure 2.3. Numbers in parentheses represent columns in the table and the letter superscripts represent the corresponding footnotes.

Sample A:

Since 91% of the soil is coarser than a No. 200 sieve (0.075), from (1) it is a coarse-grained soil. Since > 50% is coarser than a No. 4 sieve (4.75 mm), from (2) it is gravel (G). Next we see that LL = 53 and PI = 28 plots above A line in Figure 2.2, hence clay (C) fines exist. From Figure 2.3, we find $D_{10} = 0.085$ mm, $D_{30} = 0.9$ mm, and $D_{60} = 10$ mm, and from (4)[e]

Figure 2.2
Plasticity chart

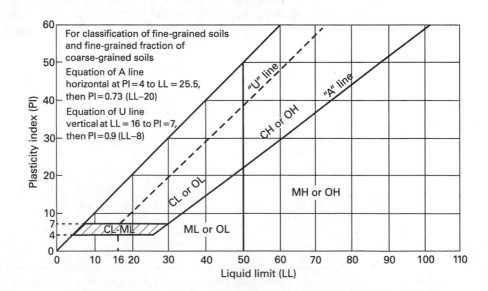

Table 2.4 Soil classification chart

				Group symbol (5)	Group name[b] (6)
CRITERIA FOR ASSIGNING GROUP SYMBOLS AND GROUP NAMES USING LABORATORY TESTS[a]					
(1)	(2)	(3)	(4)		
Coarse-grained soils (>50% retained on No. 200 sieve)	Gravels (>50% of coarse fraction retained on No. 4 sieve)	Clean gravels (<5% fines[c])	$C_u \geq 4$ and $1 \leq C_c \leq 3^e$	GW	Well-graded gravel[f]
			$C_u < 4$ and/or $1 > C_c > 3^e$	GP	Poorly graded gravel[f]
		Gravels with fines (>12% fines[c])	Fines classify as ML or MH	GM	Silty gravel[f,g,h]
			Fines classify as CL or CH	GC	Clayey gravel[f,g,h]
	Sands (≥50% of coarse fraction passes No. 4 sieve)	Clean sands (<5% fines[d])	$C_u \geq 6$ and $1 \leq C_c \leq 3^e$	SW	Well-graded sand[i]
			$C_u < 6$ and/or $1 > C_c > 3^e$	SP	Poorly graded sand[i]
		Sands with fines (>12% fines[d])	Fines classify as ML or MH	SM	Silty sand[g,h,i]
			Fines classify as CL or CH	SC	Clayey sand[g,h,i]
Fine-grained soils (≥50% passes the No. 200 sieve)	Silts and clays (liquid limit <50)	Inorganic	$PI > 7$ and plots on or above A line[j]	CL	Lean clay[k,l,m]
			$PI < 4$ or plots below A line[j]	ML	Silt[k,l,m]
		Organic	$\dfrac{\text{Liquid limit—oven dried}}{\text{Liquid limit—not dried}} < 0.75$	OL	Organic clay[k,l,m,n] Organic silt[k,l,m,o]
	Silts and clays (liquid limit ≥50)	Inorganic	PI plots on or above A line	CH	Fat clay[k,l,m]
			PI plots below A line	MH	Elastic silt[k,l,m]
		Organic	$\dfrac{\text{Liquid limit—oven dried}}{\text{Liquid limit—not dried}} < 0.75$	OH	Organic clay[k,l,m,p] Organic silt[k,l,m,q]
Highly organic soils		Primarily organic matter, dark in color, organic odor		PT	Peat

[a] Based on the material passing the 3 in. (75 mm) sieve.
[b] If field sample contained cobbles or boulders, or both, add "with cobbles or boulders, or both" to group name.
[c] Gravels with 5–12% fines require dual symbols: GW–GM well-graded gravel with silt, GW–GC well-graded gravel with clay, GP–GM poorly graded gravel with silt, GP–GC poorly graded gravel with clay.
[d] Sands with 5–12% fines require dual symbols: SW–SM well-graded sand with silt, SW–SC well-graded sand with clay, SP–SM poorly graded sand with silt, SP–SC poorly graded sand with clay.
[e] $C_u = D_{60}/D_{10}$ $C_c = \dfrac{(D_{30})^2}{D_{10} \times D_{60}}$
[f] If soil contains ≥15% sand, add "with sand" to group name.
[g] If fines classify as CL–ML, use dual symbol GC–GM, or SC–SM.
[h] If fines are organic, add "with organic fines" to group name.
[i] If soil contains ≥15% gravel, add "with gravel" to group name.
[j] If Atterberg limits plot in hatched area, soil is a CL–ML, silty clay.
[k] If soil contains 15–29% plus No. 200, add "with sand" or "with gravel," whichever is predominant.
[l] If soil contains ≥30% plus No. 200, predominantly sand, add "sandy" to group name.
[m] If soil contains ≥30% plus No. 200, predominantly gravel, add "gravelly" to group name.
[n] PI ≥ 4 and plots on or above A line.
[o] PI < 4 or plots below A line.
[p] PI plots on or above A line.
[q] PI plots below A line.

Table 2.5
Soil index properties

Sample	LL (%)	PL (%)	PI
A	53	25	28
B	36	30	6
C (not oven dried)	70	40	30
C (oven dried)	50		

$C_u = 117.6$, $C_c = 0.95$; therefore, the soil is poorly graded (*P*). From (3)^c, its symbol is GP–GC. From (6)^f, since sand $\geq$ 15%, the group name becomes poorly graded gravel with clay and sand.
Answer: GP–GC; poorly graded gravel with clay and sand.

Sample B:
Since >50% of this material passes the No. 200 sieve, from (1) it is fine-grained soil. LL < 50 and it plots below A line in Figure 2.2, hence it is silt (ML). Since sand is 27%, from (6)^k it is silt with sand.
Answer: ML; silt with sand.

Sample C:
Since >50% is finer than 0.075 mm, it is fine-grained material. LL > 50 and it plots below A line, so from (2) and (4) the soil is MH or OH. From (4) LL (oven dried)/LL (not oven dried) 50/70 is 0.71, the group symbol is OH. Since sand > 30%, from (6)^k the group name is sandy organic silt.
Answer: OH; sandy organic silt.

2.2.2 USDA *Classification*

Symbols for various soils in the USDA classification system are given in Table 2.6, and their textural division is given in Table 2.7. As illustrated in Figure 2.4, vertices of a triangle represent the extreme limits for sands, silts,

Figure 2.3
Examples of grain size distribution curves

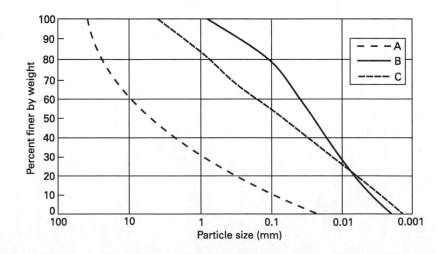

Table 2.6
Soil symbols used
in USDA

USDA soil type or state	USDA symbol
Gravel	G
Sand	S
Silt	Si
Clay	C
Loam (sand, silt, clay, and humus mixture)	L
Coarse	Co
Fine	F

Table 2.7
Textural division of
soils in USDA

Soil name	Particle sizes (mm)
Gravel	75–2
Sand	
Very coarse	2.0–1.0
Coarse	1.0–0.5
Medium	0.5–0.25
Fine	0.25–0.1
Very fine	0.1–0.05
Clays and silts	<0.05
Clay	<0.002

Figure 2.4
USDA textural
classification with
ASTM group symbols
superimposed

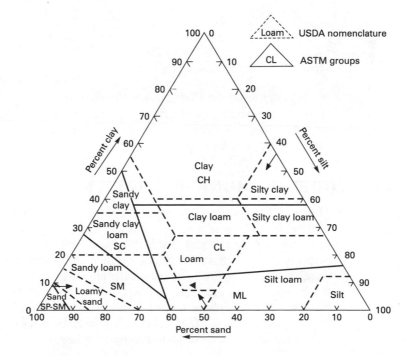

and clays. Note that only particles 2 mm in diameter and smaller are included in the chart; if larger sizes are present a correction is applied. Adjectives such as gravelly or stony are used to describe soils with significant portions of the coarse material.

Example 2.2

Classify soil C in Example 2.1 using the USDA system.

Solution: From Figure 2.3 for soil C according to Table 2.3, we obtain the following chart.

Gravel	100–92	8%
Sand	92–47	45%
Silt	47–7	40%
Clay	7–0	7%

-2 mm fraction $= 92\%$

Modified percentages are

$$sand = \frac{45 \times 100}{92} = 49\%$$

$$silt = \frac{40 \times 100}{92} = 43\%$$

$$clay = \frac{7 \times 100}{92} = 8\%$$

Hence, from Figure 2.4, it is a loam.

Superimposed on Figure 2.4 is the corresponding ASTM classification of soils for engineering purposes. The comparison between the two systems is significant because data on permeability from agricultural literature sources is often useful for engineering purposes.

2.3   UNIT WEIGHT OF MUNICIPAL WASTE

Unit weight of various components of waste are given in Table 2.8. Average unit weight of waste and its variation must be determined accurately, especially for seismic analysis. Table 2.9 shows unit weights of various waste components retrieved from landfills for unit weight determination. Those listed as durable goods are products with a lifetime of three years or more and include major appliances, furniture and furnishing, rubber tires, lead-acid automotive batteries, small appliances, and consumer electronics. The EPA lists these as "bulky and oversize" (EPA, 1992). Table 2.10 (p. 24) summarizes published values for unit weights of various types of refuse.

Table 2.8 Range of unit weight of uncompacted and compacted refuse components

Waste component	Uncompacted unit weight (lb/ft³)	Water content	RATIO OF COMPACTED TO UNCOMPACTED WEIGHT	
			Normal compaction	Well compacted
Food waste	8–30	50–80	2.9	3.0
Paper and paper board	2–8	4–10	4.5	6.2
Plastics	2–8	1–4	6.7	10
Textiles	2–6	6–15	5.6	6.7
Rubber and leather	6–16	1–12	3.3	3.3
Yard waste	4–14	30–80	4	5
Wood	8–20	15–40	3.3	3.3
Glass	10–30	1–4	1.7	2.5
Metals	3–70	2–6	4.3	5.3
Ash, brick, dirt	20–60	6–12	1.2	1.3

Source: After Tchobanoglous et al., 1977

Table 2.9
Field values of unit weight of individual waste components in municipal landfills (EPA, 1992)

Component	Unit weight (lb/ft³)	Unit weight (kN/m³)
DURABLE GOODS	19.26	2.80
NONDURABLE GOODS		
Nondurable paper	29.63	4.31
Nondurable plastic	11.67	1.70
Disposable diapers		
Diaper materials	29.44	4.28
Urine and feces	50.00	7.28
Rubber	12.78	1.86
Textiles	16.11	2.34
Miscellaneous nondurables (mostly plastics)	14.44	2.10
PACKAGING		
Glass containers		
Beer and soft drink bottles	103.70	15.09
Other containers	103.70	15.09
Steel containers		
Beer and soft drink cans	20.74	3.02
Food cans	20.74	3.02
Other packaging	20.74	3.02
Aluminum		
Beer and soft drink cans	9.26	1.35
Other packaging	20.37	2.96
Paper and paperboard		
Corrugated	27.78	4.04
Other paperboard	30.37	4.42
Paper packaging	27.41	3.99
Plastics		
Film	24.81	3.61
Rigid containers	13.15	1.91
Other packaging	6.85	1.00
Wood packaging	29.63	4.31
Other miscellaneous packaging	37.59	5.47
BIODEGRADABLE		
Food waste	74.07	10.78
Yard trimmings	55.56	8.08

Table 2.10
Unit weights for
different types of
landfill materials

Description and state	TOTAL UNIT WEIGHT	
	lb/ft^3	kN/m^3
MUNICIPAL WASTE		
Poor compaction	18–30	2.8–4.7
Moderate to good compaction	30–45	4.7–7.1
Good to excellent compaction	45–60	7.1–9.4
Baled waste	37–67	5.5–10.5
Shredded and compacted	41–67	6.4–10.5
In situ density	35–44	5.5–6.9
Municipal waste from Canada	43–89	6.8–14
Active landfill with leachate mound	42	6.6
Northeast U.S. active landfill	30–40	4.6–6.3
INCINERATOR RESIDUE		
Poorly burnt	46	7.2
Intermediately burnt	75	11.8
Well burnt	81	12.6
Ashes	41–52	6.4–8.2
HAZARDOUS WASTE LANDFILL SITE		
75-ft deep waste with soil cover	101	15.9
40- to 50-ft deep dry dust and soil	30–110	4.6–17.3
62-ft deep waste average		
Kiln dust, sludge tar, creosote, and soil (average)	73	11.5
Dusts	46	7.2
Tars	104	16.3
Contaminated soils	69	10.8
75-ft deep chemical solutions and scrap metals mixed with contaminated soil	63–74	9.9–11.6
30- to 40-ft deep landfill with 90 to 95% waste in metal drums	90	14.1

Source: Bromwell, 1978; Collins, 1977; Landva and Clark, 1990; Oweis and Khera, 1986; Peirce and Witter, 1986; Sargunan et al., 1986; Shoemaker, 1972; Sowers, 1973; Tchobanoglous et al., 1977

2.3.1 Unit Weight Determination

Municipal waste is a multiphase material. Its unit weight can be determined using a phase diagram, as shown in Figure 2.5. Let the weights of various components be represented by $W_1, W_2, \ldots, W_i, \ldots, W_n$, and the corresponding unit weights be given by $\gamma_1, \gamma_2, \ldots, \gamma_i, \ldots, \gamma_n$. The volume of each of the components will then be $V_1 = \dfrac{W_1}{\gamma_1}, V_2 = \dfrac{W_2}{\gamma_2}, \ldots, V_i = \dfrac{W_i}{\gamma_i}, \ldots, V_n = \dfrac{W_n}{\gamma_n}$.

Total unit weight
$$\gamma = \frac{W_1 + W_2 + \cdots + W_n}{V_1 + V_2 + \cdots + V_n} = \frac{W}{\dfrac{W_1}{\gamma_1} + \dfrac{W_2}{\gamma_2} + \cdots + \dfrac{W_n}{\gamma_n}}$$

$$\gamma = \frac{1}{\displaystyle\sum_1^n \frac{W_i}{W \cdot \gamma_i}} \tag{2.16}$$

Figure 2.5
Phase diagram for
multicomponent waste

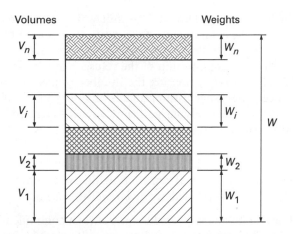

The unit weight of refuse varies widely because of the large variations in the waste constituents, state of decomposition, degree of control during placement (such as thickness of daily cover or its absence), amount of compaction, total depth of landfill, the depth from which a sample is taken, amount of water or leachate present, etc. Household refuse (garbage) usually weighs more than industrial waste. Table 2.8 shows the range of uncompacted and compacted unit weights for municipal solid waste components.

The waste components play a major role in the unit weight. However, the range of unit weights for a given waste component in uncompacted state is also quite large, as seen in Table 2.8, because these components may have voids within them. The extreme case is that of metals, where unit weight varies from 3 to 70 lb. For example, the unit weight of a soda can is about 3 lb/ft^3; if the same can is crushed flat, its unit weight would be much higher.

Since most of the components of the municipal waste stream are porous, compaction processes will reduce the voids within an individual component (intraparticle voids) as well as voids between various components (interparticle voids). For well-operated landfills, the layer thickness would usually be around 2 feet, yielding a typical unit weight of about 35 lb/ft^3 (5.4 kN/m^3). For larger thickness, the unit weight drops substantially.

The volume reduction achieved and the resulting unit weights from compaction are dictated by the intraparticle and interparticle voids and the degree to which they are saturated. The unit weight of the refuse, as placed, may be estimated if the components of the waste stream are known.

The effect of the number of passes through the compaction process depends on the composition of the refuse. For a waste layer composed mainly of old rubber tires, increasing the number of passes rarely makes a significant difference. The usual assumption is that beyond four to five passes further compaction is generally insignificant. In well-operated landfills, an experienced operator can judge the optimum number of passes.

Example 2.3

Estimate the unit weights of refuse for the 1960 U.S. average (excluding cover) for uncompacted, normal compaction, and well-compacted states. The percentage of materials generated are given in Table 2.11.

Solution: The uncompacted weight of each component is determined using Eq. 2.16. The computations are shown in Table 2.11, with values for uncompacted state in column 4. The ratios of compacted weight to uncompacted weight from Table 2.8 were used to determine the weights for the normal compaction and well-compacted states; these are tabulated in columns 5 and 6, respectively, of Table 2.11.

The calculated unit weight for well-compacted refuse (excluding soil cover) as shown in Table 2.11 is 33.4 lb/ft^3 (5.25 kN/m^3). If 10% of the volume of the landfill is assumed to be made up of soil cover material with a density of 100 pounds per cubic foot, then the unit weight will be 40.06 lb/ft^3 (6.29 kN/m^3). Assuming a 15% average moisture content of the landfill material, the moist unit weight will be 46.07 lb/ft^3 (7.23 kN/m^3). According to Table 2.10, this

Table 2.11   Estimating unit weights of refuse for the 1960 U.S. average

Waste component (1)	Percent total waste (2)	Assumed uncompacted unit weight (lb/ft^3) (3)	$\dfrac{W_i}{W \cdot \gamma_i}$ Uncompacted (4)	Normal compaction (5)	Well-compacted (6)
Food waste	13.9	18	0.0077	0.0027	0.0026
Paper and paper board	34.1	4	0.0853	0.0189	0.0138
Plastics	0.5	4	0.0013	0.0002	0.0001
Textiles	1.9	4	0.0048	0.0008	0.0007
Rubber and leather	2.3	9	0.0026	0.0008	0.0008
Yard waste	22.8	7	0.0326	0.0081	0.0065
Wood	3.4	15	0.0023	0.0007	0.0007
Glass	7.6	12	0.0063	0.0037	0.0025
Metals	12	12	0.0100	0.0023	0.0019
Ash, brick, dirt	1.5	30	0.0005	0.0004	0.0004
$\sum_1^n \dfrac{W_i}{W \cdot \gamma_i}$			0.1534	0.0386	0.0299
Average unit weight Eq. 2.16, lb/ft^3 (kN/m^3)			6.53 (1.03)	25.83 (4.06)	33.40 (5.25)

represents good to excellent compaction. If there is a leachate buildup within the landfill, the saturated unit weight below the leachate level may reach as high as 80 lb/ft^3 (12.6 kN/m^3).

Landfills with large percentages of rubble (bricks, broken concrete, etc.) have high unit weights. For typical mixed domestic and industrial sources, the use of 40 lb/ft^3 (6.28 kN/m^3) is reasonable. The refuse has also the capability to absorb water in addition to its original moisture content. This additional water has been estimated at 1.8 in./ft (Fenn et al., 1975) or about 9.4 lb/ft (136.6 N/m). Thus for the waste depicted in Table 2.11, the wet unit weight may range from 37 lb/ft^3 (5.8 kN/m^3) to 43 lb/ft^3 (6.8 kN/m^3) depending on the degree of compaction. In some old landfills underlain by clayey soils, leachate mound could develop inside the landfill. Beneath the leachate mound a unit weight of refuse of 65 lb/ft^3 (10.2 kN/m^3) to 70 lb/ft^3 (11 kN/m^3) is usually used. Another method for determining unit weight of MSW based on the consolidation characteristic of underlying soils is given in Chapter 4.

2.3.1.1 Test Pits Test pits are typically dug 10 feet deep unless the presence of free water or leachate makes that depth impractical. About 10 to 12 cubic yards of material are excavated from each test pit, and the weight and volume of the waste removed are determined. The volume of the test pit is determined either by approximate measurement or by lining and filling it with water. Because of the large size of the excavated specimen, it is difficult to determine its water content. Landva and Clark (1990) suggest using a properly vented pottery type of kiln.

A large number of test pits may be needed because of the variation in landfill composition. The results usually show considerable scatter due to the extreme heterogeneity of the waste components, the difficulty in computing the volume of test pits (which usually have very jagged sides and bottom), and the inaccuracies in determining the weight of the garbage removed. One of the limitations of the test pit is that only the materials near the surface of the landfill can be investigated.

2.3.1.2 Field Test Sections In this technique, an area about 300 feet long and 150 feet wide is filled with refuse, using usual daily field parameters (layer thickness, compaction equipment and passes). The weight of refuse is determined from the number of trucks delivering the refuse and the volume by optical surveys of dimensions. This technique is reasonably reliable for assessing the unit weight prior to advanced decomposition. The results are typically 35 to 45 lb/ft^3 (5.5 to 7.1 kN/m^3) for average compaction.

2.3.1.3 Refuse Inventories For a landfill where good records are kept, the weight of refuse dumped can be reasonably assessed from refuse inventories. The volume is determined from optical surveys and approximate estimates of the settlement of the foundation where it consists of compressible materials.

2.3.1.4 Sampling   For refuse in an advanced state of decomposition, tube samples could be obtained in a manner similar to soil sampling and the unit weight determined using standard procedures. However, the presence of materials that do not decompose, such as bricks, rocks, glass, and metals, may make it impractical to obtain specimens in this manner.

2.3.1.5 Drilling   Under suitable conditions, large-diameter holes may be drilled using coring buckets. The unit weight is determined by weighing the refuse removed and estimating the volume of the hole. Figure 2.6 shows some total unit weights determined using this method. The in-place unit weights were established using a coring bucket 12 in. (304 mm) in diameter. Immediately after removal from the hole, the materials were weighed. Volumes were estimated by assuming a diameter slightly larger than the bucket to account for imperfections in the diameter of the hole and the height of the bulk sample. Thus the indicated densities may be slightly higher than those actually existing. The conclusions derived from the data in Figure 2.6 are as follows.

1. The unit weight of the refuse material increased with increasing depth.

2. The wet unit weight of the newer fill was slightly higher than the wet unit weight of the older fill, mainly because of the higher moisture content of the newer fill.

3. The dry unit weight of the newer fill and the older fill were approximately equal.

Large-diameter augers have also proved effective in advancing the hole. Landva and Clark (1990) reported a 130-mm-diameter solid stem auger with a 140 mm bit to be most effective. However, such an auger must be withdrawn completely to obtain samples.

Figure 2.6
Total unit weight of landfill materials from drilling

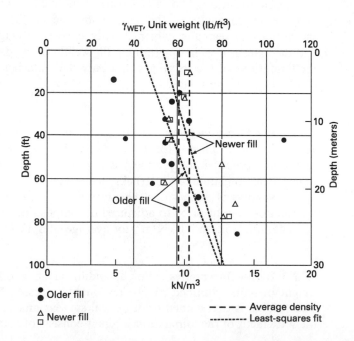

2.3.1.6 Moisture Content   The range of moisture content for municipal waste is 15 to 40% of dry weight. Landva and Clark (1990) reported a moisture content range of about 15 to 120% for landfills across Canada, noting that the moisture content increased with increasing organic contents. The moisture content is usually determined by heating a sample at 40°C to 60°C for an hour or more to avoid burning of waste. However, when the organic contents of the refuse are to be determined, a much higher temperature (440°C to 750°C) may be needed. Typical ranges for refuse components are provided in Table 2.8.

2.4   MINERAL AND INDUSTRIAL WASTE

The index properties of mineral and industrial wastes are affected by many factors, such as the nature of the ore, the chemical process used, the age of the fill, groundwater conditions, organic materials present and their nature, chemical and biological degradation, leaching, and climatic conditions. For example, the properties of fly ash vary from plant to plant, and even in the same plant they may change from time to time. Several factors—such as the type of coal, type of furnace, burning temperature, etc.—contribute to these variations.

Particle size distribution curves for some of these materials are given in Figures 2.7 and 2.8, and index properties are shown in Table 2.12. As shown in Figure 2.7, the particle sizes of coarse coal waste, incinerator residue, and bottom ash are similar to sand and gravel mixes. Similarly, as seen in Figure 2.8, fly ash, fine coal, and by-product gypsum have grain-size-distribution characteristics similar to sand and silt and plasticity characteristics of silt (as shown in Table 2.12).

Only the average grain size distribution curve is shown for each of the materials. Note that the actual grain size curves may vary considerably from those shown. Even for a given material, the curves may vary from source to source; see the two curves for the bottom ash in Figure 2.7. As an example, some statistical data for grain size for Class F and Class C fly ashes are shown in Table 2.13 (p. 31).

Figure 2.7
Particle size distribution of mineral waste

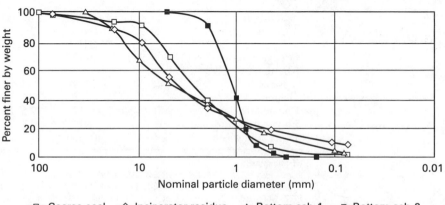

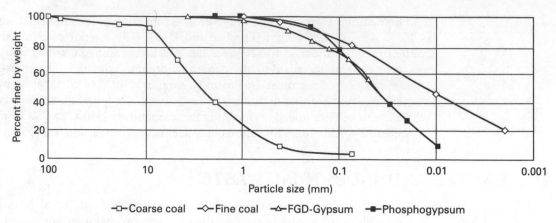

Figure 2.8 Particle size distribution of mineral waste

The sand tailings from mining have a void ratio ranging between 0.5 and 1.5. Their relative density may range from 10 to 70%. Slimes from bauxite, phosphate, and tar sand have much higher void ratio that often exceeds 8. Alberta oil sands are medium to fine, but during extraction 8 to 25% clays and silts become part of the tailings. Such fine in the tailings are of primary concern because they occupy large volume. Volume reduction is achieved

Table 2.12
Index properties
of mineral waste

Description	LL	PI	G_s	pH
Combined coal waste			2.1–2.6	
Coarse coal waste	2–20	NP	1.75–2.7	
Fine coal waste	20–65	NP	1.4–2.22	
Fly ash		NP	2.1–2.65	5.2–12.3
Boiler slag		NP	2.5–2.75	
Bottom ash		NP	2.30–2.80	
Municipal incinerator residue		NP	2.5–2.9	9–12
FGD				
Sulfite-rich	24–65	8–20	2.46–2.72	>12
Sulfate-rich	20–22	NP	2.2–2.5	>10
Gypsum		NP	2.3	7
Dredged waste	0–270	0–185	2.4–2.8	
Alumina red mud				
Fines	41–46	7–39	2.84–3.27	
Sand			3.16–3.27	
Phosphatic clay	76–245	41–175	2.5–2.9	7–8
Phosphogypsum		NP	2.3–2.53	2.8–6
Paper mill waste	70–413	40–380	1.9–2.3	

NP: nonplastic.

Source: Acar et al., 1990; Bromwell and Oxford, 1977; Chae and Gurdziel, 1976; Charlie, 1977; Hagerty et al., 1977; Holubec, 1976; Krizek et al., 1987; Lum and Tay, 1992; Mabes et al., 1977; Maynard, 1977; Moulton et al., 1976; Rollings, 1994; Somogyi and Gray, 1977; Taha and Seals, 1992; Ullrich and Hagerty, 1987

Table 2.13 Statistical grain-size data for fly ash

Particle size (mm)	CLASS F FLY ASH			CLASS C FLY ASH		
	Mean value	Standard deviation	Coefficient of variation %	Mean value	Standard deviation	Coefficient of variation %
D_{85}	0.079	0.063	80.0	0.063	0.020	31.7
D_{50}	0.023	0.015	66.9	0.022	0.011	50.0
D_{15}	0.0057	0.0048	64.8	0.0084	0.0082	97.6

Source: McLaren and DiGioia, 1987

through various means, such as reducing the amount of fines being processed, modification of extracting process to yield higher solid contents, use of additives such as lime to change segregating mixtures to nonsegregating mixtures, and reclaiming accumulated fines (Morgenstern and Scott, 1995).

Organic contents of paper mill waste may range from 40 to 90% with water content ranging between 45 and 740%. The liquidity index from field sludge samples has been reported to range from 1 to 1000. The deposits with the higher fiber contents showed the largest values (Charlie, 1977).

2.5 DREDGED WASTE

About 80 to 85% of dredged materials are particles smaller than sand (Boyd et al., 1972), and 4 to 8% of the fines may be organic silts and clays (Krizek and Salem, 1977). Where dredged waste is pumped into containment areas in the vicinity of disposal pipes, sands dominate with little or no clay-size particles. The effective size (D_{10}) decreases as the distance from the inlet pipes increases. Index properties of these materials are shown in Table 2.14.

The dry unit weight for dredged waste ranges from 54 lb/ft^3 (8.5 kN/m^3) to 72 lb/ft^3 (11.3 kN/m^3) (Krizek and Salem, 1977). Atterberg limits test data by Bromwell (1978) showed that on the plasticity chart all the points plot close to A line, except for in a brackish water environment where the liquid limit was

Table 2.14 Index properties of dredged material

Description	Sand %	Silt %	Clay %	Organic %	LL	PI	G_s
Toledo, OH	14–19	46–50	31–40	4–8	55–95	12–57	
Mobile, AL	7	18	75	5	100	35	2.72
Average U.S.	16	50	33		80–120	15–55	2.65
West Germany	0–5	35–60	10–40				
Seawater	5–95	5–70	0–40		45–212	25–65	
Newark Bay, NJ	30	48	22				
Arthur Kill, NY	24–54	17–56	14–20				

Source: Bromwell, 1978; Carrier III et al., 1984; Haliburton, 1977; Krizek and Salem, 1977; Lacasse et al., 1977; Long and Demars, 1987; Rizkallah, 1987

somewhat higher. Murdock and Zeman (1975) reported a considerable decrease in the plastic properties when a soil was either air dried, oven dried, or freeze dried. The collapse of swelling minerals, the decomposition of organics, and oxidation of iron was believed to be responsible for the reduced plasticity. The drop in the plastic properties has been reported for natural organic soils as well (Terzaghi and Peck, 1967).

2.6   SUMMARY

In this chapter we defined weight-volume relationship and index properties for natural soils and waste materials. We described classification of soils for engineering purposes and the USDA system. We examined several methods of determining unit weight of municipal wastes, their advantages, and their shortcomings. For municipal waste, we saw that unit weights may vary from 18 to 90 lb/ft^3 (2.8 to 14.1 kN/m^3) depending on the constituents of the waste, proportion of cover soil present, moisture content, and relative compaction. We also tabulated index properties of mineral, industrial, and dredged waste, noting that considerable variations in these properties exist due to the large number of factors that affect these properties.

NOTATIONS

θ	volumetric moisture content	γ'	buoyant unit weight of a material	V	total volume of material
A	activity			V_i	volume of waste component i
C_c	coefficient of curvature	γ_d	dry unit weight	V_s	volume of solids
C_u	coefficient of uniformity	γ_i	unit weight of waste component i	V_v	volume of voids
D_{10}	nominal particle diameter for 10% finer by weight, or effective particle diameter	G_s	specific gravity of solids	V_w	volume of water
				W	total weight of components
		γ_s	unit weight of solids	w	water content (%)
D_{30}	nominal particle diameter for 30% finer by weight	γ_w	unit weight of water	W_a	weight of air ≈ 0
		h_1	height (depth) of landfill	W_i	weight of waste component i
D_{60}	nominal particle diameter for 60% finer by weight	LL	liquid limit	w_n	natural water content
		n	porosity	W_s	weight of solids
e	void ratio	PI	plasticity index	W_w	weight of water
γ	total unit weight of a material (F/L^3)	PL	plastic limit		
		S	degree of saturation		

PROBLEMS

2.1 MSW materials discarded in the year 2000 are projected in Table 2.15. Determine the total dry unit weight for both the normal compaction and well-compacted states. What will be its unit weight if its moisture content were 12%?

Table 2.15
Projected MSW
for 2000

Waste component	Percent of total waste
Food waste	8.5
Paper and paper board	32.8
Plastics	14.2
Textiles	3.9
Rubber and leather	3.9
Yard waste	11.0
Wood	9.3
Glass	5.7
Metals	6.9
Ash, brick, dirt	3.7

Source: EPA, 1992

2.2 Using the ASTM soil classification system and Table 2.12, Figure 2.7, and Figure 2.8, assign group symbols and group names to the following soils.
 (a) Incinerator residue
 (b) Poorly graded gravel with sand and silt
 (c) Bottom ash 1
 (d) FGD gypsum

2.3 Laboratory analysis of FGD waste indicated the following data. For sulfite-rich waste, 100% of the material passed a No. 40 sieve, 98% passed a No. 200 sieve, its liquid limit was 38, and its plasticity index was 9. For sulfate-rich waste (i.e., gypsum), 100% of the material passed a No. 4 sieve, 83% passed a No. 200 sieve, and it was nonplastic. What are the appropriate group names and symbols for these wastes?

2.4 Grain size distributions for two municipal waste incinerator residues are given in Table 2.16. Draw their grain size distribution curves and determine their group names and symbols. What is the major difference between these two materials?

2.5 Describe how to determine the unit weight of municipal solid waste which is (a) yet to be placed in the landfill, (b) less than a year old since placement, (c) believed to be completely decomposed?

Table 2.16
Grain size distributions

Sieve	Opening (mm)	Residue 1	Residue 2
1	25.0		100
3/4	19.0	100	
3/8	9.5	76	44
No. 4	4.75	48	35
No. 10	2.00	29	25
30	0.85	22	18
40	0.425	15	12
140	0.106	4	7
200	0.075	3	6

2.6 Which of these two processes would you recommend for flue gas desulfurization? Give reasons for your recommendation.

(a) A process where sulfide-rich sludge waste is produced

(b) A process where sulfate-rich waste (gypsum) is generated

2.7 Index properties of two clays are given in Table 2.17. Determine the activity of these materials and comment on their reactivity.

2.8 Reclassify the materials of Problem 2.2 using the USDA system.

Table 2.17
Index properties

Soil	LL	PL	Percent $<2 \mu m$
Kaolinite	60	35	88
Montmorillonite	430	60	91

3 Clay Minerals

3.1 INTRODUCTION

Clays play a primary role in reducing hydraulic conductivity of soils used in the construction of liners and slurry walls for containment of new or existing waste disposal facilities. The engineering behavior of a clay mass is determined by the types of clay minerals present, the nature of the pore fluid, the interparticle forces, and applied stresses.

Fine-grained soils such as silts and clays have particle sizes less than 75 μm and plastic characteristics, due to the presence of clay minerals. Clay minerals have a crystalline structure, an equivalent diameter of approximately 2 μm, are generally platelike in shape (although some are tubular), carry a net negative charge, behave like colloids, and have a large surface area with a high percent of constituting molecules distributed on the surface. The *specific surface* is defined as the surface area per gram of mass, and it ranges from 5 to 800 m^2/g for clays, 1 to 0.4 m^2/g for silt, and 0.04 to 0.001 m^2/g for sand. Particles where the specific surface is large and the electrical forces dominate the mass forces are classified as colloids.

3.2 BONDS

The two types of bonds that are important in the study of clay minerals are *primary bonds*, which hold together the atoms, and *secondary bonds*, which hold together water molecules of adjacent sheets of crystalline lattice and affect its mineral characteristics.

3.2.1 *Primary Bonds*

Primary bonds can be either ionic bonds or covalent bonds. In *ionic bonds*, elements release or gain electrons in the outer shell of their atoms. In *covalent bonds*, two atoms share the valence electrons in their outer shells. If the centers of the bonded ions in a molecule do not coincide, the bonds are known as *polar covalent bonds*. Such molecules have both a positive and a negative charge and are known as polar molecules, or *dipoles*. Water is an example of a dipole. If the arrangement within the molecule is symmetric, the polar covalent bonds yield nonpolar molecules. Carbon tetrachloride is one such example.

3.2.2 *Secondary Bonds*

The force of attraction between dipoles is called the *dipole force*. If the positively charged end of the participating dipole is hydrogen, then the attracting force is called a *hydrogen bond* and is stronger than an ordinary dipole bond.

Properties of clay minerals and their interaction with water are significantly influenced by the hydrogen bond. *Van der Waals* bonds are due to fluctuating dipolar bonds and are relatively weak and nondirectional. However, they may contribute to cohesion. Secondary bonds and electrostatic forces are greatly affected by applied stresses.

3.3   CLAY MINERALS

In its idealized form, the crystalline structure of a clay mineral is composed of two basic building blocks: a silica tetrahedron and an aluminum octahedron. The silica tetrahedron (SiO_4) consists of a silicon atom (Si^{4+}) surrounded by four oxygen atoms (O^{2-}). Silica tetrahedra share their oxygen atoms to form either chains or sheets. A silica sheet is formed when the shared oxygen atoms in the tetrahedra lie in a plane and have hexagonal holes, as shown in Figure 3.1.

The aluminum octahedron consists of aluminum (Al^{3+}) or magnesium (Mg^{2+}) surrounded by hydroxyls, as shown in Figure 3.2. Sheet structure of

Figure 3.1
A silica tetrahedron
and a silica sheet

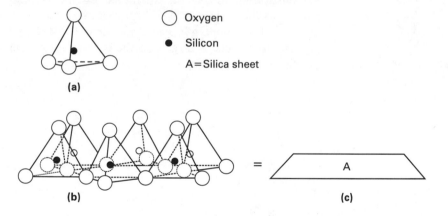

Figure 3.2
An octahedron
and an octahedron
sheet

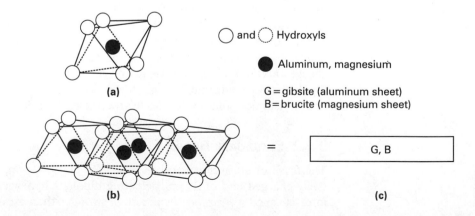

octahedra units is formed by the sharing of hydroxyls and is represented by rectangles in the figure. If only aluminum is present, it is called gibbsite $[Al_2(OH)_6]$; if only magnesium is present, it is called brucite $[Mg_3(OH)_6]$. Various clay minerals are formed as these sheets stack on top of each other with different ions bonding them together. The ions on the surface of a clay particle may be O^{2-} or $(OH)^-$, thus giving it a net negative charge. According to Bailey (1980), "Clay minerals belong to the family of phyllosilicates and contain continuous two-dimensional tetrahedron sheets linked by sharing three corners of each and the fourth corner pointing in any direction. The tetrahedron sheets are linked in the unit structure to octahedron sheets, or to groups of coordinated cations, or individual cations."

If clay particles fracture, other electrical forces may develop at the edges such that the internal ions, which are usually positive in charge, are exposed. The edges may then attract dipoles, anions, or the negatively charged faces of other clay particles. Different clay minerals are formed if during this interaction the normal locations of aluminum, magnesium, or silicon ions are occupied partially or wholly by other ions. This substitution of one kind of ion with another without a change in the crystal structure is known as *isomorphous substitution*. The isomorphous substitution causes a change in the net negative charge and may result in some distortion of crystal lattice. This charge deficiency is balanced by the external adsorption of cations at the surface of clay particles and between the layers.

The properties of clay minerals can be altered by exchanging the externally adsorbed ions or by water adsorbed on particle surface. Externally adsorbed cations can be replaced by other cations, which is known as the *ion exchange*. The replacing power for some of the commonly occurring cations in soil minerals increases from sodium to iron:

$$Na^+ < K^+ < NH_4^+ < Mg^{2+} < Ca^{2+} < Al^{3+} < Fe^{3+}$$

The ability of a mineral to adsorb external cations is known as *cation exchange capacity* (CEC). The CEC of a soil is the number of cations in milliequivalents that neutralize one hundred grams of dry clay (meq/100 g). One milliequivalent is one milligram of hydrogen or any ion that will combine with one milligram of hydrogen or displace it.

Some of the characteristics and index properties of clay minerals are given in Table 3.1. Note that the values reported in the literature vary widely even for the same exchangeable cation. This is due to a variety of factors that contribute toward the cation exchange capacity — broken bonds at particle edges, electrolyte concentration, temperature, organic contents, and replacement of hydrogen from hydroxyl. Therefore, theoretical CEC often varies from the measured or actual value.

3.3.1 *Kaolinite*

Kaolinite is formed in the weathering of orthoclase feldspar under good drainage conditions with low pH and is found in the southern United States. It has

Table 3.1
General characteristics of clay minerals with inorganic salts

Mineral	CEC meq/100 g	Specific gravity	Specific surface m^2/g	Liquid limit	Plastic limit
Kaolinite	3–15	2.6–2.68	10–20	30–60	25–35
Na				53	21
Ca				38	11
Illite	10–40	2.6–3.0	65–100	60–120	35–60
Na				61	34
Ca				90	40
Montmorillonite	80–150	2.35–2.7	700–840	100–900	50–100
Na				700	97
Ca				177	63

Source: Grim, 1968; Lambe and Whitman, 1969

a low swelling potential, low liquid limit, low activity, and yields hydraulic conductivity of 10^{-6} cm/s or higher.

Kaolinite crystal layers consist of tetrahedron and octahedron sheets that are held together by the strong hydrogen bond between oxygen and hydroxyl. A schematic of kaolinite is shown in Figure 3.3. When kaolinite sheets are stacked on each other, the hydroxyl of octahedron sheets are attracted to the oxygen of the silica tetrahedron sheet by means of oxygen bonds. Such bonds (which are ionic and covalent) are strong, but not as strong as the primary bonds and, therefore, cleavage occurs. These sheets can extend greatly in two directions and typically such crystals are 70 to 100 layers thick.

Since kaolinite contains a tetrahedron sheet, or sheet A, and an octahedron sheet, or sheet B, it is called a two-layer sheet or a one-to-one (1:1) mineral. The kaolinite platelets are about 0.05 micron thick, have a diameter-to-thickness ratio of about 20 and are usually hexagonal in shape. The structural formula is $(OH)_8Si_4Al_4O_{10}$, where one Si in 400 is substituted by Al.

3.3.2 *Illite*

Illite is very stable and is common in soils and sediments. It has a moderate swelling potential, a liquid limit higher than kaolinite, yields hydraulic conductivity of 10^{-7} cm/s or lower, and is considered suitable material for landfill liners.

Figure 3.3
Kaolinite structure.
(a) Basic building block,
(b) stacked blocks
forming particles

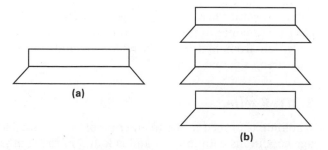
(a)

(b)

Figure 3.4
Illite structure.
(a) Basic building block,
(b) stacked blocks
forming particles

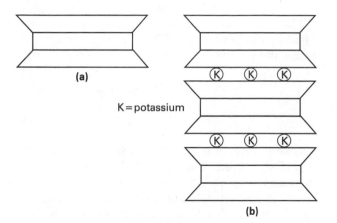

Illite crystals, as shown in Figure 3.4, consist of an aluminum sheet between two silica sheets; thus illite is a 2 : 1 mineral. Potassium ions occupy the space between the adjacent sheets and bind them together with a strong bond. There is considerable isomorphous substitution of silica with Al, Fe, and Mg, which is partly balanced by interlayer potassium. The illite particles are about 10 nm thick with a diameter-to-thickness ratio of about 50. The structural formula is $K(Al, Mg, Fe)_{2,3}(Si)_4O_{10}(OH)_2$, where the amount of potassium present varies.

3.3.3 Smectite

Montmorillonite, which is a member of the smectite group, is formed from weathering of volcanic ash under poor drainage conditions or in marine waters. The basic building sheets for smectite, another 2 : 1 mineral, are the same as for illite (i.e., sheets ABA) except there is no potassium ion present (see Figure 3.5). The bond holding the sheets is due to Van der Waals forces and exchangeable ions. It is a very weak bond and easily broken by water or other polar or cationic organic fluids entering between the sheets. There is extensive substitution of silica and alumina, resulting in considerable charge deficiency.

Figure 3.5
Smectite structure.
(a) Basic building block,
(b) stacked blocks
forming particles

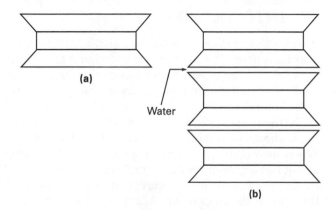

In montmorillonite, one trivalent aluminum ion in six is replaced with a bivalent magnesium ion. The resulting charge deficiency is balanced by an externally adsorbed sodium ion. The structural formula is

$$(OH)_4Si_8(Al_{3.34}Mg_{0.66})O_{20}.$$
$$\downarrow$$
$$Na_{0.66}$$

The large amount of unbalanced substitution in the montmorillonite results in high CEC. It has a particle thickness of about 0.1 nm with a diameter-to-thickness ratio of 100–400. Water molecules enter easily between the layers, which expand considerably and yield much smaller particles with a very large specific surface. Swelling potential, activity, and liquid limit are the highest in this group of clays. Sodium ions can readily be replaced by other ions. Montmorillonite where calcium is the externally adsorbed ion has lower cation exchange capacity and swelling potential. Bentonite, commonly used for advancing boreholes in subsurface investigation and slurry wall construction, contains both sodium bentonite and calcium bentonite but the proportion of the former is higher (Khera, 1995).

3.3.4 Attapulgite

Attapulgite has a double chain structure formed from bonds of silica tetrahedron sharing two of the four oxygens. Its particles are lathlike in shape, 4 to 5 μm long, and 1 to 5 nm in diameter. Its CEC is less than 5 meq/g. Sepolite (meerschaum) differs from attapulgite due to replacement within its structure.

3.3.5 Allophane

Poorly crystalline clay materials are known as allophanes and have no definite composition.

3.4 CLAY-WATER INTERACTION

When a clay particle is placed in water, a high concentration of cations occurs near its surface. This may be due to one or more of the following:

1. the release of adsorbed cations from the clay surface

2. the dipolar water molecules interacting with the surface oxygen of the clay particles

3. the release of hydrogen from the ionization of a hydroxyl group

4. the dissolving of precipitated salts, if present

When the pH (pH = $1/\log_{10}H^+$) value of a clay particle is high, the hydroxyl group has a greater tendency to form hydrogen ions. The released cations try to move away from the clay particle to equalize the charge distribution throughout the suspension. Attraction from the negatively charged surface of

Figure 3.6
Clay particle surface showing the diffused double layer and particle interaction

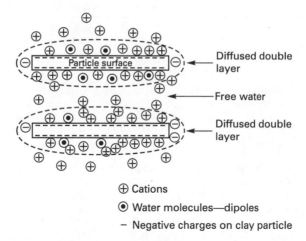

⊕ Cations

◉ Water molecules—dipoles

− Negative charges on clay particle

the particle restricts this movement, resulting in only cations adjacent to the clay surface, with a few anions near the edge (Figure 3.6).

The negative charges on the clay surface and the balancing cations surrounding it are together called the *diffuse double layer* (also called *bound water*). The thickness of the diffuse double layer depends on the cation valence, electrolyte concentration, temperature, and dielectric constant (the measure of ease with which molecules can be polarized and orientated in an electric field) of the medium.

The theoretical expression for distribution of ions adjacent to negatively charged surfaces is due to Gouy-Chapman. The thickness of the double layer is given by

$$t_{dl} = \sqrt{\frac{\varepsilon kT}{8\pi\eta e^2 v^2}} \tag{3.1}$$

where ε is the dielectric constant of the medium (i.e., the ratio of electrostatic capacity of condenser plates separated by the given medium to that of the same condenser with vacuum as the medium), k is Boltzmann's constant, T is the absolute temperature, η is the electrolyte concentration, e is the unit electronic charge, and v is the cation valence.

Values of the dielectric constant for some organics are given in Table 3.2. Figure 3.7 shows the free expansion of some clay minerals as influenced by the dielectric constant of the medium in which they are suspended. Sodium montmorillonite with an activity of 4.5 shows the highest free expansion with water, which has a dielectric constant of 80. However, the drop in free expansion is also the largest for this mineral as the dielectric constant drops to zero.

The change in free expansion of calcium montmorillonite with an activity of 2.8 is much less for the corresponding changes in dielectric constant. However, both montmorillonite clays show shrinkage with decreasing dielectric constant and it is noteworthy that sodium montmorillonite, which is most

Table 3.2
Properties of organic
liquids and water (20°C)

Description	Unit weight (g/cm³)	Water solubility (g/l)	Dielectric constant	Dipole moment (debye)
Water	0.98	∝	80.4	1.83
Hydrocarbons and related compounds				
Heptane	0.68	<0.03	2.0	0
Benzene	0.88	0.7	2.28	0
Xylene	0.88	<0.3	2.4	0
Trichloroethylene	1.46	1.0	3.4	0.9
Tetrachloromethane	1.59	0.8	2.2	0
Nitrobenzene	1.20	2	35.7	4.22
Alcohols and phenols				
Methanol	0.79	∝	31.2	1.66
Ethanol	0.79	∝	25.0	1.69
Phenol	1.06	∝	13.13	1.45
Aldehydes and ketones				
Acetone	0.79	∝	21.4	2.74
Acids				
Acetic acid	1.05	∝	6.2	1.04
Base				
Aniline	1.02	36	6.9	1.55

commonly used in containment structures, is also the most susceptible to shrinkage.

On the other hand, for a kaolinite with an activity of 0.32, free expansion increases with decreasing dielectric constant, which indicates that it would be less affected by waste chemicals. In addition to lower activity, kaolinite has low CEC and low swelling potential compared with montmorillonites.

The thickness of the double layer will also decrease as the cation valence and electrolyte concentration increase, as depicted in Figure 3.6. In a waste environment, if Ca^{2+} or Mg^{2+} replace some of the Na^+ and K^+ present in impervious soil barriers, the double layer thickness will decrease and the hydraulic conductivity will increase, with virtually no change in void ratio.

Figure 3.7
Effect of dielectric
constant on free
expansion (after
Acar et al., 1985;
Acar and Olivieri, 1989)

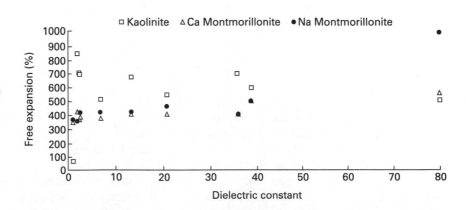

Although the theory helps us understand the interparticle response, it has limitations. First, it assumes that all the particles are of a single mineral and of the same size; this of course is not the case for natural soils. Also, the average particle spacing in a clay mass is at the lower limit of colloidal theory and the other forces in play are not considered. In addition, the assumed concentration of mineral matter in a soil fluid is much less than in a natural soil deposit. Consequently, there is a greater amount of physical and electrical interference between neighboring particles.

3.5 CLAY STRUCTURE

The arrangement of particle groups and pore spaces in a soil is generally referred to as *fabric*. The term *structure* is used to include fabric, composition, and interparticle forces. The clay particles interact through the layers of adsorbed water (i.e., the water strongly held by the soil mineral) and through the diffused layer of exchangeable ions (i.e., through attraction of dipolar water to the electrically charged soil and to the cations in the double layer, which are themselves attracted to the clay surface as shown in Figure 3.6).

The forces between adjacent particles are (a) repulsive forces between the double layers, (b) attractive Van der Waals forces and the electrical forces, and (c) bonding due to organic and inorganic materials present on particle surfaces (see Figure 3.8). The geometric arrangement of particles depends on whether

Figure 3.8
Attractive, repulsive, and net interparticle forces double layers of different thickness (Adopted from Scott, 1963)

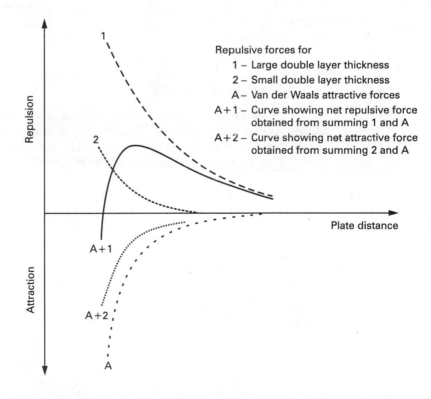

Repulsive forces for
1 – Large double layer thickness
2 – Small double layer thickness
A – Van der Waals attractive forces
A+1 – Curve showing net repulsive force obtained from summing 1 and A
A+2 – Curve showing net attractive force obtained from summing 2 and A

the net result of all the forces is repulsion or attraction. Changes in the thickness of the double layer subsequent to soil formation can alter the net particle forces, which will change soil properties such as strength, compressibility, and hydraulic conductivity without any change in void ratio.

The two elementary structures of clay particles are the dispersed structure—formed when the net particle force is repulsive—and the flocculated structure—formed when the net force is attractive. The actual structure of a natural clay deposit is much more complex. For further details, see Mitchell (1993).

3.5.1 Dispersed Structure

Repulsion is the result of interpenetration of a diffuse double layer. The repulsion is large with distilled water or monovalent exchangeable ions as pore fluid since the diffuse double layer is thicker. In high pH the broken edges of the clay particles ionize negatively, giving further impetus to the repulsive forces. The net force, which is the sum total of Van der Waals attractive forces and the repulsive forces, is repulsive, as represented by the curve (A + 1) in Figure 3.8. Under dominant repulsive forces clay platelets align themselves in a parallel, or face-to-face, orientation forming a *dispersed structure*. Figure 3.9a shows an idealized dispersed structure. Soils deposited in fresh water environment yield dispersed structure. Such deposits exhibit greater swelling and lower hydraulic conductivity. Soils, which are compacted at moisture content 2 to 3% above that corresponding to their dry unit weight, show a greater tendency toward parallel orientation of particles.

3.5.2 Flocculated Structure

With increased ionic concentration or higher-valence ions, the thickness of the diffuse double layer is reduced. Thus the clay particles can come closer to each other before their double layers interact. The net force is then attractive, as depicted by curve (A + 2) in Figure 3.8. If the pH is low as well, the edges

Figure 3.9
Basic soil structures

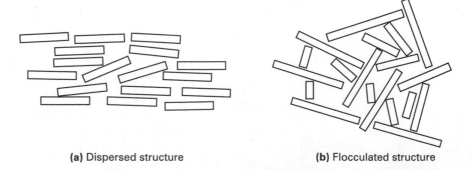

(a) Dispersed structure   (b) Flocculated structure

ionize positively, thus encouraging edge-to-face contacts. The clay particles flock to each other in a random fashion with edge-to-face orientation yielding the *flocculated* structure shown in Figure 3.9b.

3.6   THIXOTROPY

Thixotropy is the property of a clay whereby it gains shear strength at a constant volume with time, the strength is partially or totally lost upon disturbance, and the process is reversible. For example, the formation of gel is attributed to the reorientation of clay particles with time and the fabrication of a framework where there may be some edge-to-face alignment and some edge-to-edge orientation. In clays exhibiting thixotropy, the net force between particles lies somewhere between the curves (A + 1) and (A + 2) shown in Figure 3.8.

The thixotropic gain in strength of a suspension, such as montmorillonite in a slurry, is called *gelation*. The gelation ability of sodium montmorillonite makes it invaluable in sustaining trench spoils in slurry trench construction. Usually, 3 to 7% sodium bentonite suspension is adequate to stabilize a slurry trench and provide the necessary gel strength. However, for calcium-rich bentonite three to four times as much bentonite may be needed.

3.7   ORGANIC COMPOUNDS IN WASTE

Some of the properties of organic compounds were given in Table 3.2. Very few of the organic compounds available today have been investigated for their effect on properties of soils. Some information may be gained, however, by considering each class of organic chemicals rather than each chemical itself. A brief description is given here; for details, see Hart and Schuetz, 1972 and Madsen and Mitchell, 1987.

The simplest form of organic compounds are the saturated hydrocarbons, which are called *alkanes*. Methane, CH_4, is the simplest form of alkane. Increasing the number of carbon atoms yields other compounds in the form of chains (e.g., propane) or rings (e.g., cyclopropane). Removing one of the hydrogen atoms from the parent compound yields fragments of saturated hydrocarbons (the *alkyl* group). This class of compounds is relatively inert, practically insoluble in water, and shows no reaction with dilute acids or bases.

Unsaturated hydrocarbons contain a carbon-carbon double bond (e.g., ethene) in the *alkenes* group and a carbon-carbon triple bond (e.g., acetylene) in the *alkynes* group. They are highly reactive and form many of the compounds that pose environmental hazard. Unsaturated hydrocarbons built around a benzene ring (C_6H_6) are called *aromatic* compounds and exhibit low reactivity.

Alcohols are derived when a hydroxyl group replaces a hydrogen atom in saturated or unsaturated hydrocarbons (e.g., ethyl alcohol), and *phenols* are obtained if a hydroxyl group replaces a hydrogen atom in an aromatic ring (e.g., p-chlorophenol). Alcohols and phenols may be considered as an organic equivalent of water with one of the hydrogen atoms replaced by an organic group. Alcohols with a short carbon chain (e.g., methyl) have high water solubility; this solubility decreases with increasing carbon chain length. Alcohols with more than one hydroxyl group per molecule are called polyhydric alcohols. An example is ethylene glycol, which is a permanent antifreeze.

Ethers are isomers of alcohol. They have low water solubility, but they are good solvents for many organic compounds. Their reactivity is low.

Aldehydes and *ketones* contain a *carbonyl* group and are highly reactive. In aldehydes, one of the two groups attached to carbonyl is hydrogen; in ketones, both groups attached to carbonyl are organics. Formaldehyde and acetone belong to these groups.

Carboxylic acids (*organic acids*) contain a carboxyl group and are acidic because of their ability to donate a proton to more basic substances. These acids are weaker than inorganic acids. Those with lower carbon or molecular weight are soluble in water (e.g., acetic acid); this solubility decreases with increasing molecular weight.

Amines or *organic bases* have structure akin to inorganic base ammonia. They are called primary, secondary, or tertiary amines as one, two, or three of the hydrogen atoms of ammonia are replaced by organic groups. The replacing organic group may be aliphatic (compounds related to saturated or unsaturated open-chain or cyclic compounds, but which contain no benzene group), aromatic, or a combination of the two. *Aromatic amines* often have *aniline* in their names.

3.8 SOIL PORE FLUID INTERACTION

When the properties of the soil pore fluid change due to interaction with pollutants, soil structure and properties will change as well. Changes such as decreasing particle size and increasing specific surface increase the amount of exposed molecules available to interact with the pollutants and thus the changes in the soil structure and properties.

For example, the effect of pH on the water uptake capacity of sodium and calcium bentonites is shown in Figure 3.10. Note that sodium bentonite, which has the smallest particle sizes among clays, shows the most variation in water uptake with pH changes. Thus in slurry wall construction it is recommended that the pH of the slurry be maintained around 8 (Xanthakos, 1979), at which the water uptake capacity is the highest. As the pH value rises above or falls below 8, the water uptake capacity decreases significantly, which results in a decrease in diffuse double layer thickness and, therefore, an increase in shrinkage of soil. Since the pH in the waste environment can vary over a wide range, the use of sodium bentonite presents greater risk for unwanted changes in the properties of impervious barriers.

Another type of soil pore fluid interaction is the replacement of inorganic cations by organic cations. When water-soluble organics invade the clay-water system, water and other cations in the diffused double layer eventually may be

Figure 3.10
Effect of pH on water uptake for two montmorillonites (After Hermanns et al., 1987)

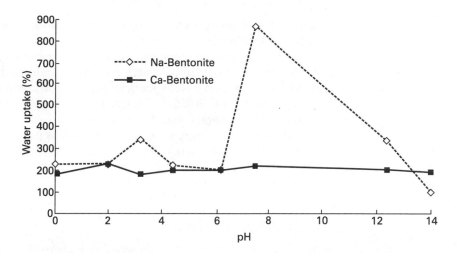

replaced by the organic cations. Clay can also take ions from insoluble substances, such as many of the organics, by exchange and adsorption even in the presence of very small amount of water.

The bond strength of an organic molecule is quite different from that of water. The dielectric constant of the organics is much less than that of water (as seen from Table 3.2), which causes certain changes in the fabric of the soil. Organic molecules can enter the interlayer spaces, causing clay layers to swell or shrink, which in turn changes shear strength and hydraulic conductivity. The rate of exchange is very rapid in kaolinite, but may take a very long time in smectites where much of the exchange capacity resides in the interlayer domain.

Clays of high activity (such as sodium bentonite) repeatedly have shown their susceptibility to changes in hydraulic conductivity when interfacing with waste chemicals. For data where clays other than sodium bentonite have been investigated, see Hermanns et al. (1987), Ryan (1987), Khera and Tirumala (1992), and Khera (1995). The effect of various permeants on different clay minerals will be described extensively in Chapter 6.

Organics with low water solubility, such as hydrocarbons and the related compounds, are not able to displace the pore water and will have little effect on soil properties. Dry clay minerals when mixed with some of the organic compounds may behave like nonplastic material. This is attributed to their low dielectric constant and lack of polar bonds.

Organic acids can dissolve clay minerals. If carbonates and iron oxides are present, they will also be dissolved. The rate of dissolution increases with increasing acid concentration, clay surface area, time of exposure, amount of MgO in the clay mineral, and decreasing particle size. Even if the organic acids are not initially present in the given waste, they may be produced by anaerobic decomposition and eventually will damage the clay minerals in liners and containment systems. Organic bases may dissolve silica sheets and/or change the chemistry of clay by replacing the adsorbed water, by interaction with adsorbed cation, by exchanging interlayer cations, or by interaction with surface oxygen of the clay mineral.

Both organic and inorganic chemicals will have a much greater effect on structures where the initial water content is high, such as in soil bentonite backfill used in slurry walls (compared with the compacted clay liners commonly used in new landfills). In soil liners with low density and high water content, clay minerals and their structure are readily intruded upon and altered. Montmorillonite, which has the smallest particles among all the clays, is most affected by changes in environmental fluids. Organically prepared clays, which are capable of absorbing other organics encountered in various types of landfills, may be a viable alternative (Mitchell, 1993).

3.8.1 *Atterberg Limits*

As discussed in Chapter 2, Atterberg limits are used as index of a soil's consistency. The effect of pollutants on Atterberg limits has not yielded any reliable results for predicting their effect on hydraulic conductivity of soils used in liners and containment structures.

3.9   CRACKING TEST

One indication of soil chemical interaction with contaminants is the *cracking test*. In a cracking test, the soil is mixed with the given contaminant, the slurry is allowed to stand, and the appearance of cracks or lack thereof is observed. Ryan (1987) reported extensive cracking of slurries made from both a treated and a nontreated bentonite with leachate containing hydrocarbons, phenol, acetone, and other organic compounds having individual concentrations of less than 75 ppm. Based on this observation and verification through hydraulic conductivity tests, it was concluded that the hydraulic conductivity of a slurry wall constructed with any of the two bentonites would increase with time. Attapulgite, which has a high specific surface but comparatively lower swelling potential and cation exchange capacity, was tested with the same leachate. The results showed no cracking of the slurry upon drying and no change in hydraulic conductivity.

3.10   SUMMARY

Clay minerals have crystalline structure with silicon tetrahedron and aluminium octahedron as their basic building blocks. When a clay particle is placed in water, a high concentration of cations occurs near its negatively charged surface giving rise to the diffuse double layer. The thickness of the double layer varies directly as the temperature and the dielectric constant, and inversely as the valence and the square root of electrolyte concentration. The thickness also has a strong influence on the soil behavior. Water-soluble organics may replace the cations in the diffuse double layer. The differences in the bond strength of organic molecules and their lower dielectric constant cause changes in the clay fabric. These changes are reflected in the water uptake of clay minerals. Because Atterberg limit tests on contaminated soils do not yield consistent results with respect to engineering behavior of the soil, some researchers have used cracking tests to assess a soil's chemical interaction with contaminants.

NOTATIONS

CEC	cation exchange capacity CEC (meq/g)	k	Boltzmann's constant $(1.38 \times 10^{-16} \text{ erg/K})$	ε	dielectric constant of the medium
e	unit electronic charge $(16.0 \times 10^{-20} \text{ coulomb or } 4.8 \times 10^{-10} \text{ esu})$	η	electrolyte concentration (ions/cm^3)	t_{dl}	thickness of diffuse double layer
PI	plasticity index	T	absolute temperature (K)		
		v	cation valence		

PROBLEMS

3.1 Define the following terms.
(a) ionic bond (b) covalent bond (c) polar bond (d) hydrogen bond
Which of the above are the primary bonds?

3.2 (a) Define diffused double layer.
(b) Can the size of a diffused double layer be changed? Explain!
(c) How are the properties of a soil mass affected
 (i) if the diffused double layer shrinks;
 (ii) if the diffused double layer expands?

3.3 (a) What are the basic types of clay structure?
(b) Under what conditions are these structures formed?

3.4 Is activity of a soil an indication of its susceptibility to changes in its properties such as free swelling?

3.5 In your judgment, which soil is more likely to be stable in a waste environment:
(a) sodium-montmorillonite-rich soil or
(b) illite-rich soil?
Explain.

4 Compressibility and Settlement

4.1 GENERAL

In this section, we present several principles of settlement for soils that can be applied to solid waste. The unit weight, γ, of a deposit usually increases with depth. The *overburden pressure*, σ, at a given depth, z, is

$$\sigma_o = \int_0^z \gamma \cdot dz \qquad (4.1)$$

If there is more than one layer, the overburden pressure is

$$\sigma_o = \sum_1^n \gamma_n z_n \qquad (4.2)$$

where n is the number of layers.

If groundwater is present, *pore water pressure*, or *neutral pressure*, u_o, is

$$u_o = \gamma_w h_w \qquad (4.3)$$

where γ_w is the unit weight of water.

Total stress minus pore pressure is called *effective stress* and is given by

$$\sigma' = \sigma - u \qquad (4.4)$$

Pore pressures may also develop when an external load is applied to a soil or refuse mass. In unsaturated materials, there may be a pressure increase in air voids as well. A thorough treatment of effective stress in fine-grained soils is given by Mitchell (1993).

4.2 SETTLEMENT

Peck et al. (1974), Schmertmann et al. (1978), and NAVFAC DM 7.1 (1982) provide details for estimating settlement in coarse-grained soils. Their methods rely on either standard penetration test (SPT) values, N, cone penetration test (CPT) values, q_c (see Chapter 7 for test details), or the elastic modulus. The settlement in saturated fine-grained soil is determined using the one-dimensional consolidation theory, which also may be applied to municipal solid waste.

4.2.1 Settlement Components

The total settlement is assumed to have three components (Figure 4.1): initial settlement, δ_i; settlement due to primary consolidation, δ_c; and settlement due to secondary compression, δ_s. Total settlement is thus

$$\delta_t = \delta_i + \delta_c + \delta_s \qquad (4.5)$$

Given the low intensity of stress increase due to the self-weight of a landfill, the initial settlement contributes little to the total settlement. For unsaturated soil, there would be a somewhat greater contribution due to rapid compression of air or gases in the voids. The settlement due to primary consolidation is determined from the consolidation theory. Secondary compression comes from creep and other causes.

4.2.2 Modulus of Soft Materials

The undrained modulus, E_u, for soft clay at applied stresses close to its strength is about 100 s_u to 200 s_u, where s_u is the undrained shear strength. For stresses less than one-third the failure stress, E_u may vary from 500 s_u to 1000 s_u. Because there is a large scatter in N or q_c values, it is not possible to select representative values of E_u for municipal waste. There is no published laboratory modulus data for municipal waste; however, based on field dynamic test data at a municipal refuse site, Sharma et al. (1990) reported E_u

Figure 4.1
Typical laboratory strain–time (log scale) curve for a saturated compressible soil under a constant load

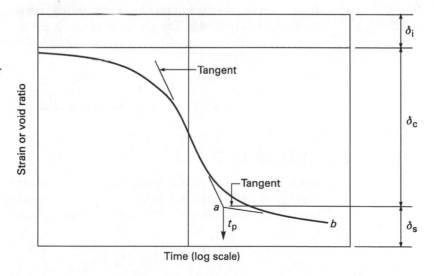

Figure 4.2
Secant modulus

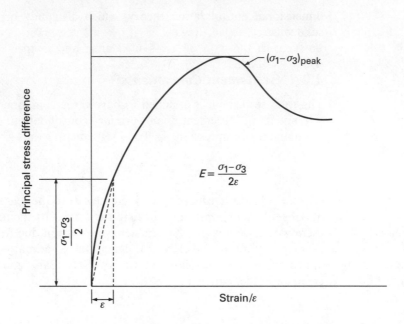

to be 84.4 MPa and Poisson's ratio, v, to be 0.49. Pressuremeter tests have also been used to determine the modulus of municipal waste (see Chapter 7).

4.2.3 *Modulus of Coarse-grained Materials*

For coarse-grained waste materials (such as sand, bottom ash, etc.), it is difficult to determine the drained elastic modulus, E. However, drained triaxial tests performed on specimens compacted to the field relative density and subjected to hydrostatic confining stress will aid in determining the values of the elastic modulus, E, and Poisson's ratio, v. The modulus is computed at a stress level of one-third to one-half the principal stress difference at peak value, as shown in Figure 4.2. The elastic modulus is assumed to be a function of q_c or N.

$$E = 2q_c \tag{4.6}$$

$$E = 7N \qquad \text{for normally consolidated fine sands} \tag{4.7}$$

4.3   INITIAL SETTLEMENT

Assuming that municipal waste behaves like peat, the initial settlement of municipal waste that has been stabilized by dynamic compaction or preloading is based on a pressuremeter tests and is given by

$$\delta_i = \frac{pb}{9}\left(\frac{2\lambda_d}{E_d} + \frac{\lambda_c}{E_c}\right) \tag{4.8}$$

where E_c is the pressuremeter modulus within a distance of $b/2$ beneath the base of the loaded area, E_d is the equivalent modulus within a distance of $8b$ beneath the base of the loaded area, and factors λ_d and λ_c are given in Table 4.1 (Baguelin et al., 1978).

The profile is divided into sublayers having thickness $b/2$. Within each sublayer, the modulus may vary. The average modulus within each sublayer is calculated using the harmonic mean as follows:

$$\frac{n}{E_{av}} = \frac{1}{E_1} + \frac{1}{E_2} + \cdots + \frac{1}{E_n} \tag{4.9}$$

where E_{av} is the equivalent modulus within a sublayer of a multilayer profile. E_d is taken as the equivalent modulus within 16 layers (each $b/2$ thick) and calculated as follows:

$$\frac{1}{E_d} = \frac{1}{4}\left(\frac{1}{E_1} + \frac{1}{0.85E_2} + \frac{1}{E_{3/4/5}} + \frac{1}{2.5E_{6/7/8}} + \frac{1}{2.5E_{9/16n}}\right) \tag{4.10}$$

Example 4.1

Consider a footing 8 ft wide and 16 ft long with a net design load of 1 tsf. The site is an old landfill that was densified by heavy compaction. Following dynamic compaction, the moduli were determined based on pressuremeter testing; the results are shown in the following chart.

Depth below bottom of footing	E (tsf)
0–8	120
16–32	100
32–64	40

Estimate the settlements under the design load.

Solution: $E_1, E_2 = 120$ tsf; $E_3, E_4 = 100$ tsf; $E_5/E_8 = 60$ tsf; $E_9/E_{16} = 40$ tsf; $E_c = 120$ tsf

Table 4.1
Pressuremeter factors λ_d and λ_c for loaded areas of length l and width b

	CIRCLE	SQUARE				
Length/b	1	1	2	3	5	20
λ_d	1	1.12	1.53	1.78	2.14	2.65
λ_c	1	1.1	1.2	1.3	1.4	1.5

From Eq. 4.9,

$$\frac{3}{E_{3/4/5}} = \frac{1}{100} + \frac{1}{100} + \frac{1}{60} \qquad \text{or } E_{3/4/5} = 81.8 \text{ tsf}$$

$$\frac{3}{E_{6/7/8}} = \frac{1}{60} + \frac{1}{60} + \frac{1}{60} \qquad \text{or } E_{6/7/8} = 60 \text{ tsf}$$

$$E_9/E_{16} = 40 \text{ tsf}$$

From Eq. 4.10,

$$\frac{1}{E_d} = \frac{1}{4}\left(\frac{1}{120} + \frac{1}{0.85 \times 120} + \frac{1}{81.8} + \frac{1}{2.5 \times 60} + \frac{1}{2.5 \times 40}\right)$$

$$E_d = 85.1 \text{ tsf}$$

From Table 4.1 for $1/b = 2.0$, $\lambda_d = 1.53$ and $\lambda_c = 1.2$. The settlement, from Eq. 4.8, is given by

$$\delta_i = \frac{pbI(1 - v^2)}{E_u}$$

$$= 0.041 \text{ ft, or } 12.45 \text{ mm}$$

This elastic compression for young refuse is usually a small fraction of the long-term creep (see Section 4.6). It should be noted that this analysis, when applied to settlement on refuse, does not account for time-dependent settlement.

4.4   PRIMARY COMPRESSION

In the one-dimensional consolidation theory, it is assumed that the soil is saturated and homogeneous, the flow is one-dimensional, the coefficient of consolidation is constant for a given load increment, the vertical displacement is small, and the load is applied instantaneously. Field time–settlement data for a landfill under a surcharge load is shown in Figure 4.3. This refuse was not saturated for its entire depth. Note the striking similarity between this curve and that shown in Figure 4.1. Since a surcharge load cannot be applied instantaneously, we cannot isolate the initial settlement. The contributions

Figure 4.3
Typical time–settlement curve for a landfill

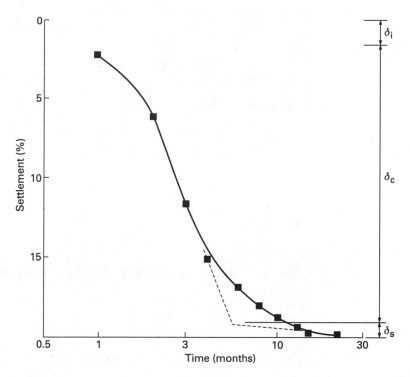

from primary compression and secondary compression are easily discernible, and in this case the latter appears to have contributed only a small proportion of the total settlement. However, decomposition of the refuse will contribute further to the secondary compression.

4.4.1 *Compressibility and Constrained Modulus*

The coefficient of volume compressibility m_v is defined as

$$m_v = \frac{\Delta \varepsilon}{\Delta \sigma'} \tag{4.11}$$

where $\Delta \varepsilon$ is the change in strain due to the stress increment $\Delta \sigma'$. The value of m_v is not constant through the entire stress range because the stress–strain curves are nonlinear.

The constrained modulus D is defined as

$$D = \frac{1}{m_v} \tag{4.12}$$

Secant constrained moduli for coal waste are given in Table 4.2. The slightly lower values of moduli for bottom ash compared to sand are attributed to

Table 4.2
Secant constrained
moduli for coal waste

Materials	Stress range (kPa)	Modulus (MPa)
Bottom ash	45–99	26.5–51.5
	770–1010	73.1–145.8
Uniform medium sand	45–99	53.9
	770–1010	178.4
Coal mine waste spoil	<100	2.4
	>100	7.2*

* Increased linearly with increasing consolidation stress.
Source: After Huang and Lovell, 1990; Koutsoftas and Kiefer, 1990

crushing of ash grains, especially at higher stress levels (Huang and Lovell, 1990).

4.4.2 *Settlement Computations Based on* m_v *and* D

For small load increments, m_v can be used for computing settlement:

$$\delta_c = m_v \cdot H \cdot \Delta\sigma' \qquad (4.13)$$

Alternatively, settlement can be given as

$$\delta_c = \sum \frac{H \cdot \Delta\sigma'}{D} \qquad (4.14)$$

where H is the thickness of the compressible layer.

Example 4.2

The thickness of a waste pile is 10 m, and its unit weight is 14 kN/m³. A 4-m-thick fill with average unit weight of 15 kN/m³ is placed over the entire area. Compute the consolidation settlement due to the fill. Use the estimated values of the constrained modulus for the mine waste spoil as given in Table 4.3.

Solution

$$\sigma'_o \text{ at mid-height} = 14 \times 5 = 70 \text{ kPa}$$

Since D varies with the stress level as the fill is placed, confining stress will increase. Therefore, we first use the thickness of the fill that will increase the stress level to 100 kPa, i.e., $\Delta\sigma = 30$ kPa.

Table 4.3
Constrained modulus
for a mine waste spoil

Stress level (kPa)	50–100	100–150
Secant constrained modulus, D (MPa)	2.5	30

$$\delta = \frac{10 \cdot 30}{2.5 \cdot 10^3} = 0.12 \text{ m}$$

$$\text{Fill height} = \frac{30}{15} = 2 \text{ m}$$

Effective stress at mid-height after placement of 2-m fill:

$$\sigma'_o = 70 + 30 = 100 \text{ kPa}$$

Additional stress from the remaining 2-m fill:

$$\Delta\sigma = 2 \times 15 = 30 \text{ kPa}$$

Additional settlement:

$$\delta = \frac{10 \cdot 30}{30 \cdot 10^3} = 0.01 \text{ m}$$

Thus, the total settlement is $0.12 + 0.01 = 0.13$ m.

4.4.3 Strain Versus Log-Pressure

The stress–strain curve of a compressible material on a semilogarithmic scale is shown in Figure 4.4. It can be approximated essentially by two straight lines. Slope of the line $B'C$ is known as *compression ratio, CR*.

$$CR = \frac{\varepsilon_u - \varepsilon_b}{\log \sigma_b - \log \sigma_a} \tag{4.15}$$

The slope of line CC' is the *recompression ratio*, or *swelling ratio*, (RR).

If the maximum past consolidation stress, σ'_p, is greater than the effective overburden stress, σ_o (as shown in Figure 4.4), then the soil is overconsolidated. The *overconsolidation ratio, OCR*, is defined as

$$OCR = \frac{\sigma'_p}{\sigma'_o} \tag{4.16}$$

Normally consolidated (NC) soils have $OCR = 1$.

4.4.4 Settlement from Strain Log–Pressure Curve

Settlement of a compressible layer of thickness H, with an average effective overburden stress of σ'_o, under a stress increase of $\Delta\sigma$ due to primary

Figure 4.4
Stress–strain curve and
graphical construction
for preconsolidation
stress determination

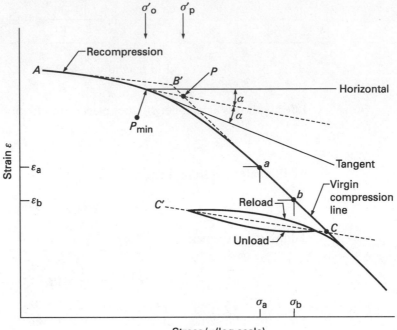

consolidation is given as

$$\delta_c = H \left[RR \cdot \log \frac{\sigma'_p}{\sigma'_o} + CR \cdot \log \frac{\sigma'_o + \Delta\sigma}{\sigma'_p} \right] \qquad (4.17)$$

Settlement for an NC soil, i.e., where $\sigma'_p = \sigma'_o$, is

$$\delta_c = H \cdot CR \cdot \log \frac{\sigma'_o + \Delta\sigma}{\sigma'_o} \qquad (4.18)$$

If Figure 4.4 is redrawn after converting the strain-to-void ratio, the shape
of the curve *ABCC'* remains unchanged. The slopes of lines *B'C* and *CC'* are
then called the *compression index* (C_c) and the *recompression* or *swelling index*
(C_r), respectively, and are given by

$$C_c = CR(1 + e_0) \qquad (4.19)$$

$$C_r = RR(1 + e_0) \qquad (4.20)$$

where e_0 is the initial void ratio of the material corresponding to σ'_o.

Table 4.4 shows these settlement parameters for various waste materials.
Where laboratory tests are performed to determine values of *CR*, considerable
variations are observed for municipal landfills because of the extreme heterogeneous nature of the refuse. Thus for construction on landfills, the use of test
fills is recommended for evaluating its settlement characteristics. An additional

Table 4.4 Settlement parameters for waste materials

Material	C_c	CR	C_α	C_α'
Peat	$0.75e_0$			
Fifteen-year-old landfill, Boston, MA		0.26		0.24
Lab test on simulated material		0.20		0.30
Old landfill, WV		0.15		0.04
Low organic contents and conditions unfavorable to decomposition	$0.15e_0$	0.15	$0.03e_0$	0.024
High organic contents and conditions favorable to decomposition	$0.55e_0$	0.41	$0.09e_0$	0.072
Municipal waste, Melbourne, Australia	$0.1e_0$			0.06
Fifteen- to twenty-year-old landfill, MI		0.08		0.02
Ten-year-old landfill, Elizabeth, NJ		0.21		0.02
Landfill, Harrison, NJ	$0.25e_0$			0.04
Recompacted municipal waste–soil mix				0.14–0.034
Old landfill with high soil contents		0.01–0.04		0.001–0.005
Laboratory data for landfills in Canada		0.17–0.36		
Dewatered leather manufacturing sludge, Italy		0.22–0.32		
Mixed sludge (steel mill, chrome plating, painting plant), Italy		0.32–0.47		
Gold mine tailings	0.37			
Gold mine tailings from centrifuge	0.75			
Compacted fly ash	0.02–0.25			
Ponded fly ash	0.32–0.65			0.07
Flue gas desulfurization sludge				
Sulfite	0.2–0.8			
Sulfite-lime and double alkali	0.5–1.0			
Sulfite with fly ash	0.02–0.35			
Sulfate	0.02–0.30			
Paper mill sludge				
Lab specimen				0.015
Field values				0.056
Dredged waste	0.3–0.7			0.002–0.013

Source: Belfiore et al., 1990; Bromwell, 1978; Dodt et al., 1987; Keene, 1977; Landva and Clark, 1990; Moore and Pedler, 1977; Oweis and Khera, 1986; Sheurs and Khera, 1980; Sowers, 1973; Stone et al., 1994; Yen and Scanlon, 1975; York et al., 1977

benefit of such test fills is that the test sections get precompressed in the process and the future settlements are reduced. The effect of decomposition, which is difficult to estimate, must not be overlooked.

Example 4.3

Laboratory tests were conducted on an undisturbed specimen of a sludge. Its natural water content = 117%, liquid limit = 163, plastic limit = 58, initial void ratio = 4.7, and dry unit weight = 4.8 kN/m³. Consolidation test results are given in Table 4.5. Determine both the compression index and the recompression index of the sludge.

Solution: First, plot the consolidation data are shown in Figure 4.5.

Table 4.5   Consolidation test data

Stress (kPa)	20	40	70	100	200	400	800	1200	400	100	20
Strain (%)	1.1	4	6.8	10	17.5	27.5	37	42	41.8	41.5	41.2

Source: Data from Belfiore et al., 1990

Figure 4.5
Consolidation test
data plotted on semilog
paper

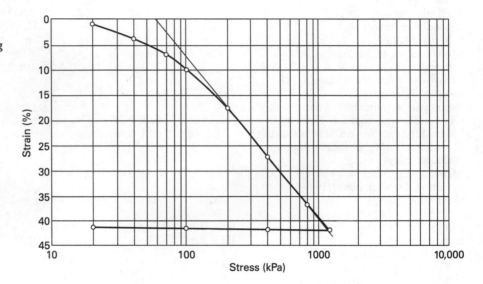

$$CR = \frac{1}{100} \times \frac{40 - 8}{\log 1000 - \log 100} = 0.32$$

$$C_c = 0.32 \times (1 + 4.7) = 1.82$$

$$RR = \frac{1}{100} \times \frac{42 - 41.5}{\log 1000 - \log 100} = 0.005$$

$$C_r = 0.005 \times (1 + 4.7) = 0.0285 \qquad \text{(from Eq. 4.20)}$$

4.5   SETTLEMENT RATE

When a saturated soil of low permeability is subjected to an external load, an excess pore pressure, u_e, develops. The equation governing the dissipation and distribution of the excess pore pressure is

$$\frac{\partial u_e}{\partial_t} = c_v \left(\frac{\partial^2 u_e}{\partial z^2} \right) \qquad (4.21)$$

where t is the time, z is the depth, and c_v is the coefficient of consolidation for vertical flow. Eq. 4.21 is known as Terzaghi's one-dimensional consolidation equation.

$$c_v = \frac{k}{\gamma_w m_v} \qquad (4.22)$$

where k is the vertical hydraulic conductivity and γ_w is the unit weight of water. (See Das (1994) for laboratory procedures.)

Assuming that the initial excess pore pressure (u_i) is equal to the stress increment ($\Delta\sigma$) and that there is complete drainage at both the top and bottom of the clay layer, we can depict the solution to Eq. 4.21 as shown in Figure 4.6. The dimensionless parameters Z (depth factor), U_z (consolidation ratio), and T (time factor) are, respectively,

$$Z = \frac{z}{H_d} \qquad (4.23)$$

$$U_z = 1 - \left(\frac{u}{u_i} \right) \qquad (4.24)$$

$$T = \frac{c_v t}{H_d^2} \qquad (4.25)$$

where H_d is the distance to the drainage boundary.

If a landfill is located on a compressible soil, the values for the consolidation ratio and the excess pore water pressure in the compressible layer can be used to indirectly estimate the average bulk unit weight of the landfill. If the foundation soils are normally or lightly overconsolidated, the preconsolidation pressure σ'_p can be determined from the laboratory tests and excess pore pressure u_e at the sampling point can be obtained from the field measurements. Then

$$\sigma'_p + u_e = \sigma'_o + \gamma_1 h_1 \qquad (4.26)$$

where σ'_o is the effective overburden pressure or consolidation pressure prior to the placement of the landfill at the sampling point, h_1 is the height (depth), and γ_1 is the average unit weight of the landfill.

$$\gamma_1 = \frac{(\sigma'_p - \sigma'_o + u_e)}{h_1} \qquad (4.27)$$

Figure 4.6
Dimensionless
parameter for
consolidation ratio
S*ource :* Reproduced
from NAVFAC DM 7.1,
1982

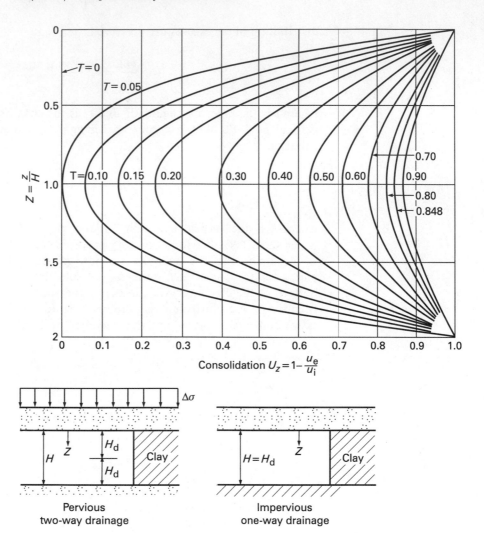

Example 4.4   At an existing landfill site the following data were obtained.

Soil profile within the landfill	Elevation (ft)
Top of organic silt and peat	0
Top of medium to fine sand	−5
Top of varved silt and clay	−10
Top of shale	−52

The piezometer head at point A (which was located under the landfill at
elevation −50 ft) was +54.6 ft. The top of landfill was at elevation +111.6 ft.
For a test specimen taken from point A, the maximum past consolidation

pressure (based on a laboratory consolidation test) was estimated as 2.23 t/ft^2. The soil profile outside the landfill was: ground level elevation (ft) = 0, water table elevation (ft) = 0. Determine the unit weight of the overlying landfill in lb/ft^3.

Solution: The effective stress outside the landfill is shown in the table.

Formation	Assumed thickness (ft)	Assumed submerged unit wt (pcf)	Effective stress, $\Delta\sigma'$ (psf)
Silt and peat	5	40	200
Medium to fine sand	5	60	300
Varved silt and clay	40	57	2280
Total			2780

$$u_e = 54.6 \times 62.4 = 3407 \text{ psf}$$

$$\sigma'_p = 2.23 \times 2000 = 4460 \text{ psf}$$

From Eq. 4.27, the unit weight of the overlying landfill is

$$= (4460 - 2780 + 3407)/111.6$$

$$= 45.6 \text{ lb/ft}^3$$

4.5.1 Average Consolidation

The average consolidation, U, is the summation of U_z over the depth of the stratum. Table 4.6 shows the relationship between the average consolidation and the time factor. Equations 4.28 and 4.29 show an empirical relationship between U and T (Leonards, 1962).

$$T = \frac{\pi(U)^2}{4} \quad \text{for } U \leq 53\% \tag{4.28}$$

$$T = 1.781 - 0.9332 \log(100 - U\%) \quad \text{for } U > 53\% \tag{4.29}$$

Table 4.6 Relationship between average consolidation, U, and time factor, T

U	0	0.1	0.2	0.3	0.4	0.5	0.6	0.7	0.8	0.9	0.95	1
T	0	0.007	0.031	0.071	0.126	0.197	0.286	0.403	0.567	0.848	1.128	∞

Example 4.5

A soil profile at a given site consists of a 20-m-thick layer of a compressible material between an upper and a lower layer of sand. The water table is at the top of the compressible layer. The tip of a piezometer is installed 6.5 m below the top. A surface load causes a stress increase of 100 kPa. After 15 days the pressure head at the tip of the piezometer is 14.97 m and after 75 days it is 13.01 m. Determine the value of c_v.

Solution: Use the given data for the piezometer tip. The initial total head at the piezometer tip upon load application is $h = [6.5 + (100/9.81)]$, or 16.69 m. The depth factor at the tip is $Z = z/H_d = 6.5/10$, or 0.65. Set up a chart as shown.

Time (days)	h (m)	Excess head $(h - 6.5)$ (m)	Consolidation ratio $(1 - u/u_i)$	T from Figure 4.6
0	16.69	10.2	$1 - \dfrac{10.2 \times 9.81}{10.2 \times 9.81} = 0$	0
15	14.97	8.47	$1 - \dfrac{8.47 \times 9.81}{10.2 \times 9.81} = 0.17$	0.11
75	13.01	6.51	$1 - \dfrac{6.51 \times 9.81}{10.2 \times 9.81} = 0.361$	0.21

$$c_v = H_d^2 \frac{T_2 - T_1}{t_2 - t_1}$$

$$= (10)^2 (0.21 - 0.11)/(75 - 15)$$

$$= 0.2 \ \text{m}^2/\text{day} \tag{4.30}$$

4.6   SECONDARY COMPRESSION

A typical plot for strain with time on a semilogarithmic scale was shown in Figure 4.1. The secondary compression ratio is the slope of line *ab* and is given by

$$C'_\alpha = \frac{\Delta\varepsilon}{\log t_2 - \log t_1} \tag{4.31}$$

where $\Delta\varepsilon$ is the change in strain between time t_1 and t_2. The secondary compression index is defined as

$$C_\alpha = C'_\alpha(1 + e_p) \tag{4.32}$$

where e_p is the void ratio corresponding to point *a* in Figure 4.1. Values of C_α for some types of waste materials were given in Table 4.4. The settlement from secondary compression is

$$\delta_s = C'_\alpha H \log \frac{t}{t_p} \qquad (4.33)$$

where t is the time at which settlement due to secondary compression is required ($t > t_p$) and t_p is the time for completion of primary consolidation.

4.7   SETTLEMENT OF MUNICIPAL LANDFILLS

The settlement of a landfill stems from one or more of the following:

1. Reduction in void space and compression of loose materials from the self-weight of the waste and the weight of the cover materials.

2. Occasional movement of smaller particles into larger voids resulting from the collapse of larger bodies, seepage, an abrupt drop in the water table, a shock wave, or vibration. This type of movement may cause unexpected depression on the surface.

3. Volume changes from biological decomposition and chemical reactions. These accelerate at high moisture content, at warmer temperatures, with a poorer state of compaction, and with a larger proportion of organic contents.

4. Dissolution from percolating water and leachate.

5. Settlement of the soft compressible soils underlying the landfill.

4.7.1 *Basic Considerations for Municipal Landfill Settlement*

The theories developed for determining settlement of fine-grained materials are not strictly applicable to landfill. Laboratory consolidation tests are not feasible because of the difficulty of obtaining reliable and representative undisturbed test specimens from landfills. Thus, field measurements are essential for developing parameters for estimating landfill settlement. Chemical and biological decomposition influences secondary compression but cannot be readily evaluated from conventional tests. Furthermore, in designing and constructing foundations on old landfill sites, one also must consider problems such as venting of gases, corrosion of buried foundation elements, leachate control, and other construction difficulties.

The deformation under surface loads occurs very rapidly and is treated like primary consolidation. Settlement under self-weight and long-term settlement from the applied loads are similar to secondary compression.

4.7.2 *Landfill Settlement from Self-weight*

Landfill settlement under self-weight will occur during the construction of a landfill. For a landfill of unit weight γ and height H, the stress increment $\Delta\sigma$ is the average vertical effective stress at mid-height, or $\gamma H/2$. The settlement is determined from the method given in Section 4.4. The constrained modulus D is $\Delta\sigma/0.435CR$, or approximately $\gamma H/2CR$. The compression ratio CR or compression index C_c may be estimated from Table 4.4. The settlement during construction is computed from Eq. 4.14.

With increasing age and height, a landfill's rate of settlement from self-weight decreases. The anaerobic environment at greater depths will reduce the rate of decomposition. In fact, the rate of settlement may eventually become constant once a landfill thickness of about 100 ft (30 m) is reached (Yen and Scanlon, 1975).

4.7.3 *Landfill Primary Compression*

A typical time–settlement curve for a refuse landfill under an applied load was shown in Figure 4.3. Initially, the settlement occurs rapidly as in primary consolidation, and then it tapers off as in secondary compression. The settlement from primary compression may be computed using Eq. 4.14 or Eq. 4.17. Various parameters for municipal landfill materials are given in Table 4.4.

Since the refuse is not always saturated, the early part of the settlement is denoted as primary compression, not consolidation; consolidation implies full saturation and dissipation of excess pore water pressure. Sowers (1973) stated that this settlement usually occurs in less than a month. The field measured data at several locations from Sheurs and Khera (1980) show that about 70 to 80% of the settlement took place within the first three months. The value of c_v ranged between 0.15 ft^2/day and 5 ft^2/day.

4.7.4 *Landfill Secondary Compression*

As previously noted, the settlement of a landfill continues after the primary compression (Figure 4.3). The long-term settlement appears to be linear on a log–time scale and can be determined by Eq. 4.33. The average value of C'_α is 0.2, with upper and lower limits of 0.32 and 0.13, respectively. After about 10 years, C'_α reaches a constant value of 0.01 to 0.02. Where soil contents are high, C'_α drops to <0.01.

If the conditions for biological decomposition and chemical reactions are favorable, the rate of secondary compression will be high. To estimate settlement contribution of a landfill from biodegradation, Stulgis et al. (1995) determined the total volatile solids (TVS) as a percentage of dry weight of total solids for a landfill and the amount of lignin in TVS. The proportion of biodegradable matter was obtained by subtracting the relative amount of lignin that is nonbiodegradable from the TVS. The computed amount of strain from biodegradation was estimated to be consistent with C'_α values based on test fill settlement data.

The mass of the total volatile solids minus lignin can be used to establish an upper bound on decomposition settlement. At initial placement, the TVS of municipal refuse is about 78% of the dry weight of the refuse and lignin is about 15% (Barlaz and Ham, 1993). Assuming the total difference is converted to gas, a column of fresh refuse will shrink by about 63%. The assumption is that the dry unit weight remains the same (loss of mass is compensated by reduction in volume). In general,

$$s_{\text{d-max}} = h_1 C_{t_1} \tag{4.34}$$

where $s_{\text{d-max}}$ is upper bound on decomposition settlement from time t_1 to the end of decomposition, h_1 is thickness of refuse at time t_1, and C_{t_1} is the fraction by dry weight of (TVS − lignin) at time t_1.

If the decay relationship for gas (see Chapter 14) is the same as for the decay of biodegradable material, then

$$C_{t_2}/C_{t_1} = \exp - \zeta(t_2 - t_1) \tag{4.35}$$

where ζ is the decay constant. Also,

$$s_{dt} = s_{\text{d-max}}(1 - e^{-\zeta}) \tag{4.36}$$

where s_{dt} is the decomposition settlement at time t after the date the refuse is tested. Methane generation may be used as an indication for potential future decomposition. A landfill with a low methane generation rate or concentration will not be expected to undergo significant decomposition settlement.

Example 4.6

A landfill 50 ft thick was found to contain 19% by dry weight TVS and 14% lignin. Assuming a 10% cover by volume and a decay constant of 0.06 yr^{-1}, determine the decomposition settlement after 10 years.

Solution

$$s_{\text{d-max}} = (1 - 0.1)(50)(0.05) = 2.25 \text{ feet}$$

$$s_{\text{d-10 yrs}} = 2.25(1 - \exp - 0.06 \times 10) = 1.015 \text{ feet}$$

4.7.5 Stress History Effect

Both stress history and the load increment ratio influence settlements. This was studied during the construction of an interstate highway, where part of an old landfill was excavated prior to surcharging. Where the stress increase from the surcharge load was less than the stress before the excavation, the settlement was from 5 to 7%. Where the surcharge stress was over 40% above the preexcavation stress, the settlement ranged from 11.4 to 16.8% (Sheurs and Khera, 1980).

4.7.6 Total Settlement

Actual computations of settlement can be quite complex. For example, to estimate the total settlement for a recently closed landfill, the following considerations will be necessary for each of the layers in the landfill.

1. Settlement from self-weight (Eq. 4.14).

2. Settlement from the weight of each of the layers that overlie the given layer (Eq. 4.18). Weight of the final cover will further add to this settlement.

3. Settlement due to secondary compression (Eq. 4.33), taking into account the fact that C'_α decreases with age.

4. Settlement of the daily cover by itself, which is typically about 20% by volume of a landfill, is not considered. However, as a landfill settles, 75% of the cover material migrates into the voids thereby increasing the density of the fill (Morris and Woods, 1990) and adding to the overall settlement.

5. Settlement of the clay liner (if present) due to primary and secondary compression.

Total settlement is the sum of the settlements of each of the layers, see for an example Morris and Woods, 1990.

Morris and Woods (1990) presented a FORTRAN program for computing time-dependent landfill settlements and densities. The program requires a knowledge of the void ratio of the landfill materials, which may vary from 3 to 6. The void ratio value of 4.0 assumed by Yen and Scanlon (1975) may be used as a starting point.

Because of the heterogeneous nature of landfill constituents and their varied rates of decomposition, differential settlements occur. The problem is further complicated by the facts that different cells are completed at different times and that the total settlement and its rate are time dependent. These differential settlements jeopardize the stability of the final cap required for each closed landfill. (See Chapter 12 for cap design.)

4.8   SETTLEMENT OF MINERAL AND INDUSTRIAL WASTE

4.8.1 Fly Ash

The compression index, C_c, of fly ash depends on the type of fly ash and the period of curing. Tests on compacted fly ash show a low C_c value (0.02–0.25). Field data indicate that the actual settlements are usually much smaller than those computed in the laboratory tests (Seals et al., 1977). To explain this, Gatti and Tripiciano (1981) presented data from consolidation tests on a virgin fly ash where load increments were applied at 1-hour intervals and 24-hour intervals. Where the load increments were applied at 24-hour intervals, the settlements were considerably less than that where load increments were applied at 1-hour intervals. This was attributed to the extra curing time, which allowed the 24-hour specimen to become stiffer. Thus, since most laboratory tests are carried out in a relatively short period of time when compared with loading in the field, the discrepancy in the data between field and laboratory tests may be partially due to the differences in curing time.

For undisturbed specimens of a ponded fly ash from Virginia, Newman et al. (1987) reported a void ratio of 1.6 and C_c and C'_α values of 0.65 and 0.07, respectively, with a coefficient of consolidation of 2.25 cm^2/min. From a site in Illinois (Cunnigham et al., 1977) for void ratios of 1.26 to 1.40, the C_c values were 0.46 to 0.32. Considering the relative void ratios for the two cases, these values seem reasonable. The value of precompression stress was determined to be larger at shallower depth than at greater depth. Note that these C_c values are much larger than those for compacted fly ash; this indicates a distinct advantage of disposal by compacting. Also, the dry unit weight of slurried fly ash, which is generally less than 70 lb/ft^3 (11 kN/m^3), is considerably lower than that of compacted fly ash, which is 82.5–133.5 lb/ft^3 (13–21 kN/m^3).

After heavy rainfalls in South and Southeast Asia (Indraratna, 1992), failures of fly ash dumps were reported as "mudflows" and not as conventional slope failures. The collapse potential of the ash is used to determine critical height of the dump. To determine the collapse potential (CP), test specimens at their natural water content are loaded in a consolidometer to different stress levels and then saturated. CP is then based on the change in void ratio upon saturation and is expressed as

$$CP = \frac{\Delta e}{1 + e_0} \cdot 100\% \tag{4.37}$$

where Δe is the change in void ratio upon saturation and e_0 is the initial void ratio of the specimen. A collapse potential of 5% is cause for concern, and a value greater than 10% is considered very undesirable (Geological Society of London, 1990).

Because of the difficulty of predicting settlements from the laboratory tests, field techniques such as standard penetration resistance (N values) and cone penetration resistance (q_c, assuming fly ash to be cohesionless) have been studied for settlement predictions. The computed settlements based on N values and q_c did not agree with the measured values but those based on the plate load tests did (Leonards and Bailey, 1982).

4.8.2 *Flue Gas Desulfurization* (FGD) *Sludge*

The compression index of FGD sludges at a stress level of 500 kPa is about two to four times greater than at a stress level of 25 kPa (Hagerty et al., 1977; Ullrich and Hagerty, 1987). At higher stress levels, the compression index for sulfite sludges was considerably larger (0.2–0.8) than for sulfate sludges (0.02–0.07). The corresponding values reported by Krizek et al. (1987) are larger: 0.5–1.0 for sulfite (lime and double alkali) sludges and 0.2–0.3 for sulfate sludges. The time for completion of primary consolidation is very short for sulfite sludge.

Fly ash is frequency added to FGD sludges to reduce their moisture content and enhance their stability. The addition of small amounts of lime to FGD sludge–fly ash mixture may add further to its stability. However, the results are not always predictable. A sulfite sludge–fly ash–lime mixture designed to yield maximum strength showed a higher compression index at a consolidation stress of 35 kPa after 28 days of curing than after 7 days of curing. A similar mixture with sulfate sludges consistently showed a lower compression index with increasing curing time. When sulfite sludge was mixed with 60 to 70% fly ash and was compacted to its maximum dry density, its C_c value was 0.02–0.25, which is more like fly ash (Ullrich and Hagerty, 1987). Krizek et al. (1976) reported higher values for the compression index (0.35) with similar amounts of fly ash.

4.8.3 *Pulp and Paper Mill Waste*

The compression index of pulp and paper mill sludges is approximately $0.4e_0$ (Wardwell and Nelson, 1981). For a given solid contents, C_c increases linearly with increasing organic contents. In a laboratory consolidation test, pore pressure dissipated within 15 minutes and the coefficient of consolidation decreased from 0.11 to 0.03 cm^2/min as the stresses increased from 10 to 320 kPa. These values were about four times smaller than the field values. Similarly, the laboratory values of the secondary compression ratio (0.015) were also about four times smaller than the field values (0.056) and were nonlinear with log time (Vallee and Andersland, 1974). It is clear that predictions based on the laboratory tests would be grossly in error.

The use of a horizontal drainage blanket and application of small earth surcharge can result in appreciable decrease in volume and increase in strength (Vallee and Andersland, 1974; Charlie, 1977).

4.8.4 *Phosphatic Clays*

Settlement of phosphatic clays under self-weight may require several decades before the disposal area can be reused. The Terzaghi theory, which assumes linear infinitesimal strain, overestimates the pore pressure dissipation at the top of a layer and underestimates it at the bottom (Gibson et al., 1981). McVay et al. (1986) compared several large strain theories (McNabb, 1960; Gibson et al., 1967, 1981; and Somogyi, 1980) and concluded that they all predicted similar settlement but the settlement rates varied. Good agreement was reported between predicted and measured values for permeability and void ratio relationships of

$$k = 4.0 \times 10^{-6} \times e^{4.11} \text{ cm/s} \qquad \textbf{(4.38)}$$

$$e = 29.43(\sigma')^{-0.29} \qquad \textbf{(4.39)}$$

The relationships of void ratio with permeability and effective stress usually hold for solid contents about 15% or higher (Scully et al., 1984). Addition of sand to clay results in reduced settlements and improved engineering properties (Bromwell, 1978).

4.8.5 *Mining and Dredged Wastes*

Wastes included here are mine processing waste, phosphatic clay slurries, tar sand tailings, and dredged wastes. The compressibility range for some of these

Figure 4.7
Range of void ratio–pressure relationships for fine tailings and other compressible materials (Rollings, 1994, Figure 1)

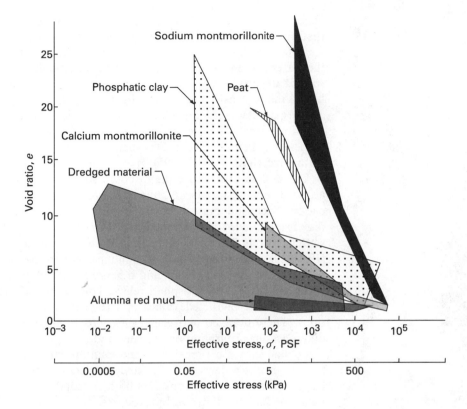

materials is shown in Figure 4.7. Note the high void ratio for phosphatic clays and dredged materials. Slurries from phosphatic clays and from dredged waste where coarse-grained particles have been segregated contain a small percentage of solids. These slurries (slimes) take many years to sediment and consolidate under self-weight.

Dredging is a periodic operation that allows desiccation as a viable alternative for reducing slurry volume in many regions. Experiences with dredged materials in the United States and Germany indicate surface drying to be an efficient means of stabilization and volume reduction (Rizkallah, 1987). Ericson and Carrier III (1995) describe the advantages of dewatering and desiccation for improving equipment access for reclamation and increasing the storage volume of fine tailings from phosphate mining and aluminum processing.

To improve the effectiveness of desiccation, surface water from the disposal and rainfall runoff should be drained quickly and effectively. In the application of desiccation, the slurry is deposited in thicknesses from about 50 cm to 1.5 m. As the materials dry, cracks a few millimeters to several centimeters wide form at the surface and extend to depths up to a meter or so. The resulting larger exposed surface areas aid in rapid drying. High capillary forces resulting from drying cause large reduction in soil volume, and the dewatering increases overburden stresses on the underlying soils. Thus the subsequent volume reduction is lower when compared with the uncracked underlying materials.

When more waste is placed on partially dry material, the cracks in the crust that provide additional drainage paths should be considered.

Alternatively, to accelerate the rate of consolidation or volume reduction, the water content of slurries can be reduced sufficiently or solids can be added so that the coarse particles remain suspended before disposal. Because of the higher initial density and greater hydraulic conductivity due to the presence of coarse particles, settlement occurs more rapidly, thereby freeing up space for additional waste. The same principle of keeping the fine and coarse particles in suspension before disposal is being applied to slurries from tar sands (Morgenstern and Scott, 1995).

Regardless of whether the disposed slurries contain only fine or both fine and coarse materials, large finite strains occur due to consolidation and desiccation. Such soil mass becomes inhomogeneous. Therefore, Terzaghi's one-dimensional consolidation theory cannot be applied for settlement and settlement rate computations. Large strain theories developed by Gibson et al. (1967, 1981) require solution to nonlinear equations. Several researchers such as Yong and Ludwig (1984), and others have developed computer solutions to these equations. For phosphatic clays, Townsend and McVay (1990) presented measured and predicted values of settlement based on such computer solutions. The difference between the various predictions were attributed to the differences in programming schemes and/or input selection since McVay et al. (1986) had shown earlier that all the solutions should produce identical results. Schroeder and Palermo (1990) provide a personal computer program based on large strain model that considers both consolidation and desiccation for managing dredging and disposal procedures.

At old dredged disposal sites where soils have consolidated to support loads, Salem and Krizek (1976) reported that the compression index for undisturbed samples ranges from 0.3 to 0.7. These values were lower than those obtained for specimens prepared in the laboratory from slurry. A relationship between liquid limit, natural water content, and compression index was given as

$$C_c = 0.04(\text{LL} + 2w_n - 50) \tag{4.40}$$

From the standard consolidation tests on undisturbed samples, the coefficient of consolidation was 6×10^{-4} cm^2/sec. The specimens consolidated from slurry in a 150-mm-deep consolidation ring had a c_v of 10^{-4} cm^2/sec. The secondary compression ratio ranged between 0.002 and 0.013 for a natural water content range of 45 to 65%. C'_α showed a linear increase as C_c increased from 0.32 to 0.55.

4.9   SUMMARY

Principles developed for saturated compressible materials that are extensively used in soil mechanics can be applied to waste materials such as municipal landfills, industrial sludges, fly ash, etc., to determine settlement. Although the assumptions of the basic theories are not always fully satisfied, reasonable estimates of settlements are possible. No reliable methods for estimating long-term settlements from biological and chemical actions are available.

NOTATIONS

ε strain (L/L)

γ unit weight (F/L^3)

δ settlement (L)

σ' effective stress (F/L^2)

ζ decay constant

ϕ' angle of internal friction for effective stress (°)

λ_b, λ_c pressuremeter factors

δ_{ct} consolidation settlement at time t (L)

σ'_h effective horizontal stress (F/L^2)

σ_z stress increase at depth z due to surface load (F/L^2)

δ_c primary consolidation/compression settlement (L)

C_α secondary compression index

C'_α secondary compression ratio

C_c compression index

CP collapse potential

CR compression ratio

C_r recompression index

c_v coefficient of consolidation for vertical flow (L^2/T)

D constrained modulus (F/L^2)

γ_d dry unit weight (F/L^3)

e void ratio

E drained modulus (F/L^2)

e_0 initial void ratio

e_p void ratio at the end of primary consolidation

E_u undrained modulus (F/L^2)

H thickness of compressible layer (L)

H_d distance to the drainage boundary (L)

h_1 landfill height (L)

H_p thickness of compressible layer at the end of primary consolidation (L)

h_w static head of water (L)

δ_i initial, immediate, or distortional settlement (L)

k vertical hydraulic conductivity (L/T)

LL liquid limit

m_v coefficient of volume compressibility (L^2/F)

n number of layers

N standard penetration resistance

γ_n unit weight in layer n

NC normally consolidated

σ_o total overburden stress

σ'_o effective overburden stress

OC overconsolidated

OCR overconsolidation ratio

σ'_p maximum past consolidation stress

q_c cone point resistance (F/L^2)

RR recompression ratio

δ_s secondary compression settlement (L)

s_d decomposition settlement (L)

$s_{d\text{-max}}$ upper bound on decomposition settlement (L)

s_u undrained shear strength (F/L^2)

t time (T)

T time factor

δ_t total settlement (L)

t_p time for completion of primary consolidation (T)

TVS total volatile solids

u pore pressure

U average degree of consolidation

u_e excess pore water pressure

u_i initial excess pore water pressure

u_o neutral pressure, or pore water pressure

U_z consolidation ratio

v Poisson's ratio

σ_v total vertical stress (F/L^2)

σ'_v effective vertical stress (F/L^2)

w water content

γ_w unit weight of water (F/L^3)

w_n natural water content

Z dimensionless depth parameter

z depth (L)

PROBLEMS

4.1 The settlement record from a test embankment on a coal mine waste spoil about 11 m deep yielded the data shown in Table 4.7. Determine the secondary compression ratio.

4.2 For a silty clay, LL = 38 and PL = 19. Data for an unconsolidated undrained triaxial test is given in Table 4.8. The confining stress was 100 kPa. Determine the undrained modulus for stress that is one-third the failure stress.

4.3 A waste pile 5 m thick is subjected to an average increase in stress of 20 kPa. The unit weight of the waste pile is 12 kN/m³. The coefficient of volume compressibility is given in Table 4.9. Determine the settlement due to the applied stress.

4.4 A 4-m-thick deposit of dredged waste has undergone desiccation over a period of time. Its unit weight is 10 kN/m³. Consolidation data indicates an *OCR* of 1.3. The *CR* is 0.35 and the *RR* is 0.08. A 2.5-m-thick fill with a unit weight of 16 kN/m³ is placed on the surface of the dredged deposit. Determine its total settlement.

4.5 The thickness of a municipal landfill is 15.5 ft. Primary compression of 1.7 ft occurred in 70 days. The secondary compression ratio is 0.019. What would be the total settlement of the landfill in 2 years?

Table 4.7
Settlement record

Time (days)	2	4	5	8	10	15	20	60	100	200
Settlement (min)	3	15	45	105	128	140	145	165	170	178

Table 4.8
Triaxial test data

Strain (%)	1	2	3	4	6	8	10	12	14
σ_1(kPa)	135	150	165	175	188	197	200	198	193

Table 4.9
Coefficient of volume compressibility

Stress level (kPa)	m_v (1/MPa)
2–40	0.15
40–80	0.02
> 80	0.015

CHAPTER 5   Shear Strength

The shear strength parameters of soils or waste material are the most important parameters for assessing the stability of structures containing such materials. Shear strength parameters may be determined by tests in the laboratory or in the field and by back-computations based on observed failures of structures. Some of the methods of strength determination are described here.

5.1   STRESS–STRAIN BEHAVIOR

In soils and waste materials, the relationship between stress and strain is nonlinear and time dependent. Volume changes or pore water pressures develop from the applied normal and shearing stresses. Figure 5.1 shows test results for Class F fly ash.

5.2   FAILURE THEORIES

According to Coulomb, the shear strength, τ_f, is expressed in terms of the cohesion, c, and the angle of friction, ϕ.

$$\tau_f = c + \sigma_f \tan \phi \qquad (5.1)$$

where σ_f is the normal stress at failure. In terms of effective stress, the strength is given as

$$\tau_f = c' + (\sigma - u) \tan \phi' \qquad (5.2)$$

Figure 5.1
Response of a Class F fly ash specimen in axial compression

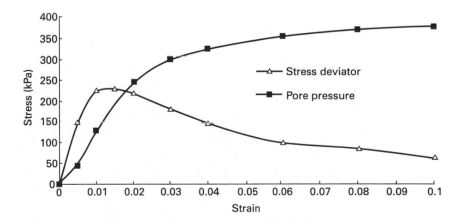

Mohr defined failure based on a critical combination of normal and shearing stresses at failure on the failure plane, as shown in Figure 5.2. The Mohr failure envelope when approximated by a straight line is known as the Mohr–Coulomb rupture line and is given as

$$\sigma_{1f} = \sigma_{3f} \tan^2\left(45 + \frac{\phi}{2}\right) + 2c \tan^2\left(45 + \frac{\phi}{2}\right) \qquad (5.3)$$

where σ_{1f} and σ_{3f} are the major and minor principal stresses at failure. The effective stress strength parameters c' and ϕ' can be determined from drained tests or tests with pore water pressure measurements. Effective stresses may be used in Eq. 5.3.

5.3 STRENGTH TESTS

In most laboratory strength tests, a specimen is subjected to a constant rate of deformation while the resulting load changes are measured and, if needed, volume changes or pore water pressure are determined. Because some waste may contain corrosive substances that damage the usual geotechnical testing equipment, apparatuses made of noncorrosive materials such as stainless steel

Figure 5.2
(a) Coulomb failure line
(b) principal stresses at failure
(c) Mohr–Coulomb failure envelope

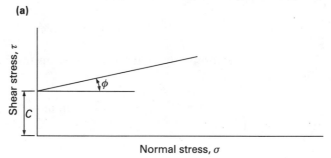

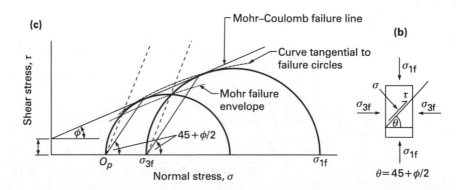

are often used instead. When hazardous materials are to be tested, the safety of laboratory personnel is paramount and proper training is a must.

5.3.1 Direct Shear Test

Direct shear tests can be used in determining the shear strength of a material (ASTM D 3080) as well as the interface resistance of different materials (ASTM D 5321). Large shear boxes similar to that shown in Figure 5.3 are used in the field for municipal waste materials and for geosynthetic interface resistance in the laboratory. Results of shearing resistance between bales of municipal waste indicate that both friction angle and cohesion intercept increase as the waste density increases (Del Greco and Oggeri, 1994).

5.3.2 Direct Shear Test for Geosynthetic Interface Resistance

Geosynthetics are extensively used in modern landfill liners and covers (see Chapter 10). A direct shear test is the most common test for measuring the shear resistance of soil–geosynthetic interface; a typical setup is shown in Figure 5.4. These tests are expansive to conduct, large homogeneous specimens of soil are difficult to prepare, and there is generally no information on vertical displacement during consolidation and/or shearing. Current procedure (ASTM D 5321) requires a shear box with a minimum dimension of 12 in. (300 mm), 15 times the d_{85} of the coarser soil used in the test, or a minimum of 5 times the maximum opening size of the geosynthetic tested. The depth of the box must be the larger of 2 in. (50 mm) or 6 times the maximum particle size of the coarser soil tested. The minimum specimen width-to-thickness ratio is 2 : 1. A flat, jawlike clamping device is used for fixing the geosynthetic to a wood, metal, or plastic block, which may be placed in either or both of the soil containers. The geosynthetic may be affixed to the block with epoxy.

Figure 5.3
Schematic of a direct shear test for soil and waste

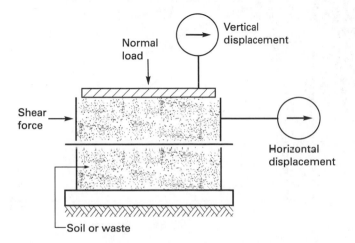

Figure 5.4
Schematic of a direct
shear test for soil–
geosynthetic interface
resistance

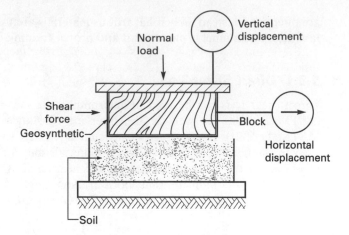

The tests are usually run at three different normal stresses representative of
the stresses in the field. Recommended strain rate is 1 mm/min for well-
drained materials. For the drained interface resistance of clay–geomembrane, a
strain rate of 0.1 mm/min or lower is suggested (Smith and Criley, 1995). The
Coulomb failure line is expressed as

$$\tau_I = c_I + \sigma \tan \phi_I \tag{5.4}$$

where τ_I is the interface shear strength, c_I is the soil–geosynthetic adhesion, σ is
the normal stress on the interface, and ϕ_I is the interface friction angle.

As seen in Figure 5.5, direct shear test curves show a peak and possibly a
residual strength. In selecting interface strength parameters, postpeak or
residual values are considered more appropriate for slope stability analysis.

Figure 5.5
Clay–geomembrane
interface resistance for
normal stresses of
150 kPa and 410 kPa
(after Byrne et al., 1992)

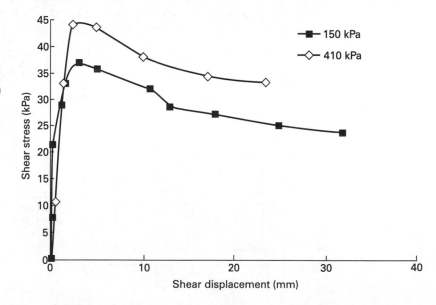

However, in the curves shown residual values have not been achieved because the interface resistance continues to drop.

Stark and Poeppel (1994) used a torsional ring shear test for evaluation of residual interface resistance between clay and a geomembrane (Figure 5.6). It is obvious that the assumption that the Coulomb line represents failure is not appropriate. At normal stresses beyond point *A*, this assumption will yield nonconservative values for interface resistance. Thus, the nonlinearity of the failure envelope should be accounted for in stability analysis.

In a sandwich test, each of the two halves of the shear box are filled with representative soils from the site, with the interfacing geosynthetic either anchored to the shear box or free floating. The effect of anchoring the geomembrane, which is considered more representative of field conditions, is shown in Figure 5.7. The advantage of sandwich tests is that soils on either side of the geosynthetic can interact as they normally do under field conditions. For example, geosynthetics used in liners and caps are frequently anchored to add stability.

Figure 5.6
Nonlinear failure envelope for geosynthetic interface resistance

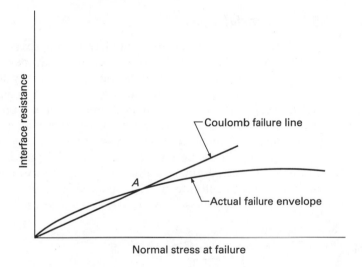

Figure 5.7
Effect of geomembrane anchorage (after Smith and Criley, 1995)

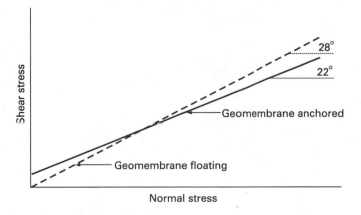

In vertically embedded geotextiles, at shallower depths failure occurred within the soil mass as a prism of soil around the embedded geotextile heaved during pullout (see Figure 5.8). With increasing depth of embedment, the boundary of the failing prism moved toward the face of the geotextile. At an embedment depth of about 3 feet, failure occurred at the soil–geotextile interface. As shown in Figure 5.9, when a geosynthetic is placed horizontally (AB) and continued vertically into the anchor trench (BC), the total resistance of length AC is more than 25% larger than the sum of the individual resistance of AB and BC (Khera et al., 1997).

5.3.3 *Unconfined Compression Test*

In an unconfined compression test, a specimen is subjected to an axial compression. This test has been used for unsaturated fly ash, stabilized fly ash, and flue gas desulfurization sludges.

5.3.4 *Triaxial Test*

All three types of triaxial test stimulate in situ stress conditions better than does a direct shear test. In particular, the measurements of stresses due to imposed deformations, volume changes, and pore water pressure are more accurate. In an unconsolidated undrained (UU) triaxial test, the specimen is not allowed to drain during application of confining pressure or axial load. In a consolidated undrained (CU) test, the test specimen is allowed to consolidate fully under the applied confining pressure but drainage is not permitted during axial loading and the pore water pressure may be measured. In a consolidated drained (CD) test, drainage is ensured during the entire test. Results of two CU tests on cement bentonite mixes are plotted in Figure 5.10.

Figure 5.8
Observed heaving of sand prism during vertical pullout test

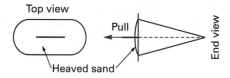

Figure 5.9
Anchor trench for geosynthetics

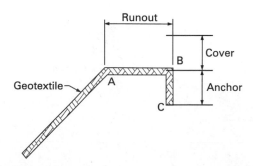

Figure 5.10
Stress–strain and pore water pressure in cement bentonite mixes (after Khera, 1995)

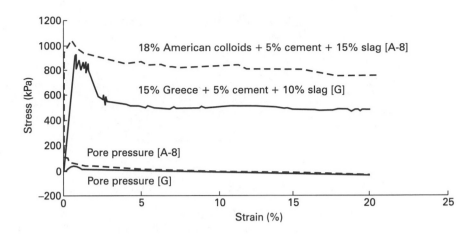

Example 5.1

In a series of CU tests on a fly ash the following data were obtained.

Test no.	σ_3 kPa	$\sigma_1 - \sigma_3$	u
1	150	140	40
2	300	310	70
3	450	450	120

Determine the cohesion and the angle of friction with respect to both total and effective stress.

Solution: Make a chart and a Mohr circle plot (as shown in Figure 5.11). The cohesion and angle of friction with respect to total stress are 0 and 19,

Figure 5.11
Mohr circle plot from data of Example 5.1

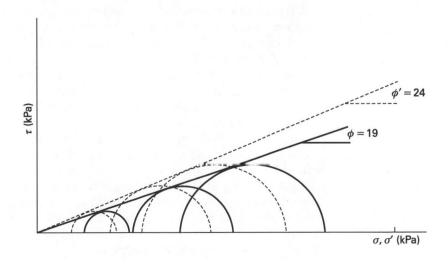

respectively. For effective stress, the cohesion is zero and the angle of friction is 24.

Test	σ_1	σ_3'	σ_1'
1	290	110	250
2	610	230	540
3	900	330	780

5.4   SELECTION OF STRENGTH PARAMETERS

For fine-grained soils with low hydraulic conductivity and in cases where failure occurs so rapidly that drainage cannot materialize, the total stress strength parameters should be used. In particular, total stress analysis is used for short-term stability or during and immediately after construction conditions. For these situations, unconsolidated undrained tests (or unconfined compression tests, if appropriate) are used to determine the strength parameters. For foundations on very soft soils where progressive failure may occur, special considerations are required (Duncan and Buchignani, 1975).

If boundary conditions are such that effective drainage can take place, drained or effective stress strength parameters can be used. Such conditions will prevail in most coarse-grained soils except where the loading is rapid (such as during an earthquake).

5.5   STRENGTH OF MUNICIPAL WASTE

Because of heterogeneous nature and varying degree of decomposition in municipal waste, it is very difficult to obtain reliable strength parameters. Some of the different methods used by engineers to determine these parameters and their results of these methods are presented here.

Unconfined compression tests carried out on compacted bales of waste by Fang and Slutter (1976) showed no failure stress even at very large strain. As shown in Figure 5.12, triaxial tests also showed no sign of failure even at strains of 20 to 40%. In a report by Cooper and Clark Engineers (1982), triaxial compression test data on Shelby tube samples taken from refuse also showed strain hardening. At strains greater than 20% with load still increased, there was still no indication of specimen failing. The report assigned a cohesion value of 35 kPa to the refuse.

Siegel et al. (1990) applied direct shear tests on refuse specimens 130 mm in diameter taken from a depth of 4.6 to 25 m. They found great variability between each specimen's composition and its response to shearing, with peak strength occurring at strains of 16 to 39% of a specimen's diameter. Thus, they selected a strain value of 10% for reporting refuse strength. If the material was considered to have no cohesion, the friction angle for individual specimens was as high as 81°. When specimens containing large proportions of soil were excluded, the friction angle dropped to 39°. Therefore, Siegel et al. warned

Figure 5.12
Stress–strain behavior
of municipal solid waste
in triaxial tests (after
Jessburger, 1994)

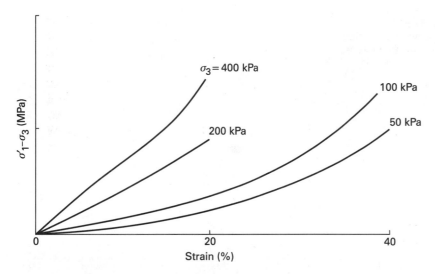

against using individual specimens for deriving the Mohr–Coulomb strength parameters.

In a summary of laboratory data on refuse from several sources, Singh and Murphy (1990) found cohesion values from 0 to 80 kPa and friction angles from 0° to 42°. In most of the reported test data the refuse was assumed to possess friction but no cohesion. Because the reported friction angles are comparable to dense sand, this assumption must be viewed critically.

Landva and Clark (1990) used direct shear apparatus with plan dimensions of 0.434 m by 0.287 m to test several kinds of waste. The rate of shear deformation was 1.5 mm/min except where pore pressure buildup indicated incomplete drainage. The waste material from old landfills in both natural and dried states exhibited a granular and fibrous nature. The shear strength parameters c and ϕ ranged from 19 to 23 kPa and 39° to 33°, respectively. For a fresh shredded waste with a large proportion of plastic sheet waste, the corresponding values were 23 kPa and 24°. For wood waste, c was 0 and ϕ was 36°. For horizontally stacked plastic bags sliding along the shear plane, c was 0 and ϕ was 9°. Although there was no clear evidence of significant change in shear strength with time where substantial decomposition occurs, regions of weakness could develop and considerably diminish the overall shear strength. Local zones of weakness may also result if waste materials with lower friction angles are concentrated in critical shear zones.

Del Greco and Oggeri (1994) interpreted their test data on municipal waste for two initial densities to postulate that the angle of friction at lower normal stress is greater than at higher normal stress, as shown in Figure 5.13. On the other hand, Kavazanjian et al. (1995) suggested $c = 24$ kPa and $\phi = 0$ for normal stress below 30 kPa and $c = 0$ and $\phi = 33°$ for higher normal stresses.

The results from in situ measurement of strength using standard penetration tests (N) and cone penetration tests (CPT) have been mixed. There are no established relationships between refuse penetration resistance and shear strength.

Hinkle (1990) reported an investigation of a landfill that was closed in 1981

Figure 5.13
Dependence of friction
angle on normal stress
for municipal waste

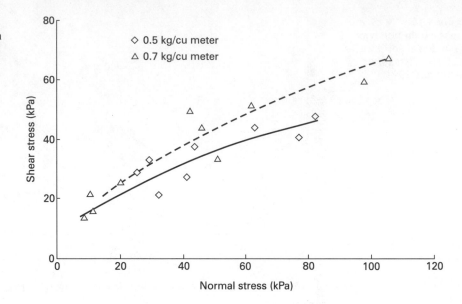

Figure 5.13
Dependence of friction
angle on normal stress
for municipal waste

near the Los Angeles Harbor. The lower 14 m of the landfill consisted of rubble and tires and the upper 15 m consisted of rubbish. The landfill had been subjected to precompression load from coke stock piles up to 12 m (40 ft) during a five-year period. The fill was randomly placed and covered each day with a thin soil cover. 40% of the cone penetrometers reached the design depth without encountering obstructions. The results were relatively consistent with tip resistance of about 5 MPa and skin friction of about 50 to 100 kPa.

Siegel et al. (1990) reported much greater difficulty penetrating a landfill in Monterey Park, California, where two or three attempts had to be made at most locations. A trend of lower-bound tip resistance of about 25 kPa (0.25 t/ft^2) was reported for most of the tests. Note this value of tip resistance is two orders of magnitude lower than that reported by Hinkle. This shows that, unlike soils, there is no correlation for cone penetration resistance for waste that can be used to determine the nature of the material penetrated.

Figure 5.14 shows the plan of a test loading conducted on a sanitary landfill in Southern California. Up to 11.6 m (38 ft) of fill with an in situ density of 19.6 kN/m^3 (125 lb/ft^3) were placed at an average rate of 0.5 m/day (1.6 ft/day). The landfill foundation is a massive conglomerate underlain by moderately cemented and thickly bedded silt stone. From the slope inclinometer data, it was seen that a movement of about 432 mm (17 in.) occurred in the upper portion and that the movement accelerated with increasing fill height. When fill placement ceased, the rate of movement decreased markedly. Evaluation of the progress of the horizontal and vertical movements indicated that the lateral movements correlated well with the vertical movements due to compression of refuse. Secondary lateral movements occurred due to spreading of the refuse material below the test fill.

The refuse slope cited was apparently stable during and after placement of the test fill. In terms of conventional slope stability analysis, the factor of safety

Figure 5.14
Plan of a test loading
(reproduced from
Oweis et al., 1985,
by permission)

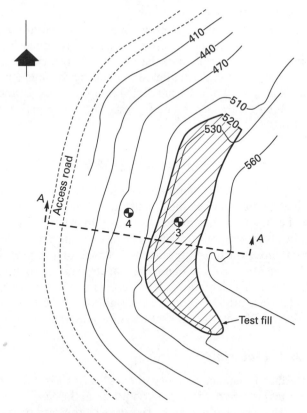

against failure would be at least 1.0. Assuming that the material could be represented by the conventional Coulomb parameters c and ϕ, their values were calculated with a factor of safety of 1.0 and the ordinary method of slices for stability analysis (Duncan and Buchignani, 1975).

Pairs of c and ϕ for the landfill are shown in Figure 5.15. Since the slope had not failed, the pairs of c and ϕ depicted by the solid line are less than the actual values and are considered conservative. In the calculations a bulk unit

Figure 5.15
Strength parameters for
landfill from various
methods (Oweis et al.,
1985; Landva and Clark,
1990; Siegel et al., 1990;
Singh and Murphy, 1990)

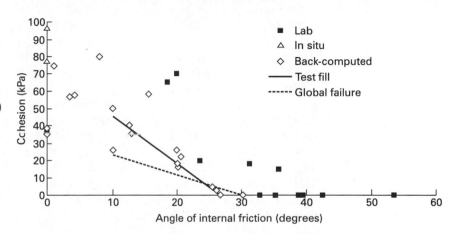

weight of 9.4 kN/m^3 (60 lb/ft^3) was assumed (Woodward-Clyde Consultants, 1984). For example, numerous "reasonable" combinations of the unit weight, cohesion, and angle of friction would produce similar safety factors. A summary of strength data based on laboratory tests, in situ field tests, and back-computations from field tests and performance records for sanitary land-fills is also shown in Figure 5.15. Note that most of the back-computed values fall along or above the line representing the ϕ and c values computed for the test fill.

5.6   INTERFACE RESISTANCE OF GEOSYNTHETICS

Long et al. (1993) summarized numerous factors that affect reported soil–geosynthetic interface shear resistance. Interface resistance test results reported in the literature are based on nonstandardized shear boxes that vary in size from 50 to 460 mm. In addition, test results on smaller specimens may differ from those on larger specimens because of the particle size of the soil, type of geosynthetics, test boundary effects, and distortion effects of the geosyn-thetics. The orientation of the geosynthetic specimen and the type of fluid (water, leachate, or others) also may affect the test results. Thus, the data in Table 5.1 for interface resistance is for guidance only.

5.6.1 *Materials*

Interface resistance of geosynthetics and soil is effected by the soil type, com-pactness, strength, drainage condition, and degree of saturation. For saturated clay, the test results may be sensitive to the rate of loading and drainage con-ditions during the test in the direct shear box. In the field, water may appear at the clay-geomembrane interface due to high compaction moisture content, wetting of clay to mitigate cracking, condensation at the bottom of membrane overlying clay ("solar still"), consolidation of clay, leachate migration, tem-perature differential, or leachate passing through the membrane. The latter will reduce interface resistance and is difficult to evaluate.

Interface resistance is also affected by the type of geosynthetic. For example, the polymer type of the geosynthetic affects the flexibility of the geo-synthetic. Geosynthetics that deform around the soil grains or get indented by the soil grains produce better frictional resistance than do stiffer types. The thickness, texture, and surface roughness of the geosynthetic also influence test results. For evaluation of the properties of geosynthetics, see Chapter 10.

In the case of geotextiles, the structure (woven or nonwoven), bonding process, and fabric openness affect the test results. Geotextiles tend to polish geomembranes, and when fully polished there may be a reduction in friction of as much as 2° (Mitchell et al., 1990). Geotextile/geonet shows higher resistance in a direction transverse to the geonet strand alignment. As shown in Table 5.1, there is considerable gain in interface resistance with textured HDPE for MH. For poorly graded sand (SP), there is only slight gain in interface resis-tance from texturing. Geogrids provide much larger resistance to pullout forces than that observed in direct shear.

Table 5.1 Average values of interface friction for geosynthetics

Interfacing materials	$\phi_1°$ PEAK		$\phi_1°$ RESIDUAL		Displacement at peak strength (mm)	Displacement at residual strength (mm)	τ_1 peak (kPa) Avg.	τ_1 residual (kPa) Avg.
	Range	Avg.	Range	Avg.				
Nonwoven (NW) (PP, PET) geotextile–geonet (HDPE)	11.9–19.1	15	10.4–27	13.5	2	4		
NW geotextile (PP, PET)/soil (SW & SP) drained	26–52	35.3	24–41	32	5	8		
Smooth geomembrane HDPE/geonet (HDPE)	7–12.4	8.7	4.7–10.9	7.5	1.5	15		
Smooth geomembrane HDPE/geonet, aligned with strand, submerged	6.3–8.9	7.6	5.4–7.2	6.3	0.7–0.9			
Smooth geomembrane HDPE/geonet, transverse to strand, subrerged	7.6–10	8.8	7.1–9.5	8.3	1–6.5			
Smooth geomembrane HDPE/NW geotextile (PET, PP)	8.1–9.8	10.9	3.7–14.5	5.2 RS, 9.5	0.75	725 RS, 35		
Unpolished HDPE/geotextile, dry	11.8–13.2	12.5	9.4–11.8	10.3	1–3			
Fully polished HDPE/geotextile, submerged	8.3–10.3	9.3	7.7–9.3	8.4	0.5–2			
Smooth geomembrane/soil (CL–CH), undrained		9		6	2	1000	7	5
Rough geomembrane (HDPE)/geonet (HDPE)		11.7		11	1.5	15		
Rough geomembrane (HDPE)/NW geotextile (PET, PP)		23.5		18	0.75	725		
Rough geomembrane HDPE/soil (CL & CH)		17		14.5	1.5	1000	10.8	9.7
Rough geomembrane HDPE/soil (SW, SM, SP)		28						
Smooth HDPE/sand (SP), saturated		19–20						
Rough HDPE/sand (SP), saturated		20–22						
Smooth HDPE/silt (MH), saturated		10–12						
Rough HDPE/silt (MH), saturated		29–32						
Smooth geomembrane HDPE/soil (CL), drained		17.6						

RS = ring shear.
Source: Mitchell et al., 1990; Long et al., 1993; Orman, 1994

Therefore, for each specific application the proposed product should be tested under expected field conditions. Great care must be taken in designing the testing program for the given field conditions, preparing the sample, performing the tests, and evaluating the results.

5.7   STRENGTH OF MINERAL AND INDUSTRIAL WASTE

Most industrial wastes—mine wastes, processes tailings, fly ash, FGD, paper mill and pulp waste, etc.—are disposed of as thin slurries in diked containment areas. Because of their high water content, they occupy large volumes and their shear strength is low. To increase the capacity or life of their storage areas, the volume of the disposed materials must be reduced.

Because it is difficult to obtain undisturbed specimens from slurry disposal sites, strength parameters for industrial waste are difficult to assess. The slurry sites may be inaccessible because the materials in the site are too weak to support sampling equipment. The sites also may contain large proportion of cohesionless materials that are difficult to sample using conventional methods, or the deposit may be weakly cemented and the sampling process may destroy the cementation.

Therefore, laboratory prepared samples are often the only kind available for testing. However, when laboratory specimens are used, the effect of disturbance, aging, and applicability of testing technique on strength properties should be included in the test assessments. Extensive laboratory research and large scale field tests are being done to utilize much of the waste in civil engineering structures without adversely affecting the environment.

5.7.1 *Mixed Coal Refuse*

Unconfined compression tests on compacted coal refuse indicate that its strength is dependent on the water content at the time of the test. With increasing water content, the strength decreases. In triaxial tests some of the loose specimens showed a sudden drop in stress level followed by a rapid increase, which was attributed to the collapse of the soil structure followed by a partial recovery due to the lateral confining pressure (Saxena et al., 1984).

The angle of friction for undrained condition ranges between 10° and 28° with cohesion ranging from 10 to 40 kPa (0.1 to 0.4 t/ft^2). For effective stress the angle of friction ranges between 25° to 43° with cohesion values similar to undrained values. The large deviation in strength is due to variation in the composition of coal waste materials. However, in designing containment areas, the angle of friction may be taken as 30° to 35° for properly compacted coal refuse while neglecting cohesion.

For uncontrolled placement of materials, the friction angle may range between 24° and 28° (Usmen, 1986). In uncontrolled loose fills, a greater surface area is exposed to chemical weathering, making the waste more susceptible to loss of strength at an accelerated rate.

5.7.2 Fly Ash

Fly ash possesses pozzolanic properties; that is, in the presence of moisture its strength increases with time (Table 5.2). This increase in strength due to curing is much more for Class C fly ash, which contains a higher proportion of lime (CaO) and other compounds, than for Class F fly ash. In addition, Class C fly ash possesses high cohesive strength whereas Class F fly ash exhibits no cohesion when in a saturated or dry state and only slight cohesion when in a partially saturated state.

5.7.2.1 Lagoon Fly Ash When fly ash is disposed of as slurry, its strength is generally lower than when it is compacted near its optimum moisture content. On old disposal sites the strength gain may be considerable, making it difficult to obtain samples even by coring. Generally, the strength decreases with increasing moisture content and increases with increasing curing time. Strain at failure decreases with increasing time.

  Cunningham et al. (1977) reported field and laboratory investigations of slurry-deposited (lagoon) fly ash and slag. In situ density had the lowest value: 8.85 kN/m^3 (56.3 lb/ft^3) with an average of about 11.7 kN/m^3 (74.4 lb/ft^3). Water content ranged between 35 and 60%. For an undisturbed sample ϕ was 27.5° and for remolded samples ϕ was 25°. At a lower elevation, ϕ was only 11.5° (very low but consistent with other data) and c was 40 kPa. Toth et al. (1988) reported ϕ of 28° to 30° from back-computations of slope failures in hydraulically placed ash. From a survey of literature, McLaren and DiGioia (1987) suggested ϕ of 32° to 33° for ash ponds and poorly compacted fly ash.

Table 5.2
Strength properties of coal and coal-burning waste

Description	Cohesion (kPa)	$\phi°$ friction angle	Unconfined compressive strength (kPa)
Coal refuse	0–25	25–40	
Undrained condition	10–40	10–28	
Effective stress	0–40	25–43	
Fly ash, Arizona 7-day			
unit wt 12.6 kN/m^3			223
unit wt 13.4 kN/m^3			331
unit wt 13.8 kN/m^3			587
Fly ash (silica 46%, aluminum 34%, calcium 7%)			
Slurry samples		37	
Compacted, undrained		41	
Effective stress			
Compacted, drained		37	
West Virginia fly ash	0–18	16–34	
Shelby tube samples (CU)		34	
Shelby tube samples (CD)		37.5	
West Virginia bottom ash		38–43	

Source: Gatti and Tripiciano, 1981; Hagerty et al., 1977; Mabes et al., 1977; Srinivasan et al., 1977

5.7.2.2 Compacted Fly Ash   Seals et al. (1977) reported results from a site where fly ash, removed primarily from an inactive lagoon, was compacted. Consolidated undrained and drained triaxial tests on compacted specimens with dry density of 14.9 kN/m^3 (94.8 lb/ft^3) and 15.9 kN/m^3 (101.1 lb/ft^3) and moisture content of 20.2% and 17.6% showed zero cohesion and ϕ of 34° and 37.5°, respectively. The final degree of saturation for these specimens was 89% and 93%, respectively. The use of a standard penetration test data to predict the angle of internal friction overestimated its value by at least 10°.

From extensive data on compacted fly ash, McLaren and DiGioia (1987) derived ϕ of 34° with standard deviation of 3.3°. However, the considerably lower ϕ value of 17° to 18° was reported by Golden and DiGioia (1991) for a compact fly ash with a dry unit weight of 10.4 kN/m^3 (66.2 lb/ft^3) to 11 kN/m^3 (70 lb/ft^3) and optimum moisture content range of 32 to 37%. For consolidated undrained tests, Wei et al. (1991) observed a decrease in strength of 20 to 25% upon immersion of a compacted Class F fly ash. Wayne et al. (1991) also reported a drop in ϕ and c upon saturation.

The addition of lime or cement, even in small amounts, may result in a substantial and rapid increase in the strength of fly ash. For ash A (Table 5.3), the addition of lime produced a higher increase than did the addition of the same proportion of cement. For ash B, cement produced the higher increase in strength. Similarly, fly ash can impart strength and stability to other materials, such as FGD, sludge, and pulp and paper mill waste.

As we will see, the strength behavior of fly ash is most significantly affected by its composition, moisture content, method of sample preparation, and age. Values of ϕ drop upon saturation, and if any value of c is observed in the unsaturated state, it drops to 0 upon saturation. Whenever fly ash is to be utilized or disposed of, representative samples must be carefully obtained or the laboratory specimen preparation must be precisely consistent with the field conditions.

The existing correlation between standard penetration resistance and angle of friction for cohesionless natural soils do not apply to combustion residues such as fly ash and bottom ash. Thus fly ash and bottom ash, which behave like silt and sand, can be used in the construction of controlled structural fill. However, upon saturation their strength shows a drop that must be accounted for in the construction design. Golden and DiGioia (1991) reported several demonstration highway projects where fly ash was used either alone or stabilized with lime and/or cement. Field monitoring data indicated the response of these structures to be excellent. However, because of the susceptibility of fly ash to frost heaving and erosion, appropriate measures such as proper drainage and soil and vegetative covers must be taken to guard against such failures. Although a major environmental concern for such uses is the impact on ground water, the leaching data reported by Golden and DiGioia (1991) and Toth et al. (1988) showed no adverse impact in this regard.

5.7.2.3 Class C Fly Ash   Indraratna et al. (1991) reported consolidated undrained and drained triaxial test data for a Class C pozzolanic fly ash. Specimens compacted at in situ water content and consolidated to a stress

Table 5.3
Effect of curing time
on stabilized ashes

Material	q_u (MPa) (7 days)	q_u (MPa) (28 days)
Michigan	0.170	0.210
New Jersey	0.310	0.425
England	0.550	0.660
West Virginia		
Ash A, with 3% lime	19.6	31.6
Ash A, with 3% cement	7.4	12.8
Ash B, with 3% lime	2.8	4.0
Ash B, with 3% cement	3.2	8.0
Lignite fly ash 10% sand 90%	2.6	4.6
Lime : fly ash : FGD sludge		
6.3 : 43.7 : 50.0	3.3	5.5
2.5 : 55.8 : 41.7	3.0	4.0
0 : 41.2 : 58.8	0.3	0.4
Fly ash with FGD cement mix		
6%	3	3.9
8%	4.48	5.24
10%	5.9	7.0
Fly ash with 10% lime	0.29	0.35
Fly ash with 5% lime	0.5	1.8
Fly ash and bottom ash with lime	1.0	2.0

Source: Chae and Gurdziel, 1976; Bowders et al., 1987; Soliman et al., 1986; Taha and Saylak, 1992; Torii and Kawamura, 1991; Ullrich and Hagerty 1987; Wayne et al., 1991

range of 0.15 to 0.6 MPa had a ϕ' of 26° and c' of 35 kPa. Compacted specimens at maximum dry density of 17.2 kN/m³ and optimum moisture content of 12% were cured for one week and consolidated to a stress range of 3.0 to 7.0 MPa. These specimens showed a ϕ' of 36° and c' of 1.8 MPa, which is a substantial increase over freshly prepared specimens.

5.7.3 Solid Waste Incinerator Residues

Unconsolidated undrained tests on partially saturated compacted specimens of fly ash from a solid waste incinerator yielded ϕ of 43° and c of 65 kPa (Poran and Ahtchi-Ali, 1989). These strength parameters are higher than generally reported for coal Class C or Class F fly ash. When the specimens were stabilized with lime, ϕ increased to 45° and with cement ϕ was 44°. Increases in c were 50 to 260% upon lime stabilization. The strength data is from unconfined compression tests conducted four days after the specimens were compacted. Because of the high concentration of some metals and other pollutants in incinerator fly ash, its use as construction material requires further research.

5.7.4 Bottom Ash

For an average void ratio ranging between 1.08 and 1.68, Seals et al. (1977) reported ϕ for wet and dry bottom ash ranging between 38° and 42.5°. Direct shear tests on bottom ash by Huang and Lovell (1990) show that for one of the bottom ashes ϕ increased from about 35° to 43° as the relative density increased from 10 to 80%. For another ash with similar relative densities, ϕ increased from about 41° to 50°. In all instances ϕ was greater for bottom ash than for sand at the corresponding relative densities.

5.7.5 Flue Gas Desulfurization (FGD) Sludge

The sludges generated by the sulfur oxide and particulate removal systems used in the United States are essentially frictional in nature. The angle of friction for compacted plain FGD sludge having dry density between 11 kN/m³ (70 lb/ft³) and 13.4 kN/m³ (85 lb/ft³) varies between 31° and 39° with negligible cohesion, as shown in Table 5.4.

In one test, the addition of fly ash to FGD dual-alkali sludge decreased the friction angle of FGD sludge under undrained conditions but showed no change under drained conditions (Krizek et al., 1987). In another test, the addition of 40 to 50% fly ash to FGD sludge showed a considerable increase in angle of friction and cohesion (Hagerty et al., 1977). The contradictory results may be due to the differences in type of fly ash or curing time.

According to Krizek et al. (1987), the compressive strength of lime-treated FGD sludge ranges from 100 to 1 MPa. When FGD sludge containing 40 to 60% fly ash was treated with 1 to 3% lime its effective friction angle increased by several degrees if the specimens were allowed a certain curing period.

Soliman (1990) reported that specimens stabilized with lime–fly ash exhibit strength by cementation not by friction because the specimens failed along a vertical plane. Also, most of the strength increase occurred during the first 20 to 30 days of curing. As the dry density of the mixes increased from 11 kN/m³ (70 lb/ft³) to 15 kN/m³ (95.4 lb/ft³), the compressive strength rose from 0.5 to 6 MPa.

Ullrich and Hagerty (1987) reported that sulfite sludges stabilized with lime–fly ash had higher initial compressive strength than similarly stabilized sulfate sludge. After curing for 28 days, a much higher strength was obtained for the sulfate sludge, although the sulfite sludge also showed a strength gain. In particular, the increased curing time resulted in greater freeze–thaw durability.

The strength of the sludge–fly ash mixture was much greater for Class C fly ash than for Class F fly ash (which lacks self-cementing characteristics). Undis-

Table 5.4
Strength parameters of FGD sludge and sludge with fly ash and lime

Sludge type	PLAIN SLUDGE		SLUDGE–FLY ASH		SLUDGE–FLY ASH–LIME	
	c (kPa)	$\phi°$	c (kPa)	$\phi°$	c (kPa)	$\phi°$
Dual-alkali	0	36	30–75	33–36	30–55	35–38
Sulfite	30	31	55–60	33–35	5–55	31–37
Sulfate	25–55	37	75–100	32–39	30–55	35–44

Source: Krizek et al., 1987; Smith, 1992

turbed samples and those compacted in the field had lower unconfined compressive strength than the corresponding laboratory prepared samples. The higher strength of the laboratory prepared specimens was attributed to better mixing and quality control. The strength increased with curing time, reaching close to the maximum value in 20 to 40 days. A curing period of 28 days appears reasonable for materials treated with fly ash and/or lime (Soliman et al., 1986).

DiGioia et al. (1991) described the construction of structural fill for a short section of an expressway where stabilized FGD sludge was used in combination with fly ash. These materials were isolated from the surrounding soils and an extensive drainage system, as shown in Figure 5.16, was used. The monitoring wells within the fly ash did not produce sufficient volume for sampling.

In evaluating the possibility of disposing of FGD sludge in the ocean, Soliman et al. (1986) used brackish water during laboratory preparation of a sample mixed with fly ash and lime. Such samples exhibited almost 90% larger unconfined compressive strength than those prepared without salt water. No relationship was reported between confining pressure and shear strength. Drained compressive strength was, however, lower than unconfined compressive strength, which may be attributed to the slower rate of deformation required for drained tests.

5.7.6 Mine Waste

As mine tailings are released in slurry form into impoundments, the particles sort themselves out, with the coarsest particles settling closest to the discharge

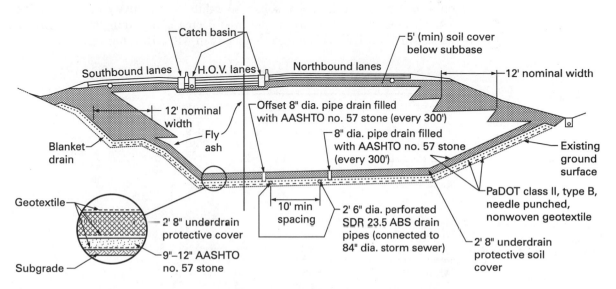

Figure 5.16 A test embankment using stabilized FGD sludge with fly ash.
Source: Reprinted from DiGioia et al., 1991

point. Such deposits are inhomogeneous and fairly loose. The behavior of tailings specimens in triaxial tests is similar to natural soils having the same range of particle sizes and densities.

In undrained tests, tailings exhibit both friction and cohesion. The angle of friction with respect to total stress ranges from 14° to 24° and cohesion, where present, may be up to 95 kPa (Table 5.5). In drained conditions, most tailings sands do not exhibit any cohesion. Compared with natural sand particles that may be subrounded or rounded, tailings particles are angular and show a somewhat higher angle of internal friction for effective stress. With increasing confining pressure, contact points get crushed and friction angle drops.

An effective stress increase from 0 to 150 kPa on tailings sand with relative density of 0 resulted in a decrease of friction angle from 41° to 29°, as shown in Figure 5.17. For dense tailings sand, Mittal and Morganstern (1975) reported a decrease in ϕ' from 37° to 31° as effective normal stress increased from 70 to 275 kPa. There was very little change in ϕ' as effective stresses were further increased up to 550 kPa.

For comparison, the dependence of ϕ' on normal effective stress for two natural sands is shown in Table 5.6. For the dry Danish sand with relative density of 95%, friction angle decreased from 52° to 42° as the effective stress was increased from about 25 to 1100 kPa. Thus the assumption of best-fit Coulomb failure line would yield nonconservative results similar to that represented in Figure 5.6 and Figure 5.13.

The range of expected stress in the field must be considered when designing a laboratory strength-testing program for tailings. Nonlinearity of the failure envelop should be accounted for in design.

For red mud, which is a bauxite residue, Somogyi and Gray (1977) reported a linear relationship between water content and undrained shear strength on a logarithmic scale. These soils also exhibited sensitivity between 3.6 and 7.4. The addition of sand to the red mud decreased its strength and increased its sensitivity. At water contents higher than 41%, leaching resulted in com-

Table 5.5
Total stress strength cohesion and friction parameters for mine waste

Material	$\phi°$	c (kPa)
Copper tailings and beach sands	19–20	0–95
Copper slimes	14–24	60
Red mud slimes	22	5

Source: Vick, 1988

Table 5.6
Dependence of friction angle on effective stress for natural sands

Material	$\phi'°$	Effective stress (kPa)
Monterey sand	52–42	50–1600
Danish normal sand	53–39	25–1100

Source: Kutter et al., 1988

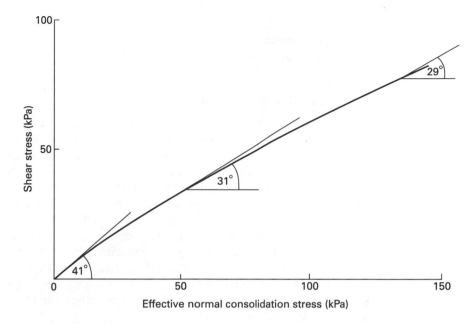

Figure 5.17
Strength dependency
on consolidation stress
(Vick, 1988)

plete loss of shear strength, but there was an increase in strength for lower initial water content.

For most tailings and slimes, ϕ' varies between 28° and 42° with majority of the values being 34° $\pm$ 3° (Vick, 1988). Because of the large variation in strength parameter, site-specific investigation must be made.

5.7.7 Pulp and Paper Mill Waste

For fresh paper mill waste, the angle of friction measured in triaxial tests decreased linearly from about 75° to 45° as the organic contents decreased from 65 to 35%. Large strains are required to fully mobilize the shear strength (Charlie, 1977). At low organic contents, the strength of the sludge approached that of its inorganic constituents. The decomposition of organics will have a similar effect. The low strength of existing sludge landfills is due to high placement water content, low permeability, long drainage distances, decomposition of a certain proportion of organics, and residual pore pressures.

The strength for the fresh waste was between 53 kPa (0.56 t/ft²) and 22 kPa (0.23 tn/ft²) for moisture contents ranging from 139 to 194%. When 10% by wet weight of sludge fly ash was added to the fresh sludge, for comparable moisture content the mixture showed a strength of only 30.5 kPa (0.32 t/ft²). The reduced strength is attributed to the lowered organic contents resulting from the addition of fly ash.

The shear strength of compacted samples obtained from the drying fields was 54.5 kPa (0.57 t/ft²) and 2 kPa (0.02 t/ft²) at moisture contents of 45 and 134%, respectively. The material for these samples was blended mechanically before compaction. Note that for comparable moisture content, the strength of 2 kPa is considerably lower than 53 kPa (which is the strength for fresh sludge). The change in shear strength with change in moisture content was reported to be much greater at lower water content than at higher water content (Jedele, 1987).

5.7.8 *Dredged Waste*

Extensive data on the shear strength of dredged materials from the vicinity of Toledo, Ohio has been reported by Krizek and Salem (1977). As one would expect, the strength decreased with increasing moisture content. The unconfined compression test showed lower strength than the vane test, probably due to unavoidable sampling disturbance. Contrary to the lower expected values from sampling disturbance, higher strength from laboratory vane is attributed partly to the confining effect of the sampling tube and partly to the difference in the height-to-diameter ratios of the two vanes. For remolded soil, there appeared to be fairly good agreement between the field and laboratory strength values. In general, the strength was found to increase with age and decrease with increasing distance from the inlet pipe. To determine the age of a deposit, the starting time was taken to correspond to one-half the final volume of dredging at that site.

Rizkallah (1987) reported laboratory vane shear tests on dredged waste stabilized with various proportions of cement, fly ash, and lime. The addition of fly ash had virtually no effect on the strength. Some gain in strength was observed with lime, but curing time of 3 to 28 days was of little consequence. The mixture with cement showed the most increase in strength and showed further improvement after 28 days of curing. Knox and Najjar (1994) reported tests on stabilizing lagoon sludge, dredged spoils, and contaminated soils at an aluminum production facility. Materials with 30% solids needed 40% cement and those with 50% solids needed 20% cement to yield the design strength for disposal in an RCRA-permitted landfill.

5.8   SUMMARY

Strength parameters for soils and waste materials are based on the validity of Mohr–Coulomb failure envelope. However, the assumption of a linear failure envelope can lead to an overestimate of shear strength. Several laboratory test methods are available that can be used for most type of materials, and appropriate testing equipment is available for corrosive material to ensure the safety of laboratory personnel. The selection of a test is determined by the type of soil, loading, and drainage conditions.

In many instances the age of a deposit can have a significant effect on the strength properties. Also, because of the extreme heterogeneous nature of the many waste materials the strength parameters can be difficult to evaluate. Thus the range of values for strength parameters given in this chapter should only be used as a guide. For actual design, site-specific investigations should be carefully conducted and field tests are highly recommended for most types of wastes.

NOTATIONS

σ	total stress (F/L^2)	u	pore water pressure (F/L^2)	c'	cohesion with respect to
ϕ	angle of friction with respect	σ_1	major principal stress (F/L^2)		effective stress (F/L^2)
	to total stress (°)	σ_{1f}	major principal stress at	c_I	soil–geosynthetic adhesion
ϕ'	angle of friction with respect		failure (F/L^2)		(F/L^2)
	to effective stress (°)	σ_3	minor principal stress (F/L^2)	τ_f	shear strength (F/L^2)
τ_I	interface shear resistance	σ_{3f}	minor principal stress at		
	(F/L^2)		failure (F/L^2)		
ϕ_I	interface friction angle (°)	c	cohesion (F/L^2)		

PROBLEMS

5.1 Refer to Example 5.1 (p. 81). Determine the magnitude of effective normal and shear stresses on the failure plane for test 3 at failure.

5.2 A specimen identical to that used in Example 5.1 is consolidated in a triaxial cell under a confining stress of 350 kPa. What would be the magnitude of σ_{1f} if the test conditions were (a) undrained and (b) drained? What would be the magnitude of pore pressure for (a)?

5.3 A drained direct shear test was performed on a coarse-grained soil with the data shown in Tables 5.7 and 5.8. Plot the horizontal displacement versus shear force

Table 5.7
Sample dimensions

Diameter (in.)	2.54
Height (in.)	0.88
Dry density (pcf)	105.65
Normal load (lb)	40

Table 5.8

Vertical dial ×0.0001″	Horizontal dial ×0.001″	Load dial (div)
0	0	0
1	10	70
12	20	105
43	30	120
73	40	125
112	50	128
128	55	124
168	70	120
200	82	115
221	95	113
241	107	110
257	120	108
270	140	104
292	170	101

Load ring constant = 0.3191 lb/div

and the horizontal displacement versus vertical displacement. Determine the angle of friction for the soil based on both peak strength and ultimate strength.

5.4 Direct shear test data for boiler slag is given in Tables 5.9 and 5.10. Plot the shear displacement versus shear stress for test 1. For test 1, the normal stress was 20 kPa. Using the results of all three tests, determine the angle of friction and cohesion.

5.5 A triaxial test was done on dry tailings. For a confining pressure of 100 kPa, failure occurred at major principal stress of 245 kPa.
  (a) Determine the friction angle with respect to effective stress.
  (b) Determine the inclination of failure plane to the horizontal plane.

5.6 The data shown in Table 5.11 were obtained from Class C fly ash. Determine the total stress strength parameters.

5.7 A consolidated undrained triaxial test was performed on a Class F fly ash, giving the data shown in Table 5.12. Effective consolidation stress was 400 kPa. Plot the strain versus principal stress difference and the strain versus pore water pressure diagrams for the specimen.

Table 5.9
Test 1

Shear displacement (mm)	Shear stress (kPa)
0.5	15
1	22
1.5	26.4
2	28
2.5	29.6
3	30.6
3.5	30.2
4	29.6
4.5	28

Table 5.10
Tests 2 and 3

Test No.	Normal stress (kPa)	Shear stress (kPa)
2	40	51
3	62	70

Table 5.11

Test	σ_3 (kPa)	$\sigma_1 - \sigma_3$ (kPa)
1	150	355
2	300	535
3	450	770

Table 5.12

Strain (%)	$\sigma_1 - \sigma_3$ (kPa)	Pore pressure (kPa)
0	0	0
0.5	150	70
1	225	130
2	220	245
3	180	300
4	145	325
6	100	355
8	85	370
10	65	375

5.8 Using the data from Problem 5.7, plot $p = \dfrac{\sigma_1 + \sigma_3}{2}$ along x-axis, $q = \dfrac{\sigma_1 - \sigma_3}{2}$ along y-axis for total stress, and p' versus q for effective stress. Determine angles ϕ and ϕ'.

6 Hydraulic Properties

6.1 BASIC HYDRAULIC PROPERTIES

We have looked at the effect of water in soil voids on the response of fine-grained soils such as clays and settlement resulting from the flow of water due to pressures from self-weight of materials and from foundations. In this chapter we will study the principles of flow of water through porous materials and apply them to the design of landfills and containment structures for controlling contamination of soils and water.

Darcy showed that for laminar flow in porous media the rate of flow (or discharge), q, is given as

$$q = -k \cdot A \frac{h}{L} \qquad (6.1)$$

where k is the *hydraulic conductivity*, A is the total cross-sectional area of soil perpendicular to the direction of flow, h is the total head loss, and L is the distance within the porous medium through which head loss occurs.

The *discharge velocity* is given by

$$v = \frac{q}{A} \qquad (6.2)$$

The *hydraulic gradient*, i, is defined as

$$i = \frac{h}{L} \qquad (6.3)$$

Laboratory values of hydraulic conductivity are reported at 20°C. Temperature corrections are based on the viscosity of water or the *permeant*, as given by

$$k_{20} = k_t \frac{\mu_t}{\mu_{20}} \qquad (6.4)$$

where k_t is the hydraulic conductivity, μ_t is the viscosity at temperature t, and μ_{20} is the viscosity at 20°C. Eq. 6.4 may be used for determining hydraulic conductivity at any other temperature. Typical values of hydraulic conductivity for some natural soils are given in Table 6.1.

Table 6.1
Hydraulic conductivity
of soils and rocks

Soil or rock formation	Range of k (cm/s)
Gravel	1–5
Clean sand	10^{-3}–10^{-2}
Clean sand and gravel mixtures	10^{-3}–10^{-1}
Medium to coarse sand	10^{-2}–10^{-1}
Very fine to fine sand	10^{-4}–10^{-3}
Silty sand	10^{-5}–10^{-2}
Glacial till	10^{-10}–10^{-4}
Homogeneous clays (unweathered)	10^{-9}–10^{-7}
Shale	10^{-11}–10^{-7}
Sandstone	10^{-8}–10^{-4}
Limestone	10^{-7}–10^{-4}
Fractured rocks	10^{-6}–10^{-2}

Source: Freeze and Cherry, 1979; Peck et al., 1974

According to Bernoulli's equation the total head is given by

$$\frac{p_1}{\gamma} + \frac{v_1^2}{2g} + z_1 = \frac{p_2}{\gamma} + \frac{v_2^2}{2g} + z_2 + h \tag{6.5}$$

where p is the pressure, γ is the unit weight of water, v is the flow velocity, g is the acceleration due to gravity, z is the elevation, and h is the loss in head. Because of the low flow velocity in soils, velocity head is neglected and Eq. 6.5 becomes

$$\frac{p_1}{\gamma} + z_1 = \frac{p_2}{\gamma} + z_2 + h \tag{6.6}$$

Head losses through a soil column are shown in Figure 6.1.

The actual velocity of flow of a permeant, which carries contaminants through a soil, is considerably higher than the discharge velocity because the flow takes place through pore space only and not the entire cross section. Seepage velocity, or average linear velocity, v_s, is defined as

$$v_s = \frac{v}{n} \tag{6.7}$$

where n is the porosity. Since the path of travel for a permeant is not linear, the microscopic velocity is generally larger than the average linear velocity.

6.1.1 Intrinsic Permeability

Included in the many factors that affect hydraulic conductivity are the permeant density, viscosity, and temperature; soil type, structure, particle shape, size, and particle size distribution; and pore size, pore size distribution, micropores, fissures, joints, stratification, and other discontinuities. The effect of

Figure 6.1
Various heads during flow through a porous medium

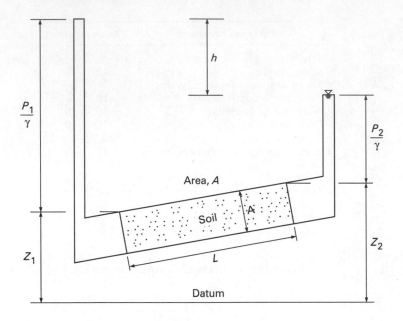

permeant density and viscosity can be accounted for by considering the *intrinsic permeability K*, which is given by

$$K = k \frac{\mu}{g \cdot \rho} = k \frac{v}{g} \qquad (6.8)$$

where μ is the dynamic viscosity of the fluid (g/cm/s = poise), ρ is its mass density (g/cm^3), and $v = \mu/\rho$ is the kinematic viscosity (cm^2/s). The unit of intrinsic permeability is cm^2.

6.1.2 *Contaminant Transport Through Soils*

Contaminated solutes are transported through soils in several ways. The most common contaminant transport mechanisms are

- *Advection*, where solutes are transported by seepage
- *Chemical diffusion*, where solutes move from regions of higher concentration to regions of lower concentration
- *Dispersion*, where velocity differences due to variation in pore sizes, pore size distribution, and tortuosity result in the mechanical mixing of chemicals.

In mathematical modeling, the combined effect of diffusion and dispersion is called *hydrodynamic dispersion* and is given as

$$D = D_e + \alpha v_s \qquad (6.9)$$

where D_e is the effective diffusion coefficient and α is dispersivity. Molecular diffusion coefficients in free water, D_o, range from 8×10^{-10} to 15×10^{-10} m^2/s for organic contaminants at 20°C (Acer and Haider, 1990). The maximum rate of migration occurs in infinite dilution or free water. Also, D_o increases with temperature. Note that the effective diffusion coefficient, D_e, in a soil mass is less than D_o. Longitudinal dispersivity in the direction of flow reported by Anderson (1979) ranged from 3 to 200 m, where the lower values are for fine-grained soils. For soils with hydraulic conductivity $<10^{-7}$ m/s and hydraulic gradient <1 (as would be the case for most soil liners), dispersivity is negligible. Therefore, D can be assumed to represent chemical diffusion alone. Chemical concentration in solutes may decrease due to

• *Adsorption*, where dissolved solids from contaminant are stripped due to the reaction between the permeant and the soil

• *Biodegradation*, where some of the organics may be broken down by bacteria.

Adsorption results in partial *removal* of solute; i.e., it is *partitioned* between the solution and the soil solids. For instance, replacement of one or more types of cations on the soil surface with other cations is a form of adsorption (see Section 3.3). The mass of solute, S, adsorbed by the soil is a function of chemical concentration, c:

$$S = K_f c^m \qquad (6.10)$$

where K_f and m are constants determined by plotting adsorption data on a log–log scale. At low concentration (as is usually the case for municipal landfills) $m \cong 1$. Then Eq. 6.10 becomes

$$S = K_d c \qquad (6.11)$$

where K_d is the distribution coefficient. For a clayey soil, Barone et al. (1992) reported a K_d of 0.19 ml/g for acetone and 11.3 ml/g for toluene. The values reported by Acer and Haider (1990) for various soils with different organic contaminants ranged from 0.02 to 17.3 ml/g.

In biodegradation, the rate of concentration reduction for first-order decay is

$$\frac{\partial c}{\partial t} = -\lambda c \qquad (6.12)$$

where λ is the first-order decay constant. Because several factors influence decay, it is difficult to determine its parameters. Both biodegradation and adsorption retard the pollutant transport process. However, if the nature of influent changes, desorption may occur and some of the contaminants may go back into the permeant.

Depending on the type of reaction between soil and contaminants, some of these transport mechanisms may have considerable effect on the fate of contaminants. According to Freeze and Cherry (1979), an equation that describes the transport of reactive solutes through saturated soil under hydraulic and concentration gradients, dispersion, and retardation is

$$R \frac{\partial c}{\partial t} = D \frac{\partial^2 c}{\partial z^2} - v_s \frac{\partial c}{\partial z} - \lambda c \tag{6.13}$$

where c is the concentration at time t and depth z, and $R = 1 + \rho_d \cdot K_d/n$ is the retardation coefficient in which ρ_d is the dry density. For reactive solutes, the higher R is, the slower the solute moves. For nonreactive contaminants, $R = 1$ and the first-order decay generally is neglected (the last term in Eq. 6.13 drops). If the soils are unsaturated, the problem becomes much more complicated as D and v_s vary with time and location.

Although some analytical solutions are available for Eq. 6.13, computer solutions are more convenient to use. Example problems for simple cases are given in Chapter 10. For a detailed presentation of contaminant transport through soil liners, modeling, computer solutions, and evaluation of various coefficients, see Rowe et al. (1995). For the simple case where a surface of an infinitely deep stratum is subjected to constant concentration c_0 such that

$$c(0, t) = c_0 \qquad t \geq 0$$

$$c(z, 0) = 0 \qquad z > 0$$

$$c(\infty, t) = 0 \qquad t \geq 0 \tag{6.14}$$

the solution is given as (Ogata, 1970; van Genuchten and Alves, 1982)

$$c(z, t) = \frac{c_0}{2} \left[\operatorname{erfc}\left\{ \frac{zR - v_s t}{2(DRt)^{0.5}} \right\} + \exp\left(\frac{v_s z}{D} \right) \operatorname{erfc}\left\{ \frac{Rz + v_s t}{2(DRt)^{0.5}} \right\} \right] \tag{6.15}$$

where $\operatorname{erfc}(z)$ is the complementary error function; i.e., $\operatorname{erfc}(z) = (1 - \operatorname{erf}(z))$.

Example 6.1

A clay liner 0.9 m thick has a small leachate head above it. For retardation coefficients of (a) 0 and (b) 5, determine the ratio of concentration of the contaminants at the base of the liner as a percent of the concentration at the top of liner after 100 years. Assume a dispersion/diffusion coefficient of 2.5×10^{-3} m^2/yr, discharge velocity of 6.3×10^{-3} m/yr, porosity of 0.5, dry density of 1.5 g/cm^3, and 100% saturation.

Solution:

(a) $R = 1$, $v_s = 6.3 \times 10^{-3}/0.5 = 0.013$

$$\frac{zR - v_s t}{2(DRt)^{0.5}} = \frac{0.9 \times 1 - 0.013 \times 100}{2(0.0025 \times 1 \times 100)^{0.5}} = -0.4$$

$$\frac{zR + v_s t}{2(DRt)^{0.5}} = 2.2$$

$$e^{0.013 \times 0.9/0.0025} = 107.77$$

$$c/c_0 = 0.5[\mathrm{erfc}(-0.4) + 107.77 \times \mathrm{erfc}(2.2)]$$

$$= 0.815 \text{ or } 81.5\%$$

(b) $R = 5$

$$\frac{zR - v_s t}{2(DRt)^{0.5}} = 1.431$$

$$\frac{zR + v_s t}{2(DRt)^{0.5}} = 2.594$$

$$c/c_0 = 0.5[\mathrm{erfc}(1.431) + 107.77 \times \mathrm{erfc}(2.5942)]$$

$$= 0.0347 \text{ or } 3.47\%$$

Example 6.2

Determine the concentration ratio in Example 6.1 if only diffusion occurs.

Solution:

$$v_s = 0, \ R = 1$$

$$\frac{zR}{2(DRt)^{0.5}} = 0.9$$

$$c/c_0 = 0.5[\mathrm{erfc}(0.9) + \mathrm{erfc}(0.9)]$$

$$= 0.2031 \text{ or } 20.31\%$$

6.2 LABORATORY MEASUREMENTS

Laboratory tests are generally easier to perform than field tests but there are certain difficulties associated with them. If a soil is inhomogeneous, stratified, fissured, or jointed, the size of the sample may not be large enough to be representative of the soil in the field. Soil may be disturbed during sampling, or it may not even be possible to obtain a sample of certain materials at waste sites.

In determining hydraulic conductivity of compacted soils, large field samples are considered more appropriate because compacted soils are often inhomogeneous. If the hydraulic conductivity is to be determined for a specimen prepared by laboratory compaction, the laboratory compacted specimen may not be representative of the material in the field. In addition, anisotropy in the material hydraulic conductivity will yield erroneous values if the laboratory flow direction does not correspond to the field flow direction.

6.3   PERMEANT

Standard methods are available for determining the hydraulic conductivity of soils with water but no such methods exist for soils interfacing with liquid wastes. Thus, to determine the effect of leachate or other waste liquids on the hydraulic conductivity of a soil the baseline hydraulic conductivity has often been obtained by some researchers with 0.01 N $CaSO_4$, which is representative of the salt concentration found in soils. ASTM recommends the use of tap water, which is now a common practice, but if the tap water is extremely brackish a solution of 0.005 N $CaSO_4$ is suggested by the ASTM.

6.4   HYDRAULIC CONDUCTIVITY TESTS

Parameters such as state of stress, hydraulic gradient, degree of saturation, and temperature influence the outcome of hydraulic conductivity tests and must be carefully controlled.

6.4.1 *Compaction Mold or Rigid Wall Permeameter*

The standard compaction mold (ASTM D 698)—with a diameter of 101.6 mm (4 in.), a height of 116.4 mm (4.5 in.), and a volume of 944 cm^3 (0.0333 ft^3)— serves as permeameter for compacted soils. The base of this rigid wall permeameter is modified to accept a porous disk, as shown in Figure 6.2. The chamber above the compacted soil is filled with the desired fluid and a constant head is applied at the upper end of the specimen. When the flow rate is constant, the hydraulic conductivity is determined from the total amount of flow, Q, in a given time, t, as

$$k = \frac{QL}{hAt}$$

(6.16)

Results of hydraulic conductivity tests from rigid wall permeameters are shown in Figure 6.3. Although this test is straightforward, it has limitations.

Figure 6.2
Rigid wall permeameter
(EPA, 1989)

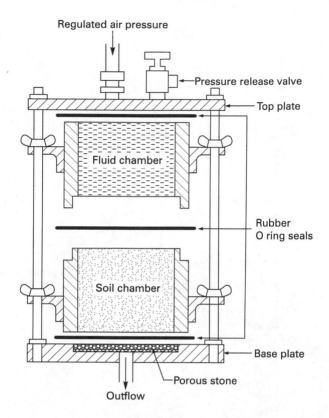

For example, it has no provision for applying vertical stress to the test specimen. In addition, if the soil shrinks due to interaction with the permeant, flow
through shrinkage cracks cannot be controlled. Large voids may develop at
the soil–cell wall interface causing leakage. Also, there is no satisfactory
method available to ensure saturation of the test specimen. Finally, the application of back-pressure adversely affects test results as it increases the formation of channels at soil–cell interface (Edil and Erickson, 1985).

A smaller compaction mold known as the Harvard miniature mold — with
a 33.3 mm diameter and 72.7 mm height — has also been used as a permeability cell. Soil is compacted by a tamper that produces a kneading action
similar to that of a sheepsfoot roller. Because only small samples can be tested,
this mold is not desirable for compacted specimens.

In selecting a rigid wall permeameter, factors that influence the soil
structure — such as the method of compaction, energy input, and the moisture
content at which a specimen is compacted — must be considered. For example,
kneading compaction at moisture content wet-of-optimum yields a more
oriented or dispersed structure (which exhibits lower hydraulic conductivity)
than does static compaction, which produces a more random structure. Therefore, the laboratory method of specimen preparation should correspond to the
proposed method of compaction in the field. For example, if sheepsfoot roller
is to be used in the field, the laboratory specimen could be compacted using a

Figure 6.3
Hydraulic conductivity
with organic permeants
in a rigid wall
permeameter
(reproduced from
Anderson, 1982,
by permission)

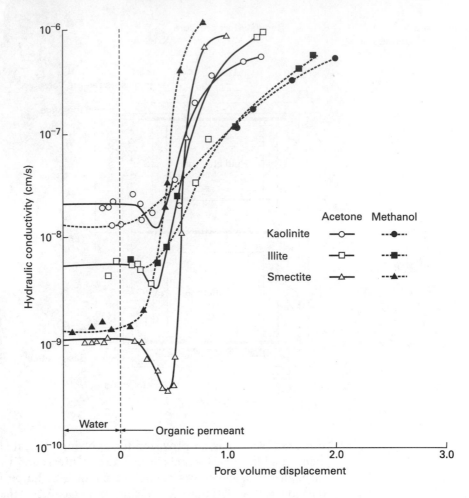

tamper to produce kneading at the appropriate moisture content. However, some researchers recommend against using kneading compaction in the laboratory.

6.4.2 *Flexible Wall Permeameter*

A schematic diagram of a flexible wall permeability cell and control panel are shown in Figure 6.4. Where toxic or corrosive liquid must be used as pore fluid, specially designed permeant interface devices are used to isolate the control panel from the cell. The test specimen diameter may range from 38 to 152 mm and its height is about 1/2 to 1 times the diameter. The specimen is enclosed in flexible membranes with filter papers covered with porous disks at each end. The membranes are sealed to the bottom pedestal and the top cap. To protect the membranes from damage by corrosive permeants, a Teflon tape (1 ml thick) or Saran Wrap is wrapped around the specimen before enclosing it in the membrane (Daniel and Liljestrand, 1984; Khera, 1995).

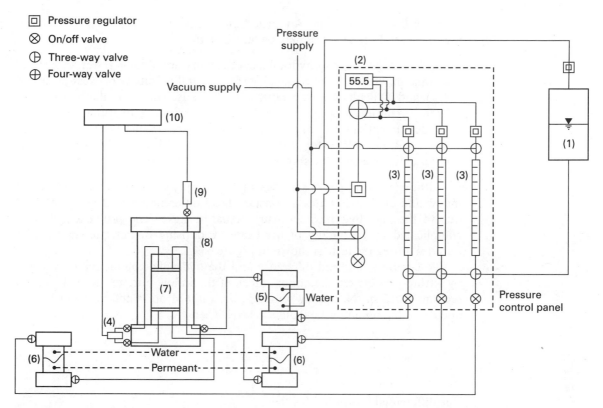

Figure 6.4 Flexible wall permeameter and control panel (1) water reservoir; (2) pressure indicator; (3) volume change burettes; (4) differential transducer; (5) bladder accumulator; (6) permeant interface device; (7) soil specimen; (8) flexible wall permeameter; (9) cell pressure transducer; (10) data logger.

Drainage lines are purged of air and the test specimen is saturated using back-pressure. A back-pressure of 200 to 500 kPa or even higher is applied in small increments (25–50 kPa) to the pore fluid while the chamber pressure is maintained at 25 to 50 kPa above the back-pressure. The value of the pore pressure parameter B is computed as

$$B = \frac{\Delta u}{\Delta \sigma_3} \tag{6.17}$$

where $\Delta \sigma_3$ is the applied increase in cell pressure and Δu is the resulting change in pore pressure. The process is repeated until a B value of 0.95 or greater is achieved. During saturation, the application of vacuum to a specimen can help achieve a higher value of B (Dunn and Mitchell, 1984). The saturation back-pressure is maintained during entire permeability test.

After a sample has stabilized under the applied pressures, flow is initiated by creating a pressure difference between two ends of the specimen. During the hydraulic conductivity test, both inflow and outflow measurements are recorded from time to time. A typical flow rate is shown in Figure 6.5. According to ASTM D 5084, a test may be terminated

if at least four values of hydraulic conductivity are obtained over an interval of time in which (1) the ratio of outflow to inflow rate is between 0.75 and 1.25, and (2) the hydraulic conductivity is steady. The hydraulic conductivity shall be considered steady if four or more consecutive hydraulic conductivity determinations fall within $\pm 25\%$ of the mean value for $k \geq 1 \times 10^{-10}$ m/s or within $\pm 50\%$ for $k < 1 \times 10^{-10}$ m/s, and a plot of the hydraulic conductivity versus time shows no significant upward or downward trend.

Although the apparatus is designed to apply constant pressures to the inflow and outflow burettes, a constant head difference cannot be maintained due to the play that most pressure actuators have. The problem is further complicated by the falling of headwater and rising of tailwater in the two burettes during the test, as shown in Figure 6.6.

To compute the head difference and the hydraulic conductivity, we begin by letting a_i = the cross-sectional areas of the inflow burettes, a_o = the cross-sectional area of the outflow burettes, dz_i = the drop in inflow burette level, and dz_o = the rise in outflow burette level. Then

$$dz_o = \frac{a_i}{a_o}\, dz_i \qquad (6.18)$$

The differential loss in head is then

$$dh = dz_i + dz_o \qquad (6.19)$$

Substituting Eq. 6.18 into Eq. 6.19, we have $dh = dz_i + \dfrac{a_i}{a_o}\, dz_i$ or

$$dz_i = \frac{a_o}{a_o + a_i}\, dh \qquad (6.20)$$

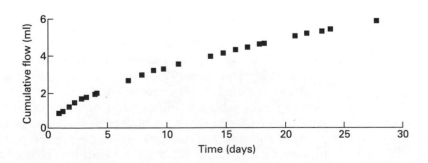

Figure 6.5
Typical time versus cumulative flow curve

Figure 6.6
Falling headwater and
rising tailwater in
a flexible wall
permeameter

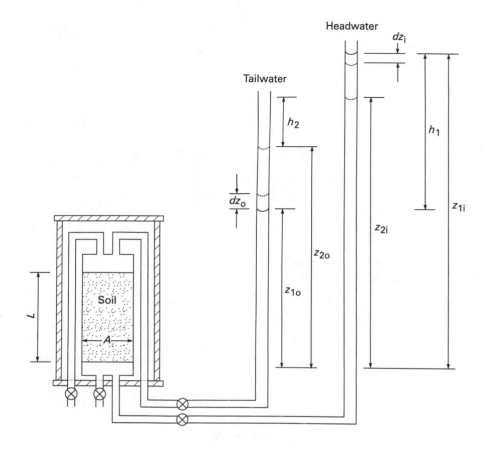

The total flow through inflow burette is

$$dq = a_i \cdot dz_i \tag{6.21}$$

Substituting Eq. 6.20 into Eq. 6.21, we get

$$dq = \frac{a_i a_o}{a_i + a_o} \, dh \tag{6.22}$$

From Eq. 6.1, the discharge in time dt is

$$dq = -kA \frac{h}{L} \, dt \tag{6.23}$$

Substituting Eq. 6.22 into Eq. 6.23, we have

$$kA \frac{h}{L} \, dt = -\frac{a_i a_o}{a_i + a_o} \, dh \tag{6.24}$$

Separating the variables and taking the limits gives us

$$-a_i a_o L \int_{h_1}^{h_2} \frac{dh}{h} = kA \cdot (a_i + a_o) \int_{t_1}^{t_2} dt \tag{6.25}$$

where h_1 is the initial total head difference between inflow and outflow at time t_1, and h_2 is the final total head difference between inflow and outflow at time t_2.

After integrating and taking the limits, the hydraulic conductivity is given by

$$k = \frac{a_i a_o L}{A(a_i + a_o)\Delta t} \ln \frac{h_1}{h_2} \tag{6.26}$$

where $\Delta t = t_1 - t_2$. If inflow and outflow burettes have the same cross-sectional area (as is generally the case), then $a_i = a_o = a$, and Eq. 6.26 yields

$$k = \frac{aL}{2A \cdot \Delta t} \ln \frac{h_1}{h_2} \tag{6.27}$$

Since additional pressures p_i and p_o are applied to the inflow and outflow burettes, respectively, the corresponding initial head difference is given by

$$h_1 = \left(z_{1i} + \frac{p_{1i}}{\gamma_w}\right) - \left(z_{1o} + \frac{p_{1o}}{\gamma_w}\right) \tag{6.28}$$

and the final head difference is given by

$$h_2 = \left(z_{2i} + \frac{p_{2i}}{\gamma_w}\right) - \left(z_{2o} + \frac{p_{2o}}{\gamma_w}\right) \tag{6.29}$$

where the definition of z is given in Figure 6.6. These values of head are substituted into Eq. 6.26 for hydraulic conductivity computations.

The plots of hydraulic conductivity with time from a flexible wall permeameter are shown in Figure 6.7. According to ASTM guidelines, the

Figure 6.7
Hydraulic conductivity time plot for water, aniline, and phenol (Khera and Tirumala, 1992. Copyright ASTM: reprinted with permission.)

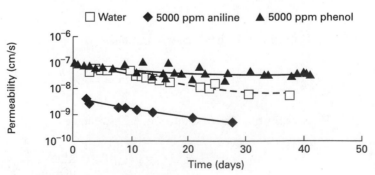

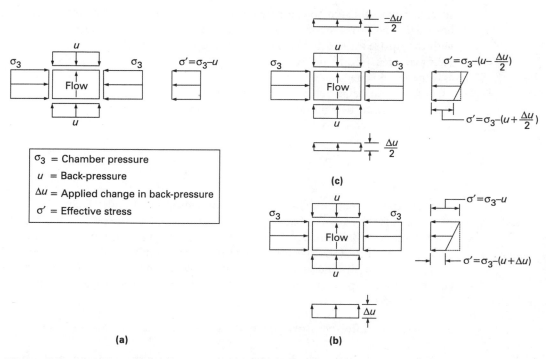

Figure 6.8 Stresses on test specimens in flexible wall cell at various loading stages

maximum hydraulic gradient may not exceed 30 for soils with hydraulic conductivity of less than 10^{-9} m/s, and it may not exceed 2 for soils with hydraulic conductivity of 10^{-5} m/s.

Total and effective stresses and pore water pressure during consolidation in a flexible wall cell are shown in Figure 6.8a. The values of Δu are based on the hydraulic gradient to be generated and are computed as

$$\Delta u = i \cdot L \cdot \gamma_w \qquad (6.30)$$

The hydraulic gradient in Figure 6.8b is produced by raising the pore pressure Δu on the upstream side by a half the amount ($\Delta u/2$) and lowering it on the downstream side by the same amount ($-\Delta u/2$). The average effective stress on the specimen remains unchanged. During the permeability test, the specimen will have a tendency to swell on the upstream side from the decrease in effective stress and consolidate on the downstream side from the increase in effective stress.

If, on the other hand, the required gradient is generated by increasing the pore pressure Δu on the upstream side as shown in Figure 6.8c, the average effective stress on the specimen will decrease by $\Delta u/2$. During the permeability test, this specimen will swell on the headwater end. In either case, until the specimen has adjusted to the new effective stress, inflow and outflow will not be equal. The cross-sectional area of the specimen will vary along its length with the upstream end being the largest. However, the difference in the cross-sectional area is difficult to determine (without disassembling the specimen) and therefore is neglected.

If specimen height changes are to be measured and field anisotropic stress conditions are to be simulated during a permeability test, a triaxial cell is used as the permeameter.

Example 6.3

A specimen for a permeability test has been consolidated under a chamber pressure of 500 kPa and a back-pressure of 300 kPa. For the desired hydraulic gradient, a head difference of 100 kPa is required between the upstream and downstream ends of the specimen. Determine the chamber pressure and pore pressure if the hydraulic gradient is obtained by (a) maintaining effective stress constant on the upstream end and (b) maintaining a (zero) average change in effective stress.

Solution: (a) Effective stress upstream and downstream $= 500 - 300 = 200$ kPa

Since the effective stress has to remain constant on the upstream end, we must reduce the pore pressure u on the downstream side by 100 kPa.

Location	Chamber pressure (kPa)	Pore pressure (kPa)	Effective stress (kPa)
Upstream	500	300	200
Downstream	500	(300 − 100) or 200	300

Therefore, the average effective stress $= \dfrac{(200 + 300)}{2}$ or 250 kPa.

(b) Average effective stress at the end of consolidation $= 200$ kPa

We must increase u upstream by $\dfrac{100}{2}$ or 50 kPa and decrease u downstream by 50 kPa.

Location	σ_3 (kPa)	Pore pressure (kPa)	σ_3' (kPa)
Upstream	500	350	150
Downstream	500	250	250

Therefore, the average effective stress $= 200$ kPa.

6.4.3 *Constant Rate of Flow*

In the constant rate of flow method of testing hydraulic conductivity a flow pump is used to inject or withdraw permeant from the specimen at a constant rate of flow (ASTM D 5084). The induced head difference across the length of the specimen is monitored with a differential pressure transducer (Olsen et al., 1985). A plot of the head differential for a test on a sand bentonite mix is shown in Figure 6.9.

Direct flow measurements are not necessary in this method, and hydraulic conductivity is determined in a relatively short period of time. The time needed for constant flow hydraulic conductivity reported by Olsen et al. (1994) was a few minutes for $k \geq 10^{-8}$ m/s, a few hours for 10^{-11} m/s $< k < 10^{-10}$ m/s, and a few days with 10^{-3} m/s $< k < 10^{-12}$ m/s. These test durations are much less than those needed for the other methods.

Low flow rates are recommended for this method. If a high flow rate is sustained, the hydraulic gradient could become very high. By maintaining a low flow rate, a small hydraulic gradient — comparable with the field values — can be applied and errors due to seepage-induced consolidation can be minimized. In tests where a certain number of pore volume is required to pass through the specimen, the equipment can be set to inject the required amount of permeant automatically.

6.4.4 *Double-ring Cell*

Anderson et al. (1985) reported that a double-ring rigid wall permeameter was effective in mitigating the leakage problem of the compaction mold cell. In this permeameter, a ring is built into the base plate so that the area within the ring is half of the total area of the base. The flow adjacent to the side walls is isolated from the flow through the area enclosed by the ring. Leakage is indicated if the flow rates from the central core and the outer annular area are unequal. Like the compaction permeameter, no provision is made for the application of confining pressure. Use of back-pressure may have an adverse

Figure 6.9
Head difference from flow pump test

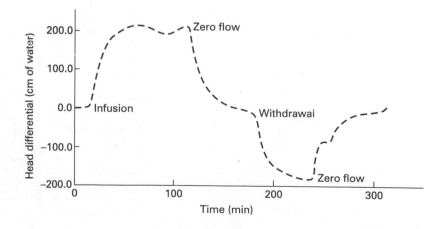

effect on the test results. Wu and Khera (1990) give a modified version of double-ring permeameter with several improvements.

6.4.5 *Oedometer Cell*

A fixed-ring consolidation cell is easily adapted to function as a permeameter, as shown in Figure 6.10. One of the two openings at the bottom is closed, and the permeant enters the soil through the other opening and exits from the top. The head is applied using a standpipe of a small cross-sectional area, a. To reduce the loss due to evaporation, a moist paper towel may be used to cover the open end of the standpipe. A constant head is maintained at the exit.

As shown in Figure 6.10, for a drop in head, dh, during the time interval dt, the change in the volume of water in the standpipe is equal to the flow of water through the soil. For a soil with hydraulic conductivity k, if the total head difference is h, then

$$a \cdot dh = -k \cdot \frac{h}{L} \cdot A \cdot dt \qquad \textbf{(6.31)}$$

where A is the cross-sectional area of the test specimen and L is the length of the specimen. If we let the total head difference decrease from h_1 to h_2 in the time interval from t_1 to t_2, we can separate the variable in Eq. 6.31 and take

Figure 6.10
Fixed-ring
consolidation
permeameter

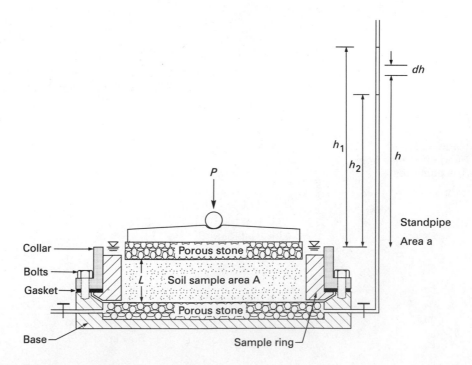

the limits to obtain

$$\int_{h_1}^{h_2} \frac{dh}{h} = -\frac{k \cdot A}{a \cdot L} \int_{t_1}^{t_2} dt \qquad (6.32)$$

Evaluation of Eq. 6.32 yields

$$k = \frac{aL}{A \cdot \Delta t} \ln\left(\frac{h_1}{h_2}\right) \qquad (6.33)$$

where $\Delta t = t_1 - t_2$. If an additional pressure p is applied to the standpipe to reduce the test duration, then

$$k = \frac{aL}{At} \ln\left(\frac{h_1 + \dfrac{dp}{\gamma_w}}{h_2 + \dfrac{dp}{\gamma_w}}\right) \qquad (6.34)$$

The hydraulic conductivity can also be computed from the rate of settlement using Eq. 6.35 based on the one-dimensional consolidation theory described in Section 4.5.

$$k = c_v \cdot m_v \cdot \gamma_w \qquad (6.35)$$

where c_v is the coefficient of consolidation, m_v is the coefficient of volume compressibility, and γ_w is the unit weight of water.

The fixed-ring consolidation cell has many advantages. First, it allows simulation of field stresses and reduces formation of shrinkage cracks and side-wall leakage. In addition, the hydraulic conductivity values can be determined at different effective vertical stresses or void ratios. Also, because of the small thickness of the specimen, a test is completed in a short time. Finally, either laboratory prepared or undisturbed specimens can be used.

However, this method also has limitations. For example, it is difficult to perform tests at very low consolidation pressure. Also, because it is an open system, permeant containing volatile organic compounds should not be used. The volatiles will evaporate during the test thereby altering the composition of the effluent, it will not be possible to compare the composition of influent and effluent, and the chemicals may pose a health hazard to the laboratory personnel.

Example 6.4

A sand mixed with 5% bentonite was consolidated in a fixed-ring consolidation permeameter 6.35 cm in diameter. Under a stress of 100 kPa, the coefficient of consolidation was determined to be 9.155×10^{-6} m^2/min and the coefficient of volume compressibility was 0.0007 m^2/kN.

(a) Determine the hydraulic conductivity of the specimen corresponding to the given consolidation stress. (b) For the falling head permeability test, the diameter of the standpipe was 7 mm, and the total head dropped from 824 to 804 mm in 12 min. Determine the hydraulic conductivity based on the falling head test if the height of the specimen was 22.07 mm.

Solution:

(a)

$$k = c_v \cdot m_v \cdot \gamma_w$$

$$= 9.155 \times 10^{-6} \times 0.0007 \times 9.81$$

$$= 6.29 \times 10^{-8} \text{ m/min}$$

(b) Use Eq. (6.33).

$$a = \left(\frac{7}{2}\right)^2 \cdot \pi = 38.49 \text{ mm}^2$$

$$A = \left(\frac{6.35 \times 10}{2}\right)^2 \cdot \pi = 3.167 \times 10^3 \text{ mm}^2$$

$$k = \frac{38.49 \times 22.07}{3.167 \times 10^3 \times 12} \times \ln\left(\frac{824}{804}\right) \times \frac{1}{1000}$$

$$= 5.49 \times 10^{-7} \text{ m/min}$$

6.4.6 API Test

The American Petroleum Institute (API) test is most commonly used for determining the hydraulic conductivity of bentonite slurry. The apparatus consists of a rigid cylinder with porous stone and outflow at the base (Figure 6.11). In a

Figure 6.11
API filter test

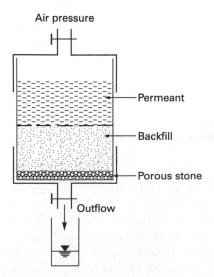

Air pressure

Permeant

Backfill

Porous stone

Outflow

test that can be completed in a few hours, the permeant is subjected to an air pressure. Hadge and Barvenik (1985) adopted it for quality control of backfill at a slurry wall construction site. The agreement between API and flexible wall hydraulic conductivity was within ±66%. If this test is adopted at other sites, its accuracy must be verified against the flexible wall permeameter.

6.5 FACTORS AFFECTING HYDRAULIC CONDUCTIVITY

Some of the factors that have significant effect on soil hydraulic conductivity are discussed in the following subsections.

6.5.1 Fine-grained Components

The addition of a small percentage of fines in a coarse-grained soil can decrease its hydraulic conductivity by several orders of magnitude. Plastic fines, which are more effective and have greater capability to adsorb contaminants than nonplastic fines, are recommended (D'Appolonia, 1980). However, fines with high plasticity may be more difficult to work and blend with coarser materials than those with low and medium plasticity. Also, the difficulty in workability may produce inhomogeneous fill, which may show considerable variation in hydraulic conductivity.

6.5.2 Freeze–Thaw

In cold regions, compacted soil containment structures may undergo freeze–thaw cycles during their lifetime. These cycles may cause more than three orders of magnitude increase in hydraulic conductivity. The changes in hydraulic conductivity take place during the first 3 to 10 freeze–thaw cycles, with most significant alteration taking place in the first one (Othman et al., 1994). Specimens frozen at comparatively fast rate show a relatively greater number of ice lenses and larger increases in hydraulic conductivity (Othman and Benson, 1992). The largest increases occur in soils with the lowest initial hydraulic conductivity.

6.5.3 Compaction Effect

In a compacted soil, the hydraulic conductivity depends on the water content during compaction and method of compaction. A soil compacted wet-of-optimum has a hydraulic conductivity two to three orders of magnitude lower than one compacted dry-of-optimum. Both kneading and impact compaction yield lower hydraulic conductivity than does static compaction (Mitchell, 1976). However, static compaction is suggested for laboratory specimens so that the test values, though higher, will be more in line with field values (Dunn and Mitchell, 1984).

Daniel (1994) reported increase in hydraulic conductivity with increasing size of compacted specimens. Broderick and Daniel (1990) reported that a higher degree of compaction reduces contaminants' ability to attack the soil. Further discussion of compaction effects will be found in Chapter 9.

6.5.4 *Sample Saturation*

If a soil sample is not completely saturated, its measured hydraulic conductivity will be lower. Hydraulic conductivity varies as third power of the degree of saturation (Mitchell, 1993).

Although none always yield complete saturation, various methods for specimen saturation have been used. Compacted samples are sometimes allowed to soak from the bottom. In some instances vacuum is applied at the top while permitting inflow at the bottom. However, vacuum should not be applied if the effective consolidation stress has to be lower than one atm. In a flexible wall permeameter, the use of back-pressure yields the best results (see Section 6.4.2).

6.5.5 *Hydraulic Gradient Effect*

In materials of soft consistency or low density, hydraulic conductivity decreases with increasing gradient. The impact on stiffer materials is less. The hydraulic gradients that exist across compacted soil liners seldom reach 20, which is about the highest one would ever find in the field. In most cases, these values are less than 1. According to EPA regulations, the height of leachate in a landfill must not exceed 300 mm. For a soil liner 3 ft (914 mm) thick, hydraulic gradient will be around 1.

High gradients result in larger effective stresses at the downstream end of the specimen (see Figure 6.8), which causes changes in the specimen volume and the nature of the void system. High gradients may also cause migration of soil particles that may cause either clogging in the larger pores or an increase in pore size and pore volume due to erosion. These changes result in hydraulic conductivity values that are not always reliable.

At low hydraulic gradients some organics may not permeate a water-saturated soil. At higher gradients, the macropores and other discontinuities that develop in the soil result in larger hydraulic conductivity, especially in rigid wall permeameter tests. In the flexible wall permeameter test, the increases in hydraulic conductivity are small. In fact, the hydraulic conductivity may be underestimated in a flexible wall permeameter because the cracks that develop will close if high confining pressure is applied. On the other hand, if the confining stresses are small the effect of increasing the gradient may increase the size of the cracks. This phenomenon is consistent with the concept of hydraulic fracture, for which the critical value for the ratio of change in pore pressure to effective stress is 0.5 to 1.0 (Bjerrum, 1972).

The use of the proper hydraulic gradient cannot be overemphasized.

Hydraulic gradients found in the field are too low to be realistic for completing laboratory tests in a reasonable time period. However, it is recommended that a low hydraulic gradient be used for the determination of hydraulic conductivity with due consideration of the effective confining stress. If higher gradients must be used, in the case of dense soils its detrimental effect may be somewhat reduced if the upstream back pressure is increased and the downstream back pressure is decreased by half the amount needed for the desired hydraulic gradient.

6.5.6 Consolidation Stress

In general, hydraulic conductivity decreases with increase in consolidation. However, the decrease in hydraulic conductivity is much greater for increases in consolidation stresses between 0 to 50 kPa than at stresses greater than 50 kPa (Wu and Khera, 1990). In laboratory soil–bentonite backfill material, the settlements under the expected field stresses were measured to be 10% by Engemon and Hensley (1986). The measured field settlements of the same backfill material in the actual slurry wall were 1/10 of 1%, indicating that the effective consolidation stress in the field was substantially lower than that computed from the self-weight of the wall.

The consolidation pressure in a flexible wall permeameter also tends to close the cracks that result from soil–chemical interaction (Khera and Thilliyar, 1990). Furthermore, hydraulic gradient generates seepage forces that augment the applied effective consolidation stresses. Therefore, for slurry wall backfill materials, the use of any consolidation pressure except the smallest practical value during the hydraulic conductivity measurements will lead to k values that are nonconservative for design.

Fernandez and Quigley (1991) showed that the application of effective stress in a fixed-ring permeameter can greatly reduce or eliminate large increases in hydraulic conductivity. For pure ethanol, which has a dielectric constant of about 32, at a static vertical stress of 20 kPa there was no increase in k, but at zero effective stress k increased a hundredfold. For dioxin with a dielectric constant of about 2, even at a vertical stress of 160 kPa there was a twofold increase in hydraulic conductivity. Postdamage application of effective stress restored the clay damaged by polar ethanol but required a static vertical stress of about 100 kPa. For nonpolar dioxin, the postdamage effective stress did not close the clay openings caused by permeation at zero static stress.

Increase in overburden pressure reverses the increase in hydraulic conductivity of liners that have undergone freeze–thaw cycles. Once again, for a given decrease in hydraulic conductivity, the required applied pressure during a freeze–thaw cycle is less than that needed after the damage has occurred (Othman and Benson, 1992).

Based on this discussion, it may be postulated that for new construction considerable benefit can be reaped if placement of waste is so arranged that the static stress in the field is high before leachate or freeze–thaw cycles can damage a compacted soil liner.

6.6   PHYSICOCHEMICAL CHANGES

The nature of permeating chemicals can change the hydraulic conductivity characteristics of clay soils by alterations in soil fabric, increase or decrease in the diffused double layer thickness, solution of soil components by strong acids and bases, blockage of soil voids due to precipitation or movement of fine particles from one area to another, and growth of microorganisms. The effect of a given chemical varies according to the composition of the clay. An extensive summary and discussion of the effect of permeants on the hydraulic conductivity of various naturally occurring and commercially available clay soils may be found in a report by the EPA (1986).

6.6.1 *Inorganic Compounds*

Inorganic compounds may impact the engineering properties of soils with clay components. A change in chemical or electrolyte concentration of the pore fluid in a soil may modify the swelling and dispersion characteristics of clay particles and thus have an effect on the hydraulic conductivity. The effect of a change in electrolyte concentration from 0.6 to 0.1 N NaCl on the hydraulic conductivity of various proportions of illite–silt mixtures formed by different sedimentation procedures are depicted in Figure 6.12 (Mitchell, 1993). The internal swelling started to increase and the hydraulic conductivity began to decrease with the flow of electrolyte of lower concentration. As the concentration of pore fluid dropped to 0.1 N, which occurred at 1.5 or greater pore volume of flow (PVF), there was no further change in relative hydraulic conductivity.

Soils with greater swelling potential, such as sodium bentonite, show larger changes in hydraulic conductivity with the exchange of the adsorbed cations. The monovalent Na^+ causes a larger interlayer spacing and more dispersion of the clay particles than does the bivalent Ca^{++}, which has smaller

Figure 6.12
Decrease in hydraulic conductivity from internal swelling.
(*Fundamentals of Soil Behavior*, 2nd Ed, James K. Mitchell, copyright © 1993 by John Wiley and Sons, Inc. Reprinted by permission of John Wiley and Sons, Inc.)

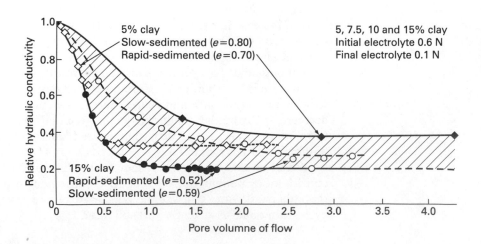

hydrated radius. In general, the affinity of clay minerals to cations increases as the ion valence increases. However, the high concentration of lower valence ions in a solution can negate the greater replacing power of higher valence ions. The exchanging of cations affects the thickness of the double layer. As the result the soil structure may alter, its volume may change, and it may develop cracks and macropores that may have considerable effect on the hydraulic conductivity.

Alther et al. (1985) reported an increase in hydraulic conductivity of more than one order of magnitude for a bentonite as the concentrations of chlorides of potassium, sodium, magnesium, and calcium were increased. For divalent cation, the hydraulic conductivity stabilized after a small increase in salt concentration. Yong (1986) reported that in an undisturbed clay distilled water removed much larger proportions of monovalent cations (up to 30 times) than divalent cations. Since sodium montmorillonite derives its low hydraulic conductivity characteristics from the presence of monovalent sodium ion, a permeant may increase hydraulic conductivity.

6.6.1.1 Inorganic Acids   Increasing concentration of inorganic acids triggers flocculation, inhibits swelling, and increases dissolution of clay minerals. Kaolinite shows the least solubility and smectite the most. Concentrated acids cause increases of several orders of magnitude in the hydraulic conductivity of clay soils, essentially due to dissolution. With dilute acids, there may be an initial tendency toward decrease in hydraulic conductivity due to clogging of the pores by precipitated salts. In some instances, this is followed by an increase in hydraulic conductivity as the precipitates leach out. Dilute hydrochloric acid with pH of 1, 3, and 5 have little effect on the hydraulic conductivity (Lentz et al., 1984)

6.6.1.2 Inorganic Bases   Inorganic bases tend to cause dispersion of soils and may even dissolve soil particles. Lentz et al. (1984) found that sodium hydroxide in tap water with pH of 9 and 11 had no effect on the hydraulic conductivity of compacted clays. A pH of 13 caused a decrease in the hydraulic conductivity by factors of about 2.5 to 13. The decrease in hydraulic conductivity was small for the monovalent ions and was attributed to the precipitation of salts from the tap water. The larger decreases, observed for magnesium montmorillonite, were believed to be from the replacement of divalent magnesium ion with monovalent sodium ion.

The effect on the water adsorption capacity of sodium and calcium bentonites was shown in Figure 3.10. A lower hydraulic conductivity is to be expected for pH between 8 and 11 with sodium bentonite because of its high swelling potential in this range of pH. Furthermore, the changes in pH drastically reduce the water adsorption capacity of sodium montmorillonite, thereby resulting in increased hydraulic conductivity.

Data on soil–bentonite backfill and slurry wall filter cake from D'Appolonia (1980), who used an API type of filter press, showed increases in hydraulic conductivity by factors of 5 to 10 with a 5% solution of sodium hydroxide. The differences in soil types, moisture content, density, and th permeameter used in the two investigations may be responsible for the divergent results.

6.6.2 *Organic Compounds*

Test data with organic compounds showing large changes in hydraulic conductivity were reported by Anderson (1982) (see Figure 6.3). His study drew the attention of many researchers to this area and a large body of data has become available on the hydraulic conductivity of compacted soils. The results are, however, conflicting; with the same chemicals, hydraulic conductivity has been reported to increase, decrease, or show no change. The divergent results have been attributed to the interaction between the soil and the chemical, the appearance of cracks, large pores, and channels in the soil specimen, and the testing procedures and the equipment used.

When a pore fluid has a low dielectric constant, it will have a tendency to curtail soil swelling and, therefore, its hydraulic conductivity will increase. Figure 6.13 shows the relationship between hydraulic conductivity and the dielectric constant for an illite-rich lean clay. The lowest hydraulic conductivity is obtained for water, which is polar and has the highest dielectric constant (80). Alcohols and acetone, which also are polar and have dielectric constants ranging from 20 to 35, show intermediate values for hydraulic conductivity. Nonpolar hydrocarbons and related compounds have the lowest dielectric constant values (less than 3) and the highest hydraulic conductivity.

6.6.2.1 Hydrocarbons Hydrocarbons (saturated, unsaturated, halogenated, and aromatic compounds) usually have a very low water solubility, a negligible dipolar moment, and a low dielectric constant. In most instances when compacted soils are permeated with pure compounds, the hydraulic conductivity increases by several orders of magnitude in rigid wall permeameters. In such cases, cracks and other discontinuities develop within the soil and flow occurs through these discontinuities. In flexible wall permeameters, the same compounds show a decrease in hydraulic conductivity. Less than 10% of the water in the soil pores is displaced by the compound (Fernandez and Quigley, 1985). The low hydraulic conductivity in a flexible wall permeameter is due to restraining of cracks and macropores. Furthermore, flow occurs only through a small proportion of the specimen.

In the study by Fernandez and Quigley (1985), complete replacement of pore water was obtained by sequential permeation of the soil, first with alcohol (ethanol), then with a hydrocarbon (benzene). This resulted in a hydraulic conductivity increase of four orders of magnitude, as seen in Figure 6.14. When the course of sequential permeation was reversed and the pore hydrocarbon was replaced by water, the hydraulic conductivity decreased by the same order of magnitude. Often the reported changes in the hydraulic conductivity are negligible for any concentration within the solubility limits of the compound.

Figure 6.13
Hydraulic conductivity
as a function of
dielectric constant
(reproduced from
Fernandez and Quigley,
*Canadian Geotechnical
Journal*, Vol. 22, 1988,
by permission)

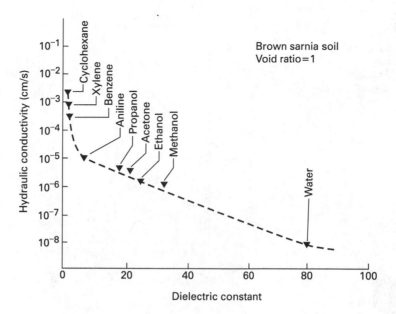

6.6.2.2 Alcohols, Phenols, and Ketones The high water solubility of acetone and phenol allows them to replace pore water easily. In a rigid wall permeameter the hydraulic conductivity may increase by as much as three orders of magnitude. These compounds may initially show decreases in hydraulic conductivity but this is followed by a two to three orders of magnitude increase as more of the pore water is replaced. If water is used again as the permeant, the hydraulic conductivity decreases.

There is not much change in hydraulic conductivity if the chemical concentrations are less than 75%. Bowders and Daniel (1987) also reported only a slight change in hydraulic conductivity in kaolinite and illite-chlorite with methanol of up to 80% concentration in water. But pure methanol increased hydraulic conductivity of kaolinite by 7.5 times and of illite-chlorite by 44 times in rigid wall permeameters.

Fernandez and Quigley (1988) studied the effect of water-soluble ethanol at various concentrations on the hydraulic conductivity of Sarnia clay compacted with a Harvard miniature tamper in a rigid wall permeameter. All but one of the tests had no vertical stress during the permeation; one test was performed under an effective vertical stress of 160 kPa. As shown in Figure 6.15a, hydraulic conductivity, k, decreased for ethanol concentration of up to 60%. With further increase in ethanol concentration, k increased rapidly to yield a value of more than two orders of magnitude higher with pure ethanol. As seen in Figure 6.15b, when intrinsic permeability K — which accounts for permeant density and viscosity (Eq. 6.8) — is used for plotting the data, all ethanol concentrations resulted in increase of K compared with water.

The greater affinity for water of double-layer cations prevented a significant collapse of the double layer at low concentrations. The much larger

Figure 6.14
Effect of pore fluid on
hydraulic conductivity
(reproduced from
Fernandez and Quigley,
1988, by permission)

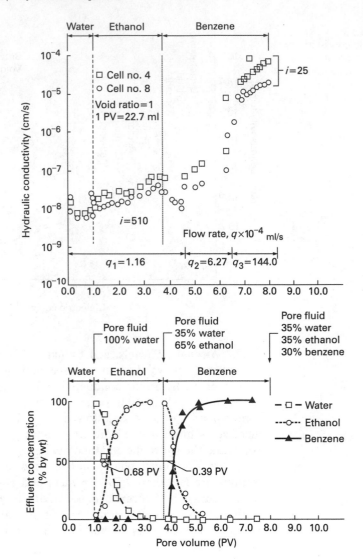

increase in intrinsic permeability at high concentration was viewed as the indication of collapse of the diffused double layer. The effluent contents with pure ethanol were almost 100% alcohol. After-test chemical analysis of the specimen showed that only 70% of the pores contained ethanol, which led to the conclusion that channel flow was occurring through the specimen.

The specimen with effective vertical stress of 160 kPa showed a slight decrease in hydraulic conductivity. Intrinsic permeability and hydraulic conductivity both showed a small increase because the confining vertical pressure

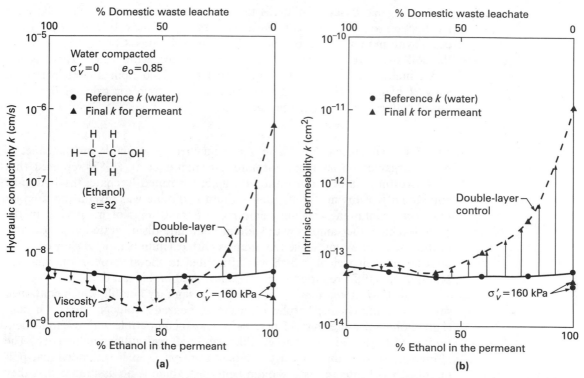

Figure 6.15 (a) Hydraulic conductivity (b) intrinsic permeability of compacted Sarnia clay permeated with leachate–ethanol solution. Tests were initially performed with water (•) and then with permeants (▲) with various ethanol concentrations (reproduced from Fernandez and Quigley, *Canadian Geot.*, Vol. 25, 1988, p. 584, by permission)

compressed the soil and prevented the large increase in hydraulic conductivity that occurred in tests without vertical stress.

6.6.2.3 Organic Acids Both increases and decreases in hydraulic conductivity have been reported for pure and dilute states. The presence of any salt in the soil with which the acid can react may result in an initial decrease in hydraulic conductivity due to clogging of the pores by the precipitates. Thus, long-duration tests are necessary to establish whether the precipitates eventually will leach out. If they do, it may lead to larger pore sizes, piping, and consequently higher hydraulic conductivity. The clay particles themselves may also be dissolved by the acid, resulting in similar effects.

Hermanns et al. (1987) used high-sulfate-resistance cement in a sodium bentonite mix and a calcium bentonite mix. Permeant consisted of 2.5% each of acetic acid and propanic acid. Over a four-month period, the hydraulic conductivity of the calcium bentonite mix decreased from 5×10^{-8} cm/s to 2.8×10^{-11} cm/s, whereas that of the sodium bentonite mix remained at 7×10^{-8} cm/s.

6.6.2.4 Organic Bases　Aniline is the most commonly tested organic base for its effect on soil hydraulic conductivity. Large hydraulic conductivity increases occur with pure chemical but not with dilute solutions. Changes occurring in the soil structure may be reversed by permeation with water. There is not much migration or dissolution of particles. A 5000 ppm aniline permeant showed hydraulic conductivity to decrease with time for calcium bentonite mixed with cement and blast furnace slag.

6.6.2.5 Mixed Chemicals　There is not much data available on the effect of mixed organics on sodium bentonite. Hermanns et al. (1987) reported that after two months of permeation with a reconstituted leachate, the hydraulic conductivity of the mix containing sodium bentonite was several times larger than that containing calcium bentonite. After 30 days of immersion in the same leachate, the samples with sodium bentonite disintegrated and lost 90% of their original weight while the samples with calcium bentonite showed a 5% increase in their weight (which was attributed to the absorption of leachate). Nußbaumer (1987) reported similar results based on his test data.

Ryan (1987) noted that a leachate from a sanitary landfill, which contained several organic chemicals (phenol, acetone, benzene, toluene, etc.) with concentrations not more than 75 ppm each, was incompatible with different bentonites and yielded an unacceptable value of hydraulic conductivity. The aforementioned findings clearly demonstrate that even if the individual pollutants do not interact with sodium bentonite, there is no assurance that they will not affect it adversely when more than one are present in the aqueous permeant.

6.7　SOIL FLUID COMPATIBILITY

Currently there are no standardized test procedures available for determining the effect of waste chemicals on soils used in waste containment structures. According to the current practice, a test specimen is first permeated with water and then with leachate. If actual samples of the waste leachate or chemicals are not available, similar samples from other sites—or specially prepared samples with the expected constituents—are used. However, it is not the leachate alone that one must consider. There is also the possibility of the presence of immiscible contaminants or nonaqueous phase liquids (NAPLs). If NAPLs are present, they should also be checked for compatibility: high concentrations of NAPLs can cause slurry walls to degrade rapidly.

Immersion tests are also used to determine the effect of permeant on the long-term integrity of soils. Test specimens are prepared from the soil, weighed, and immersed in the leachate. The specimens are then removed from the leachate at various time intervals and their weights are compared with the original weights. They are also visually examined for distress (Khera and Tirumala, 1992).

6.7.1 *Baseline Hydraulic Conductivity*

Baseline hydraulic conductivity is established with water. The test termination criteria in a flexible wall permeameter for water, given in Section 6.4.2, is dictated by ASTM D 5084. Similar termination criteria may be used with other permeameters.

6.7.2 *Hydraulic Conductivity with Leachate*

After the baseline hydraulic conductivity has been established, the permeant is switched from water to the leachate. ASTM termination criteria for hydraulic conductivity with water is considered not fully applicable to waste liquids. Termination criteria often suggested are based on the number of pore volume of flow (PVF). All pore space in a soil may not be interconnected or conductive. The effective porosity, n_e — which controls seepage velocity — can be estimated from (Edil et al., 1994)

$$n_e = \frac{k \cdot i}{L} \cdot t_{50} \tag{6.36}$$

where t_{50} is the time at which effluent concentration is 50% of the influent concentration. The number of PVF during time t, is given by

$$\text{PVF} = \frac{\text{Total flow}}{\text{Effective volume of pore space in specimen}}$$

For a test specimen of area A and length L, volume of pore space is $n_e \cdot A \cdot L$. If the rate of flow through the specimen is q, then

$$\text{PVF} = \frac{q \cdot t}{n_e \cdot A \cdot L} \tag{6.37}$$

If all the voids are interconnected and the soil is saturated, then the volume of pores through which permeant can flow is the same as the total void space, i.e., $n_e = n$. When combined with Eq. 6.2 and Eq. 6.7 it will yield

$$\text{PVF} = \frac{v_s \cdot t}{L} \tag{6.38}$$

where v_s is the seepage velocity. According to the termination criteria proposed by Peirce and Witter (1986), at least one pore volume of permeant must be passed through the soil and the slope of a regression line for hydraulic conductivity versus the time plot may not vary appreciably from zero at a 95% confidence level.

For a compacted kaolinite specimen Bowders (1988) reported that hydraulic conductivity that was stabilized after five PVF increased by 17 times with the passage of an additional quarter of a pore volume. Compacted Sarnia clay data from Fernandez and Quigley (1991) indicated increasing hydraulic conductivity until four PVF with ethanol and five PVF with dioxin. Thus a criteria based on PVF alone does not appear to be adequate. Daniel (1994) proposed that permeation be continued (a) for at least two PVF and (b) until concentration of the chemicals in effluent that may cause changes in hydraulic conductivity have similar concentrations as in influent.

Great care must be exercised while sampling the effluent for chemical analysis. Samples should be taken as close to the specimen as possible. Where permeant interface device is used (see Chapter 5), effluent samples may be obtained from it since it is a closed system. For volatile organics, precautions must be taken to mitigate evaporation during and after sampling. Daniel (1994) has described the following scheme.

A stainless steel "tee" connector is mounted to the valve on the permeameter that opens and closes the effluent line. The tee is fitted with a septum that is held in place with a drilled-out solid ferrule. A hypodermic syringe with needle is used to puncture the septum and extract a sample of liquid from the effluent line. The septum will not leak for at least 30 to 40 punctures. The same syringe is usually used to inject the effluent sample into a device for analysis of water chemistry.

Breakthrough curves that represent the ratio of effluent concentration to influent concentration (c/c_0) are shown in Figure 6.16. Advection or plug flow is depicted in curve 1 for which at PVF = 1, $c/c_0 = 1$. Curve 2 shows the effect of dispersion; for soils of low hydraulic conductivity this effect is inconsequential. Curve 3 displays the effect of chemical diffusion and adsorption (or

Figure 6.16
Breakthrough curves for a leachate or chemical permeant
(1) plug flow
(2) advection and dispersion
(3) initial adsorption
(4) retardation
(5) desorption
(6) rapid flow through macropores

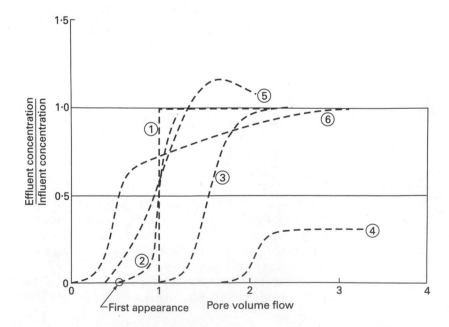
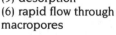

partitioning), which can have much greater influence in fine-grained soils than dispersion. Curve 4 portrays the likely effect of retardation. Curve 5 shows desorption, and curve 6 depicts flow occurring through macropores on break-through. A considerable number of PVF may be required for some of these curves to reach equilibrium.

If the hydraulic conductivity of the soil is much higher after permeation with waste liquid than with water and is greater than the design values, then the soil under investigation is considered unsuitable for containment of that waste. Other materials should be considered and evaluated for their per-formance.

6.8 VOLUME CHANGE

There are several factors that can cause changes in soil volume. For example, organic cations may destroy the swelling capacity of smectite. Grim (1968) reported a decrease in the water adsorption capacity of smectite due to coating of interlayer spaces by organics. The water adsorption capacity decreased as the size of the organic ion increased.

Volume change may also result from changes in effective stress, as when a hydraulic gradient is applied. In a constant head permeameter, once the soil has reached equilibrium under a given hydraulic gradient there is no further change in effective stress. In the variable head permeameter, however, the effec-tive stress changes with head variations. This is especially true in a flexible wall permeameter where small diameter burettes are used for inflow and outflow measurement and the level of headwater drops while that of tailwater rises. In such a case, Fernandez and Quigley (1991) observed up to 3% decrease in specimen height due to seepage force.

Changes in soil volume also occur due to alterations in clay structure. If the double layer shrinks, the dispersed structure may tend toward flocculation. Such volume changes are much greater for the smectite group than for illite or kaolinite group. These volume changes may or may not be visually observable. Their effect on the soil mass may be the development of cracks, fissures, joints, change in pore space distribution, shrinking, and swelling. After these changes, the clay may have greater or lesser hydraulic conductivity than before.

6.9 MICROORGANISM

Microorganisms have the beneficial effect of attenuation. They may produce certain organic acids that can react with trace contaminants. They can also oxidize or reduce certain metals and chemicals and deposit them in the soil voids thereby retarding the rate of flow. Beneficial as they may be, if micro-organisms cannot develop in the field, they should not be allowed to develop in the laboratory tests either. A disinfectant may be used to prevent their growth.

6.10 LABORATORY TESTS

The results of laboratory tests are used to compare the performance of different materials that might be used in liners or slurry walls. Also, laboratory tests are used for checking the hydraulic conductivity of undisturbed block or tube specimens obtained from the field. In conducting laboratory tests, the following criteria should be observed.

1. Perform soil–contaminant liquid compatibility tests carefully. Use a termination criteria that accounts for all possible mechanisms of containment structure failure with respect to hydraulic conductivity.

2. Laboratory values of moisture content and energy input should not be more than the field values.

3. If a liner is to be put to service before it is completely hydrated, allow only partial hydration of the laboratory specimens.

4. When estimating the field hydraulic conductivity from laboratory tests, use the highest laboratory values. For other test conditions such as magnitude of hydraulic gradient, type of permeameter, etc., follow the recommendations described in the previous sections.

5. Use carefully obtained undisturbed tube samples and block samples for hydraulic conductivity checking. Obtain the largest possible specimens for these tests.

6. Use consolidation stresses consistent with those most likely to occur in the field. Much lower damage occurs to the liner if static stresses are high.

6.11 LABORATORY/FIELD HYDRAULIC CONDUCTIVITY VARIATIONS

Laboratory hydraulic conductivity values are often lower than those in the field. For natural soils, the ratio of field hydraulic conductivity to laboratory hydraulic conductivity has been reported to be as high as 46,000 in some cases but generally it is larger than 1 (Olsen and Daniel, 1981). It has been shown by Day and Daniel (1985) and Reades et al. (1987) that hydraulic conductivity of the laboratory prepared samples and the undisturbed samples from liners with small thickness is several orders of magnitude lower than field samples. However, for thicker liners there is a good agreement between the laboratory, the undisturbed sample, and the field values.

Some of the factors contributing toward higher hydraulic conductivity in the field are

1. The inhomogeneous nature of the compacted soil. A few zones of high hydraulic conductivity may often govern the overall hydraulic conductivity.

2. Disparity in soil structure between the field and the laboratory samples due to differences in the method of compaction.

3. Formation of desiccation cracks, if compacted soil is not protected.

4. Disturbance of the soil samples obtained from the field (such as smearing of the surfaces).

5. Considerably higher confining pressures used in the laboratory than those existing in the field.

6. Use of much higher gradients in the laboratory than in the field.

7. Frequency and size of clods in the soil before compaction. (Larger clods yield higher hydraulic conductivity.)

8. Inadequate attention to soil moisture content during field operations.

9. Exposure to freeze-thaw cycles in the field.

Techniques for overcoming some of these difficulties and methods of determining hydraulic conductivity in the field are described in Chapters 7 and 10.

6.12 HYDRAULIC PROPERTIES OF MUNICIPAL SOLID WASTE

Laboratory or field measurement of hydraulic conductivity of municipal refuse is not a routine step in an investigation program. Estimates are usually made based on local experiences and published data. Field pumping tests may be necessary for a proper determination of the hydraulic properties of refuse.

Laboratory data on baled refuse (Fang, 1983) indicate hydraulic conductivity of 7×10^{-4} cm/s (2 ft/day) for dense specimens with unit weight of 11.2 kN/m^3 and 15×10^{-3} cm/s for loose specimens with unit weight of 5.6 kN/m^3. Intermediate values of 5×10^{-3} cm/s and 3.5×10^{-3} cm/s were measured for unit weights of 7.7 kN/m^3 and 8.2 kN/m^3, respectively. Laboratory data on shredded refuse (Fungaroli and Steiner, 1979) suggest hydraulic conductivity in the range of 10^{-2} cm/s to 10^{-4} cm/s. In the absence of field measurements, a value of 10^{-3} cm/s is appropriate (Oweis et al., 1990).

Leachate buildup occurs in old landfills that have no underdrain system and where the hydraulic conductivity of the underlying formation is substantially lower than that of refuse. For refuse on an impervious base, the leachate buildup can be estimated from the method given in Chapter 11. The leachate buildup may be measured in the field and the hydraulic conductivity of refuse k_r, can be determined from the value of percolation (using, for example, the water balance method or actual data) and the method of Chapter 11.

This technique was applied to a landfill in the Hackensack Meadows in northern New Jersey (Oweis and Khera, 1986). The back-calculated k_r was 2.6×10^{-3} cm/s. A field pumping test at the same landfill yielded a k_r value of about 10^{-3} cm/s. The critical parameter in removing leachate by pumping from refuse appears to be the specific yield or (ratio of the volume of drainable leachate to the total volume of the refuse). The Hackensack Meadows pumping test indicated a projected specific yield of about 10%, which is at the low end of the range for fine sands. The pumping rate was 12 gal/min (45 l/min).

The falling head test can be used for estimating the hydraulic conductivity of refuse. The values derived are, however, largely affected by the local composition. For example, a high concentration of plastics yields lower hydraulic conductivity. Because of the heterogeneous nature of refuse and its random channels, values of hydraulic conductivity are more difficult to determine than in soils composed of aggregate.

For example, the common assumption is that leachate is generated only after refuse reaches its field capacity. In some cases, however, leachate is generated before the landfill as a whole reaches its field capacity. In addition, not all the leachate flows to a well-designed surface drainage system; leachate has been observed to spring from the sides of a landfill despite the presence of a functional drain at the toe. The hydraulic conductivity produced by such side channeling, or "fingering," is commonly known as *secondary hydraulic conductivity*. The moisture content at field capacity is less than the saturation water content. Therefore, initially moisture transport occurs under unsaturated conditions. Above the field capacity, the hydraulic conductivity is of the same order of magnitude as at saturation.

6.13 HYDRAULIC PROPERTIES OF MINERAL WASTE

The hydraulic conductivity of coarse coal waste can vary over a wide range depending on the variability in particle size distribution and in situ unit weight. Reported laboratory values have ranged from 10^{-3} to 10^{-6} cm/s. For fresh uncompacted waste it may be as high as 10^{-1} cm/s and for well-compacted waste that has undergone weathering it may be as low as 10^{-8} cm/s. The hydraulic conductivity for fine coal waste (tailings) has been reported between 10^{-3} to 10^{-7} cm/s (Holubec, 1976).

The hydraulic conductivity of fly ash ranges from 10^{-4} to 10^{-7} cm/s. McLaren and DiGioia (1987) reported a mean value of 1.32×10^{-5} cm/s for Class F fly ash and 1.13×10^{-5} cm/s for Class C fly ash. The value of permeability decreases with time and (obviously) with increasing density. The hydraulic conductivity values for bottom ash range from 5×10^{-3} to 10^{-1} cm/s. At disposal sites of fly ash, bottom ash is frequently used as drainage material because of its excellent flow characteristics.

The hydraulic conductivity of FGD sludge varies from 5×10^{-3} to 5×10^{-6} cm/s. Hydraulic conductivity of 10^{-9} cm/s was reported for sludge from a double-alkali scrubbing system containing 30% and 50% fly ash (Krizek et al., 1976). The lower values are more common to sulfite-rich sludges and the higher to sulfate-rich sludges. When the sludge is mixed with fly ash its hydraulic conductivity increases but with increased degree of compaction the hydraulic conductivity decreases. With the addition of lime and increased curing time there may be some decrease in hydraulic conductivity. Tests by Soliman (1990) on lime-stabilized fly ash FGD specimens yielded hydraulic conductivity values ranging from 1.3×10^{-6} to 5.6×10^{-6} cm/s. For compacted solid waste incinerator ash, Poran and Ahtchi-Ali (1989) reported hydraulic conductivity of 1.6×10^{-1} cm/s for a void ratio of 1 and 1×10^{-1} cm/s for a void ratio of 2.5 with a linear correlation between hydraulic conductivity and void ratio.

Pulp and paper mill sludges have a hydraulic conductivity range of 10^{-7} to 10^{-8} cm/s. Andersland and Mathew (1973) found that the addition of 10% fly ash increased the hydraulic conductivity of these sludges by one to two orders of magnitude at low pressure but with increasing effective stress the hydraulic conductivity decreased rapidly.

6.14 SUMMARY

Solutes are transported through soils due to advection, dispersion, and diffusion. The transport process may be affected by adsorption and biodegradation. For soils with hydraulic conductivity $<10^{-9}$ m/s and hydraulic gradient <1 (as would be the case for most soil liners), dispersion is negligible and chemical diffusion can play a major role in transport. Laboratory methods of measuring hydraulic conductivity consist of flexible wall permeameter, fixed wall permeameter, double wall permeameter, consolidation ring permeameter, and constant rate of flow. For flexible wall permeameters, ASTM D 5084 is the standard method for determining hydraulic conductivity with water. Back pressure is used to ensure complete saturation.

There is no standard method for evaluating hydraulic conductivity of soils with waste chemicals. A baseline value is determined with tap water or 0.005 N $CaSO_4$ solution. Number of pore volume flow and concentration of influent and effluent is used as termination criteria to assess compatibility. The hydraulic gradient and the effective stress may show a significant effect on the test results.

In most instances, the aqueous solutions of a single organic and inorganic contaminant have less adverse effect on soils containing clays than solutions of concentrated chemicals. If several organics are mixed, the hydraulic conductivity may be higher even at low concentrations. Water, which has the highest dielectric constant, yields the lowest values for hydraulic conductivity. The results reported in the literature are often contradictory and no theory is available that can predict the response of a soil to various chemicals. Although hydraulic conductivity values for various waste types are given, because of the large range of these values site-specific tests must always be performed.

NOTATIONS

α dispersivity (L)

γ unit weight (F/L^3)

λ first-order decay constant (1/T)

ρ density (M/L^3)

μ dynamic viscosity (M/LT)

ν kinematic viscosity (L^2/T)

σ'_3 effective consolidation stress (F/L^2)

σ_3 minor principal stress, total chamber pressure (F/L^2)

ρ_d dry density (M/L^3)

μ_t dynamic viscosity at temperature t (M/LT)

γ_w unit weight of water (F/L^3)

A total cross-sectional area of soil (L^2)

a cross-sectional area of standpipe or burette (L^2)

a_i inflow burette area (L^2)

a_o outflow burette area (L^2)

B pore pressure parameter

c chemical concentration (M/L^3)

c_0 source concentration (M/L^3)

c_v coefficient of consolidation (L^2/T)

D chemical diffusion coefficient in soil/rock (L^2/T)

D_e effective diffusion coefficient (L^2/T)

D_o molecular diffusion coefficients in free water (L^2/T)

g acceleration due to gravity (L/T^2)

h	total head loss (L)		temperature t (L/T)	S	mass of solute (M)
h_1	head difference at the beginning of flow period (L)	L	the distance through which head loss occurs (L)	t	time of flow (T)
h_2	head difference at the end of flow period (L)	m_v	coefficient of volume compressibility (L^2/F)	t_{50}	time for 50% effluent concentration (T)
i	hydraulic gradient	n	porosity	u	pore water pressure (F/L^2)
K	intrinsic permeability (L^2)	n_e	effective porosity	Δu	change in pore pressure (F/L^2)
k	hydraulic conductivity (L/T)	p	pressure (F/L^2)	v	discharge velocity (L/T)
K_d	distribution coefficient (L^3/M)	PVF	pore volume flow	v_s	seepage velocity (L/T)
		q	rate of flow (L^3/T)	z	elevation or depth (L)
K_f	constant	Q	total amount of flow (L^3)		
k_t	hydraulic conductivity at	R	retardation coefficient		

PROBLEMS

6.1 Does the thickness of the double layer have any effect on hydraulic conductivity? Explain. If monovalent ions are replaced by divalent ions, how would the hydraulic conductivity be affected?

6.2 What effect does effective stress have on hydraulic conductivity? If a soil shows increase in hydraulic conductivity with a certain leachate, could the application of static stress to the soil revert the detrimental effect of the leachate?

6.3 What is a compatibility test and why is it necessary to perform one? Can index properties be used as compatibility test? If so, which ones and how?

6.4 Compare the advantages and disadvantages of flexible wall permeameters and fixed wall permeameters. Can back pressure be used to saturate test specimens in both these tests? Explain.

6.5 A test specimen 100 mm in diameter and 50 mm in length is to be tested for hydraulic conductivity. If the maximum permissible gradient is 30, determine the pressure difference that should be applied at the two ends of the specimen.

6.6 What hydraulic gradient would be acceptable for a soil with hydraulic conductivity of 3×10^{-8} cm/s? Would the acceptable hydraulic gradient for a soil with hydraulic conductivity of 10^{-3} cm/s be less or more?

6.7 After one-dimensional consolidation under 0.125 tsf, the height of a specimen was 0.9525 in. and diameter was 2.477 in. The time and dial reading data are given in Table 6.2. A falling head permeability test was performed at the end of next stress level of 0.25 tsf. The area of the standpipe was 0.387 cm², and the temperature

Table 6.2　Consolidation data under 0.25 tsf

Elapsed time (min)	0	0.1	0.25	0.5	1	2	4	8	15	30	60	120	1305	5640
Dial reading × 10^{-4} in.	1384	1410	1415	1422	1436	1452	1477	1509	1543	1574	1594	1604	1616	1620

during the test was 21°C. The flow data are given in Table 6.3. Determine the hydraulic conductivity of the soil using

a. The data from the falling head method.

b. The coefficient on consolidation (Section 4.5). All answers must be given in CGS units.

6.8 In a falling head permeameter test, the specimen length is 25 mm, diameter is 63.5 mm, and hydraulic conductivity is 3.1×10^{-6} m/s. The diameter of the standpipe is 3 mm.

a. How long would it take for the head difference to change from 1.2 to 1.01 m?

b. How long would it take for the same head to drop if an additional pressure of 20 kPa is applied to the upstream standpipe?

c. If the hydraulic conductivity was measured at 25°C in the laboratory, what would be its value at 20°C? (*Hint:* The dynamic viscosity of water is 8.94 millipoise at 25°C and 10.05 millipoise at 20°C.)

6.9 In a flexible wall permeameter, hydraulic conductivity of a compacted soil was determined to be 1.8×10^{-9} cm/s with 4% ethanol. Kinematic viscosity of the permeant at 20°C was 0.0305 cm^2/s. What is the intrinsic permeability of this soil at 20°C?

6.10 With concentrated hexane, the intrinsic permeability of kaolinite and a mixed soil was computed to be 4×10^{-15} m^2 and 8×10^{-15} m^2, respectively. At the test temperature, the density of hexane was 0.66 g/ml and its dynamic viscosity was 2.94 millipoise. What were the measured values of hydraulic conductivity for each of the two soils?

6.11 A test specimen is 70 mm in diameter and 35 mm in length. Its total density is 1.9 g/cc. Specific gravity of the solids is 2.7, and the hydraulic conductivity of the specimen is 9×10^{-8} cm/s. In a constant head permeability test, a hydraulic gradient of 50 is applied. Determine the

a. discharge velocity,

b. seepage velocity if only 89% of the pores are interconnected,

c. time to pass one pore volume of fluid through this specimen under the applied gradient.

6.12 In a certain landfill, the contaminant source concentration is 1500 mg/l, discharge velocity is 0.003 m/yr, diffusion coefficient is 0.015 m^2/yr, porosity is 0.45, $\rho_d K_d = 1.6$, and linear thickness is 1.2 m. Compute the concentration of contaminant plume after 20 years, 40 years, 80 years, and 160 years.

6.13 A permeability test is to be performed in a flexible wall permeameter on a 71-mm-diameter and 38-mm-long specimen after consolidating it at a cell pressure of 500 kPa and a back pressure of 350 kPa. For a hydraulic gradient

Table 6.3
Falling head data

Clock time	Head (cm)
10:05	86.5
10:30	83.4
11:00	80
11:45	75
12:25	71

of 30, determine the chamber pressure and pore pressure at the two ends of the specimen if the gradient is generated by allowing no change in the

a. effective stress on the upstream side,
b. effective stress on the downstream side,
c. average effective stress.

CHAPTER 7

Site Investigation

7.1 INTRODUCTION

Surface and subsurface conditions of soils and waste materials must be evaluated for proper design, operation and maintenance, and future utilization of a disposal facility. Prominent geotechnical design functions and the usually required geotechnical input are identified in Table 7.1. Some of the data needs are related to the physical and engineering characteristics of the subsoils; others relate to the characteristics of the wastes. The techniques used to collect the required data include remote sensing (air photo interpretation, infrared photography); geophysical methods (electrical methods, seismic methods); test pits; test boring and penetrometer resistance; sampling of soils and rocks; in situ tests for measuring properties of soils, rocks and waste materials, groundwater quality, and profile; and geotechnical monitoring. Several of these methods are described in this chapter.

7.2 HEALTH AND SAFETY

In any site investigation program in an environment where refuse or potentially toxic waste is expected, proper environmental safety measures should be followed. The appropriate level of protection is usually determined from the available evidence and is subsequently modified during work at the site.

A solid waste is deemed hazardous if (a) the waste is listed as hazardous waste; or (b) the waste meets specific criteria on ignitability, corrosivity, reactivity, and toxicity. The U.S. Environmental Protection Agency (EPA) lists a solid waste as hazardous if the EPA determines, based on organic and inorganic constituents, that the waste significantly contributes to mortality or serious illness or poses a substantial present or potential threat if not regulated. Wastes identified as such (based on tests of effects on rabbits and rats or the knowledge of effects on humans) are identified as "acute" hazardous waste.

If a representative sample of a solid waste is liquid, it is ruled ignitable if it has a flash point of $<60°C$ ($140°F$) determined by test methods specified in the regulations. If the waste is not liquid, it is ruled ignitable if under standard temperature and pressure it is capable of causing fire through friction, absorption of moisture, or spontaneous chemical changes and, when ignited, burns so vigorously and persistently that it creates a hazard. Ignitable compressed gas and oxidizers are hazardous under this criterion.

A solid waste is ruled corrosive if its representative sample is aqueous and has a solid waste $2 \geq \text{pH} \geq 12.5$. A solid waste is also ruled corrosive if its representative sample is liquid and it corrodes steel (SAE 1020) at a r

Table 7.1   Data needs for geotechnical design functions

Required investigation	DESIGN FUNCTION[a]												
	1	2	3	4	5	6	7	8	9	10	11	12	13
Waste characteristics													
Disposal practice									×	×			×
Physical, chemical component				×					×	×	×		×
Unit weight				×	×				×	×			×
Strength									×	×			×
Age													×
Hydraulic properties				×	×	×			×	×	×		×
Deformation characteristics			×								×		×
Leachate buildup				×	×	×			×	×	×		
Leachate quality				×	×	×					×		
Site geology/hydrology													
Seismic history		×	×	×	×	×	×	×	×	×			×
Rock geology		×	×	×	×	×	×	×		×	×		×
Rock hydraulic properties				×		×	×					×	×
Aquifer characteristics				×		×	×					×	×
Groundwater flow												×	
Recharge and discharge area												×	
Seasonal high water levels				×		×							
Groundwater quality						×						×	
Water supply wells and reservoir characteristics						×						×	
Groundwater chemistry				×		×						×	
Index properties													
Grain size, Atterberg limits, moisture content	×	×	×			×	×	×	×	×			×
Soil chemistry		×	×	×		×		×			×		×
Foundation soils													
Borrow Area Soils													
Stratigraphy				×	×	×			×	×	×	×	×
Classification	×	×	×	×	×	×	×	×	×	×	×	×	×
Engineering properties													
Strength		×	×			×			×	×	×		×
Deformation		×	×			×			×	×	×		×
Hydraulic conductivity		×	×	×		×	×	×	×	×	×		×
Erodability	×	×	×										

[a] 1-daily and intermediate cover; 2-runon/runoff system; 3-capping; 4-liner; 5-subsurface collection drain; 6-contaminant barrier; 7-gas venting; 8-leachate and gas management structures; 9-landfill slope stability; 10-landfill foundation stability; 11-geotechnical instrumentation; 12-groundwater monitoring; 13-construction on landfill

greater than 6.35 mm (0.25 in.) per year at a test temperature of 55°C (130°F). These corrosivity criteria do not cover wastes that do not contain liquids.

A solid waste is ruled reactive if any of the following apply:

1. It is normally unstable and readily undergoes violent change without detonating; it reacts violently with water; or when mixed with water it forms potentially explosive mixtures or generates toxic gases, vapors, or fumes in a quantity sufficient to present a danger to human health or the environment.

2. It is readily capable of detonation or explosive decomposition or reaction at standard temperature and pressure.

3. It is capable of detonation or explosive reaction when subjected to a strong initiating source or when heated under confinement.

4. It is a cyanide- or sulfide-bearing waste that when exposed to pH conditions between 2 and 12.5 can generate toxic gases, vapors, or fumes in a quantity sufficient to present a danger to human health or the environment.

5. It is a forbidden explosive as defined in other codes of federal regulations (49 CFR 173.53), Class A explosive (49 CFR 173.55), or Class B (40 CFR 173.88).

A waste is ruled hazardous if the concentration of any of the contaminants listed exceeds the regulatory level given in Table 7.2. The determination of concentration is based on the Toxicity Characteristic Leaching Procedure (TCLP) method. Where the waste contains less than 0.5% filterable solids, the waste itself after filtering is considered to be the extract for the purpose of the test. The liquid, if any, is separated from the solid phase and stored for later analysis (either separately or in combination with the extract from the solid phase, depending on compatibility). The solid phase is extracted with an amount of extraction fluid equal to 20 times the weight of the solid phase. Particle size reduction by crushing or other means is permitted. The extraction fluid used is a function of the alkalinity of the solid phase. If the pH is greater than 5, an extraction fluid having a pH of 2.88 ($\pm$0.05) is used. Extraction is achieved by placing the extract in a rotary agitation device, rotated at 30 ($\pm$2) rpm for 18 ($\pm$2) hours at 23 ($\pm$2) °C during the extraction period.

Four levels of protection—A, B, C, and D—have been identified by various agencies. Level D is used when sites are identified as having no toxic or hazardous waste and protection requirements consist of only work uniforms (coveralls, hard hats, boots), with splash goggles in some cases. Level C protection includes disposable outer clothing, safety shoes or boots, hard hat, rubber gloves with cotton liner, and full-face air-purifying respiratory protective equipment. Level B requires a self-contained breathing apparatus (SCBA) and disposable chemical-resistant outer clothing, overboots, and gloves. Level A protection is for the worst possible conditions and requires an SCBA and a totally encapsulating chemical-protective suit with a positive seal, safety boots or shoes, rubber gloves, and hard hats.

For drilling in a municipal landfill, Level C protection is usually sufficient, although an SCBA may be needed (especially for drillers near the borehole). Explosive gases should be periodically monitored beneath the drilling rig. For example, where methane concentrations reach 15% of the lower explosive limit (LEL), work may need to be stopped or venting may be required.

The proper selection of protective tools and clothing should be determined by a qualified individual. Table 7.3 provides a list of some of the field instruments used for monitoring potentially toxic and/or explosive gases and vapor

Table 7.2
Maximum
concentration of
contaminants for the
toxicity characteristic

Contaminant	Regulatory level (mg/L)
Arsenic	5.0
Barium	100.0
Benzene	0.5
Cadmium	1.0
Carbon tetrachloride	0.5
Chlordane	0.03
Chlorobenzene	100.0
Chloroform	6.0
Chromium	5.0
o-Cresol	200.0
m-Cresol	200.0
p-Cresol	200.0
Cresol	200.0
2,4-D	10.0
1,4-Dichlorobenzene	7.5
1,2-Dichloroethane	0.5
1,1-Dichloroethylene	0.7
2,4-Dinitrotoluene	0.13
Endrin	0.02
Heptachlor (and its epoxide)	0.008
Hexachlorobenzene	0.13
Hexachlorobutadiene	0.5
Hexachloroethane	3.0
Lead	5.0
Lindane	0.4
Mercury	0.2
Methoxychlor	10.0
Methyl ethyl ketone	200.0
Nitrobenzene	2.0
Pentachlorophenol	100.0
Pyridine	5.0
Selenium	1.0
Silver	5.0
Tetrachloroethylene	0.7
Toxaphene	0.5
Trichloroethylene	0.5
2,4,5-Trichlorophenol	400.0
2,4,6-Trichlorophenol	2.0
2,4,5-TP (Silvex)	1.0
Vinyl chloride	0.2

Table 7.3 Field instruments for health and safety

Instrument	Hazard monitored	Application	Detection method	Limitations	Ease of operation	General care and maintenance	Typical operating times
Ultraviolet (UV) photoionisation detector (PID) (Photovac 10A10) (HNU PI-101)	Many organic and some inorganic gases and vapors.	Detects total concentrations of many organic and some inorganic gases and vapors. Some identification of compounds is possible if more than one probe is used.	Ionizes molecules using UV radiation; produces a current that is proportional to the number of ions.	Does not detect methane. Does not detect a compound if the probe used has a lower energy level than the compound's ionization potential. Response may change when gases are mixed. Other voltage sources may interfere with measurements. Readings can only be reported relative to the calibration standard used. Response is affected by high humidity.	Effective use requires that the operator understands the operating principles and procedures and is competent in calibrating, reading, and interpreting the instrument.	Recharge or replace battery. Regularly clean and maintain the instrument and accessories.	10 hours; 5 hours with strip chart recorder.

Table 7.3 *Continued*

Instrument	Hazard monitored	Application	Detection method	Limitations	Ease of operation	General care and maintenance	Typical operating times
Flame ionization detector (FID) with gas chromatography option (OVA 128)	Many organic gases and vapors.	In survey mode, detects the total concentrations of many organic gases and vapors. In gas chromatography (GC) mode, identifies and measures specific compounds. In survey mode, all the organic compounds are ionized and detected at the same time. In GC mode, volatile species are separated.	Gases and vapors are ionized in a flame. A current is produced in proportion to the number of carbon atoms present.	Does not detect inorganic gases or some synthetics. Sensitivity depends on the compound. Should not be used at temperatures less than 40°F (4°C). Difficult to absolutely identify compounds High concentrations of contaminants or oxygen-deficient atmospheres require system modification In survey mode, readings can be only reported relative to the calibration standard used.	Requires experience to interpret data correctly, especially in the GC mode. Specific identification requires calibration with specific analyte of interest	Recharge or replace battery. Monitor fuel and/or combustion air supply gauges. Perform routine maintenance as described in the manual. Check for leaks.	8 hours; 3 hours with strip chart recorder.

Combustible gas indicator (CGI) (MSA explosimeter)	Combustible gases and vapors.	Measures the concentration of a combustible gas or vapor.	A filament, usually made of platinum, is heated by burning the combustible gas or vapor. The increase in heat is measured.	Accuracy depends on the difference between the calibration and sampling temperatures. Sensitivity is a function of the differences in the chemical and physical properties between the calibration gas and the gas being sampled. The filament can be damaged by certain compounds such as silicones, halides, tetraethyl lead, and oxygen-enriched atmospheres. CGI does not provide a valid reading under oxygen-deficient conditions.	Effective use requires that the operator understands the operating principles and procedures.	Recharge or replace battery. Calibrate immediately before use.	Can be used for as long as the battery lasts, or for the recommended interval between calibration, whichever is less.

Table 7.3 *Continued*

Instrument	Hazard monitored	Application	Detection method	Limitations	Ease of operation	General care and maintenance	Typical operating times
Oxygen meter (MSA oxygen meter)	Oxygen (O_2).	Measures the percentage of O_2 in air.	Uses an electrochemical sensor to measure the partial pressure of O_2 in the air and converts that reading to O_2 concentration.	Must be calibrated before use to compensate for altitude and barometric pressure. Certain gases, especially oxidants such as ozone, can affect readings. Carbon dioxide (CO_2) poisons the detector cell.	Effective use requires that operator understands the operating principles and procedures.	Replace detector cell according to manufacturer's recommendations. Recharge or replace batteries before expiration of the specified interval. If the ambient air is more than 0.5% CO_2, replace or rejuvenate the O_2 detector cell frequently.	8 to 12 hours.
Direct-reading colorimetric indicator tube (Draeger)	Specific gases and vapors.	The compound reacts with the indicator chemical in the tube, producing a stain whose length or color change is proportional to the compound's concentration.	The measured concentration of the same compound may vary among different manufacturer's tubes. Many similar chemicals interfere. Greatest sources of error are (1) how the operator judges stain's endpoint and (2) the tube's limited accuracy. Affected by high humidity.	Minimal operator training and expertise required.	Do *not* use a previously opened tube even if the indicator chemical is not stained. Check pump for leaks before and after use. Refrigerate before use to maintain shelf life of about 2 years. Check expiration date of tubes. Calibrate pump volume at least quarterly. Avoid rough handling that may cause channeling.	

Gamma radiation survey instrument (Thyac III)	Gamma radiation.	Environmental radiation monitor.	Scintillation detector.	Does not measure alpha or beta radiation.	Extremely easy to operate, but requires experience to interpret data. Rugged, good in field use.	Must be calibrated annually at a specialized facility. Can be used for as long as the battery lasts, or for the recommended interval between calibrations, whichever is less.
Portable infrared (IR) spectrophotometer	Many gases and vapors.	Measures concentration of many gases and vapors in air. Designed to quantify one- or two-component mixtures.	Passes different frequencies of IR through the sample. The frequencies adsorbed are specific for each compound.	In the field, must make repeated passes to achieve reliable results. Requires 115-volt AC power. Not approved for use in a potentially flammable or explosive atmosphere. Interference by water vapor and carbon dioxide. Certain vapors and high moisture may attack the instrument's optics, which must then be replaced.	Requires personnel with extensive experience in IR spectrophotometry.	As specified by manufacturer.

Source: Occupational Safety and Health Guidance Manual for Hazardous Waste Site Activities, Tables 7.1 and 7.2, NIOSH/OSHA/USCG/EPA, October 1985.

Even when contaminants are not suspected, it is prudent to develop a simple health and safety plan identifying the name and route to the nearest hospital and emergency phone numbers. A more extensive plan is essential for work at a contaminated site. Paragraph (1) of OSHA 29 CFR 1910.120 outlines the requirements of a health and safety plan for hazardous waste sites:

1. Provide background information on site (status, physical features, disposal practices, past monitoring data, health complaints, etc.).

2. Identify known or suspected contaminants (location, concentration, and action level that would require upgrading protective equipment).

3. Evaluate risks associated with various field operations.

4. Identify key personnel and alternates for site safety and remedial response.

5. Address levels of personnel protective equipment including a decision tree for upgrading or downgrading the level of protection.

6. Designate work areas (exclusion zone, contamination reduction zone, and support zone).

7. Establish decontamination procedures for personnel and equipment.

8. Determine safety work force in each zone during entry and subsequent operations.

9. Establish emergency procedures for evacuation and communication response to fire and explosions; emergency phone numbers (e.g., fire department, hospital, police, ambulance, medical consultant, poison control) must be included in the plan.

10. Verify individual health and safety training for use of protective gear, field instruments, and site task activity.

11. Describe procedures and record keeping and equipment required to monitor the work area for potentially harmful substances.

12. Consider weather and other conditions that affect safety. (For example, wind direction will control the direction of vapors coming out from a borehole, so it would be unsafe to stand in a downwind location.)

13. Implement procedures for controlling access to site.

14. Identify medical surveillance requirements.

7.3   INVESTIGATION PHASES

A geotechnical investigation for large sites, such as disposal facilities, is usually carried out in three phases: desk study, preliminary investigation, and detailed investigation.

7.3.1 Desk Study

This phase covers review of available topographic and geological data, aerial photographs, site reconnaissance, and data from previous investigations. In the United States, topographic and geological data, including water supply papers, are available from the United States Geologic Survey and each state's Geological Survey. Useful data on near-surface soils are available from the soil surveys published by the United States Department of Agriculture (USDA) and the Soil Conservation Service (SCS). Although soil surveys are primarily for agricultural purposes, they do contain considerable information for geotechnical engineering use. In the United Kingdom, topographic and soil maps are available from the Ordnance Survey. Geological records and maps can be obtained from the Geological Museum. Other useful data are available from highway departments, universities, public libraries, and many large cities around the world that have developed extensive geological information (Legget, 1973).

The value of topographic and geological maps is limited as they do not contain adequate details for geotechnical purposes. Greater details can be obtained from aerial photographs, which are the most common type of remote sensing technique. An experienced air photo interpreter can develop valuable information on the site, such as depth to rock, drainage patterns, site morphology, and surface soil type (Way, 1973). Aerial photographs are usually available in nine-inch frames in scales of 1 : 12,000 to 1 : 80,000. Other types of imagery (NAVFAC DM 7.1, 1982) include Skylab, NASA, SLAR, and thermal infrared. These are generally used for more detailed investigations such as fault studies and water resources.

The outcome of the desk study usually reveals the broad geological features at the site. It also helps determine the scope of the preliminary investigation, such as the number of borings needed, where they should be located, and their required depth.

7.3.2 Preliminary Investigation

The areal extent and depth of the preliminary investigation partially depends on the nature of the geological details identified in the desk study. However, for a new land disposal facility, the minimum scope of the investigation is often dictated by regulations such as those shown in Table 7.4.

Test pits may be used for visual examination of soil strata or waste materials at shallow depths. The minimum depth of borings for sites where foundation instability is not a concern is usually less than 50 ft (15 m) beneath the base of the disposal facility. Otherwise, some or all of the borings are advanced deeper to obtain data for preliminary stability assessment. The preliminary investigation defines the general site conditions, hydraulic conductivity, and groundwater flow. Sufficient data are collected to develop a preliminary design for the facility to submit for regulatory preliminary approval.

Table 7.4
Minimum number of borings (NJDEP)

Acreage	Total no. of borings	No. of deep borings
Less than 10	4	1
10–49	8	2
50–99	14	4
100–200	20	5
More than 200	24 plus 1 boring each additional 10 acres	6 plus 1 boring each additional 10 acres

7.3.3 Final Investigation

Depending on the uniformity of site conditions, the preliminary and final investigations may be combined into one phase. Various criteria have been proposed (NAVFAC, 1982; ASCE, 1976) for the extent and depth of investigations for various engineering structures. Table 7.5 presents guidance for scoping out an exploration program utilizing test borings.

7.3.4 Drilling Methods

Several methods are available for advancing boreholes to obtain samples of underlying materials (Clayton et al., 1982). For land disposal facilities the two commonly used techniques are the hollow-stem continuous flight auger and rotary drilling. Others are the reverse rotary, air rotary, and bucket auger

Table 7.5   Guidance for borings

Areas of investigations	Boring layout	Boring depth
New or expanded and disposal facility	Establish subsurface sections at most critical areas. Provide four to six borings along each section to establish data for stability analysis. Supplement by test pits for examination of near-surface soils and waste.	Advance borings into a relatively incompressible soil to depths where the increase in stress is 10% or less of the existing effective overburden stress, where settlement is not a key design factor. Advance to below active or potential failure surface or to a depth where foundation failure is unlikely.
Facility structure	Minimum of four borings at corners plus one in the interior for 5000 ft^2 (465 m^2). Supplement by test pits for examination of near-surface soils and waste.	Advance borings into a relatively incompressible soil or where vertical stress induced by the structure is less than 10% of the overburden effective stress. Advance for a minimum of 30 ft (9.1 m).
Retention ponds	Advance preliminary borings at 100–300 ft centers (30.5–91.4 m). Add intermediate borings along the center line at critical facilities (cut-off, outlet, and inlet structures). Supplement by test pits for examination of near-surface soils and waste.	Add borings to a minimum of 0.5–1 times the width of the embankment or to the relatively hard stratum.
Containment barriers	Advance preliminary borings at 500 ft (152.4 m) intervals. Advance intermediate borings for a final spacing of 300 ft. Supplement by test pits for examination of near-surface soils and waste.	Extend a minimum of 10 ft (3 m) into stratum with the design hydraulic conductivity.

methods. In extremely difficult conditions, the percussion or cable tool drilling method may be used. Descriptions of each of these methods follow.

7.3.4.1 Hollow-Stem Continuous Flight Auger The hollow-stem auger (Figure 7.1) serves as a casing and has a continuous outer spiral. It contains an inner rod with a plug at the lower end, which keeps the soil from entering the auger during the advancement of the hole. The drill is operated by power from a truck or drilling rig.

For sampling, the plug is replaced by a sampler. In soils such as sand and silt below the water table, when the plug is removed the high water pressure outside the stem may cause the soils to be washed into the casing. If representative samples are needed from such soils, the plug should not be used during augering. Instead, the casing should be cleaned from within or the hole should be stabilized by mud or by keeping a positive head of water above the groundwater level.

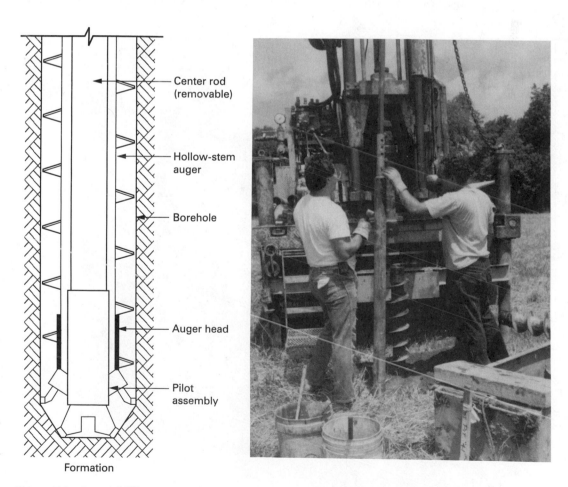

Formation

Figure 7.1 Auger drilling

7.3.4.2 Rotary Drilling
The rotary drilling method (Figure 7.2) employs power to rotate the drilling rods and the cutting bit through which water or drilling fluid is circulated to remove cuttings from the hole. The drilling fluid is prepared from bentonite and water and has a density of 68 to 72 lb/ft^3. If the stability of the hole cannot be maintained by the drilling fluid, a casing is driven in and its inside is cleaned before taking a sample. This method can be used in all soils except those containing large boulders and cobbles.

7.3.4.3 Reverse Rotary
This method differs from rotary drilling in the way mud is introduced into the hole. Instead of circulating the mud through the drill rod, mud is introduced into the annulus around the drill pipe as the rods are rotated. The mud (along with the cuttings) returns to the surface through the drilling rod by either inducing suction or air-lifting the mud and cuttings.

7.3.4.4 Air Rotary
Air rotary rigs employ air down the drill pipe to advance the hole, with cuttings discharged up the annulus around the drill pipe. Some

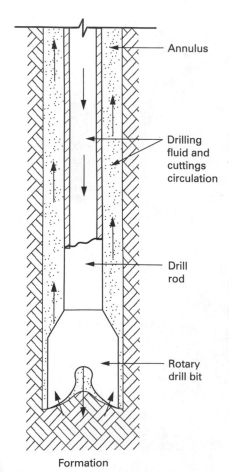

Annulus

Drilling fluid and cuttings circulation

Drill rod

Rotary drill bit

Formation

Figure 7.2 Rotary drilling

air rigs are equipped with a downhole hammer that pulverizes the material and uses air to return the cuttings to the surface. When drilling below water, a casing is required to prevent the hole from collapsing. An air rig with a down-hole hammer is suitable for advancing holes but is typically not equipped to recover samples using the split barrel sampler.

7.3.4.5 Bucket Auger The bucket auger (Figure 7.3) consists of a cylindrical bucket 10 to 72 in. in diameter, with teeth at the bottom. The bucket is attached to a kelly bar that is rotated and pushed into the ground. When the bucket is filled, it is brought to the surface and emptied. A casing needs to be advanced when drilling below water or leachate mounds in landfills.

7.3.4.6 Percussion or Cable Tool Drilling In this method, a hole 4 in. or larger in diameter is advanced by alternately raising and dropping a heavy drilling bit (Figure 7.4). A small amount of slurry is kept in the hole, which keeps the cuttings in suspension. The slurry is removed by a bailer or a pump as it reaches its soil-carrying capacity. This method can be used in most instances where obstructions are anticipated. When possible the borehole is advanced ahead of the casing.

7.3.5 Drilling in Refuse

In advancing a test boring through refuse, difficulties may be encountered that require one or more relocations of the boreholes before the refuse layer is penetrated. Augering is the preferred method of advancing a hole through refuse and is usually attempted first. However, textiles and other objects may prevent advancement of the auger. Rotary drilling is not usually successful

Figure 7.3
Bucket auger rig

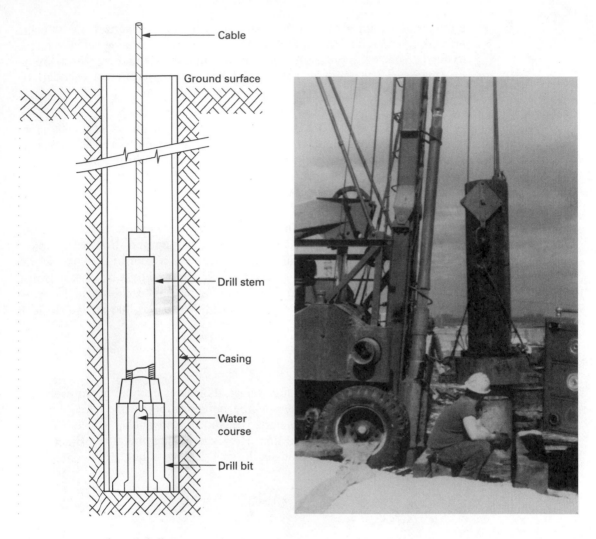

Figure 7.4 Cable tool drilling

because circulation of the drilling fluid is difficult to maintain. An air rotary with a downhole hammer is suitable in most cases. The most difficult drilling is usually below the leachate or water level where collapse of the hole becomes a problem.

For large holes (over 6 or 8 in. in diameter), the conventional drilling rigs usually do not have enough power to advance the hole either through augering or rotary action. In such cases, a large machine normally used for caisson construction may be considered. One alternative is to use a pile-driving rig to advance the casing; another is to use a bucket auger, which has proven to be successful above the leachate level. Another effective, although slow, method is cable tool drilling, which can penetrate almost all types of obstructions. Its rate of penetrating through refuse varies from 4 to 8 ft per 8 hour day.

7.3.6 Sampling

A survey of the current practice of soil sampling is contained in the 1979 Proceedings of the International Society of Soil Mechanics and Foundation Engineering. Samples from current soil sampling procedures can be broadly grouped as "disturbed" or "undisturbed" samples.

Disturbed samples are usually adequate for classification and physical property testing such as Atterberg limits (liquid limit, plastic limit, and moisture content), chemical testing (chemical corrosivity and contamination), and laboratory resistivity tests for assessing, potential for galvanic corrosion. Disturbed samples are usually recovered from test borings at (typically) 5-ft intervals using a standard split barrel (ASTM D 1586) or a split spoon sampler.

Undisturbed samples are required to conduct laboratory strength, permeability, and compressibility testing. Undisturbed samples, which are of better quality, can be obtained with techniques that minimize sample disturbance (NAVFAC DM, 7.1 ASTM D1587-74).

In areas suspected of contamination, samples should be screened using field instruments (see Table 7.3). Any positive reading may trigger health and safety procedures and decontamination protocols for sampling devices and drilling rigs. The major requirements are

1. Before entering the site, possible off-site contamination must be removed from drilling equipment by steam cleaning.

2. Between samples, samplers should be decontaminated. This usually involves:

  a. Brushing off visible dirt and mud.
  b. Scrubbing and washing with organic-free, clean water.
  c. Scrubbing and washing with trisodium sulfate.
  d. Scrubbing and washing with methanol or acetone.
  e. Rinsing with deionized or distilled water.

Between boreholes, all casings, rods, samplers, and other equipment used should be steam cleaned on a pad where contaminated fluids and materials can be collected and disposed of. Fluids and cuttings from the borehole that are contaminated should be stored in drums for eventual disposal.

7.4 IN SITU MEASUREMENT OF SOIL AND ROCK PROPERTIES

7.4.1 Standard Penetration Test

In the standard penetration test (SPT), the penetration resistance of the subsurface materials is determined by driving a split barrel sampler with a 140-lb hammer dropping 30 in. The number of blows, N, required to drive the sampler a distance of 12 in. (300 mm) after an initial penetration of 6 in. (150 mm) is referred to as the SPT N value (ASTM D 1586).

A low N value can be due to the sampler being driven through disturbed soil, improper cleaning of the borehole, or failure to keep a higher hydraulic head inside the hole than outside the hole (which causes sand boiling at the bottom of the hole). Hollow-stem augering should be avoided when drilling below the water table. High N values can be caused by restriction in the fall of the hammer, a plugged sampler, or gravel or large objects in otherwise loose soil.

The N value is affected by the energy delivered to the drill rod, which is affected by the type of rig and local practice. Seed et al. (1985) evaluated and reported on local practices in several countries (Table 7.6). In Japan, for example, a rope and pully practice delivers about 67% of the free-fall energy whereas rigs equipped with a mechanical trip hammer deliver about 78% of the free-fall energy. Thus if a trip hammer is used, the equivalent N using the rope and pully is about 16% more than that using the trip hammer. In the United States, the delivered energy using a safety hammer and two rope wraps around a pully is about 60% of free-fall energy. If a donut hammer and two wraps are used, the delivered energy is only about 45% of free-fall energy. Thus if a trip hammer is used in the United States, where the practice is a safety hammer and two wraps, the equivalent N would be about 30% more. Thus Seed et al. (1985) suggested normalizing the practice to 60% free fall as the N value.

The size of the hole and drilling methods also affect the N value. The N value has been correlated with many engineering parameters for soils. For example, the Gibbs and Holtz (1957) relationship of N and relative density D_r can be approximated as

$$D_r = 30(N')^{0.42} \qquad \text{for } N' > 5 \tag{7.1}$$

$$N' = C_n \cdot N \tag{7.2}$$

where

$$C_n = \left(\frac{1}{\sigma_v'}\right)^{1/2}$$

with σ_v' expressed in t/ft^2.

Also, the angle of internal friction for granular soils may be correlated to the SPT N value using the following relationship (Parry, 1977).

$$\phi = 25 + 28\left(\frac{N_{55}}{\sigma_v'}\right)^{1/2} \tag{7.3}$$

where N_{55} is the SPT blow count corresponding to an energy ratio of 55% and σ_v' is the vertical effective stress expressed in kPa. A crude estimate of the dynamic shear modulus (NAVFAC, 1982) is

$$G(\text{t/ft}^2) = 120 \, N^{0.8} \tag{7.4}$$

Table 7.6
Local practices
affecting N value

Country	Practice	Correction factor for 60% rod energy (N_{60})
United States	Safety hammer, rope and pully release	1.0
	Donut hammer, rope and pully release	0.75
Japan	Mechanical trip hammer	1.3
	Donut hammer, rope and pulley release	1.12
China	Donut, free-fall release	1.0
	Donut, rope and pulley release	0.83
Argentina	Donut, rope and pulley release	0.75

The N value is, in most cases, of very little value for sanitary landfills and should not be used for other than obtaining samples and classification. As shown by the typical values in Figure 7.5, the scatter is so wide that no meaningful utilization can be made.

Figure 7.5
N values in refuse

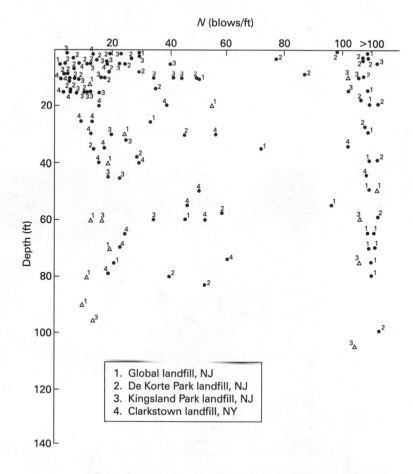

1. Global landfill, NJ
2. De Korte Park landfill, NJ
3. Kingsland Park landfill, NJ
4. Clarkstown landfill, NY

7.4.2 *Cone Penetration Test*

The static (Figure 7.6), or Dutch, cone penetration test (CPT) consists of forcing a cone into the ground at a rate of 10 to 20 mm/s and measuring the pressure needed for each increment of penetration (ASTM D 3441). Where both friction (f_s) and point resistance (q_c) are needed, a Begamann friction cone is used. The point resistance (q_c) is obtained while the cone is extended 80 mm. Then the total resistance of the cone and friction sleeve is obtained during the last 120 mm of the total 200 mm that the outer rods are advanced. The frictional resistance (f_s) of the sleeve is the total penetration resistance less the q_c.

Errors may be introduced due to friction between telescoping rods, friction between soil and the mantle immediately above the cone, excessive compression of the inner rod at high resistance (10 tons), the cone's large length at greater depths (30 m), etc. Some of these difficulties can be overcome with the use of an electric cone. An electric cone is more rapid; allows automatic data acquisition, reduction, and plotting; and has greater accuracy and repeatability.

Because no soil samples can be recovered for identification, cone resistance, q_c, and friction ratio, f_s/q_c, are used to differentiate between various soil types. Clean sands generally exhibit low friction ratios, which increases with increase in clay content. A soil classification chart based on a standard friction electric cone is given in Figure 7.7 (Olsen and Farr, 1986). To determine the true value of q_{cn}, we start with the minimum n of 0.6 and compute q_{cn}. Using the computed value of q_{cn} and the normalized value of f_{sn}, we can determine the new value of n from the graph. Each successive n will be larger than the previous one. Usually less than five repetitions are adequate for convergence.

Figure 7.6
Components of an acoustic cone (courtesy, Converse Consultants)

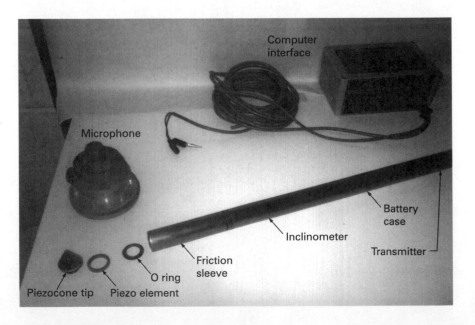

Figure 7.7
Soil classification based on standard friction electric cone tests (reproduced from Olsen and Farr, 1986, by permission)

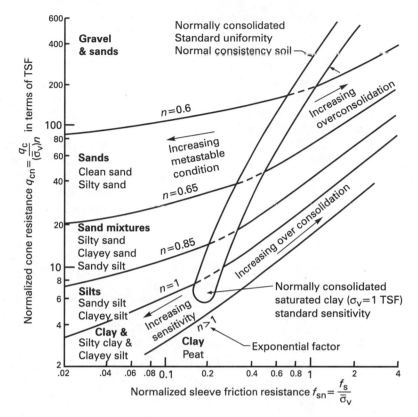

The cone penetration resistance q_c has been correlated with various engineering parameters. One useful relationship is

$$s_u = (q_c - \sigma_{vo})/N_k \qquad (7.5)$$

where s_u is the undrained shear strength, σ_{vo} is the total vertical stress at the point of measurement, and N_k is the cone factor. N_k ranges from 10 and 20. Robertson and Campanella (1983b) recommend using an N_k value of 15 for a preliminary estimate of s_u and a value of 10 for sensitive soils.

An estimate of the friction angle can be made in terms of the cone resistance q_c as follows (Robertson and Campanella, 1983a):

$$\tan \phi = \frac{1}{C_1} \ln \left(\frac{q_c/\sigma_v'}{C_2} \right) \qquad (7.6)$$

with $C_1 = 6.82$ and $C_2 = 0.266$.

The Young's modulus of sands, E, is estimated as

$$E = \alpha q_c \qquad (7.7)$$

The empirical parameter α depends on stress history and the fraction of failure stress level at which E is determined. At 25% of the failure stress level, for normally consolidated sand the value of α varies from 1.5 to 3.0 and for over-consolidated sand α is 4 to 10 times the value for normally consolidated sands. A relationship between q_c and E is given in Figure 7.8.

The ratio of cone resistance q_c (kg/cm²) to the SPT N value has been correlated with the mean grain size D_{50} (mm). The ratio q_c/N increases with increased grain size, as illustrated in Table 7.7. The value of D_{50} may be estimated from Table 7.7 if actual soil is not available for classification (Robertson and Campanella, 1983a).

The conventional cone has been modified to include a filter element for measuring pore water pressure. The filter may be located at the cone tip, cone face, or immediately behind the cone. A data logger records the point resistance, sleeve friction, and pore pressure. Because the position of the filter element affects q_c, the recorded q_c is corrected using

$$q_T = q_c + u(1 - a) \qquad (7.8)$$

Figure 7.8
Relationship between q_c and E (reproduced from Robertson and Campanella, 1983b, by permission)

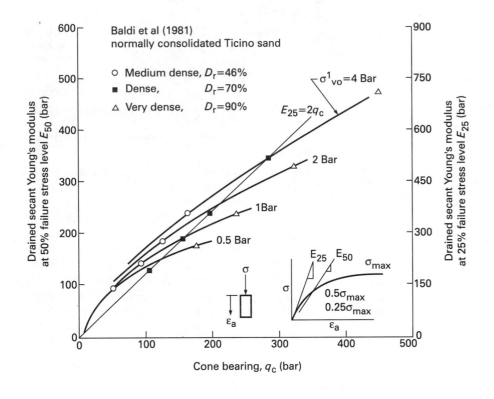

Table 7.7
Variation of q_c/N with mean grain size

Mean grain size, D_{50} (mm)	Soil	q_c/N
0.001	Clay	1
0.01	Sandy silts and silts	2
0.1	Silty sand	4
0.5	Medium sand	5–6
1.0	Gravelly sand	6–8

where q_T is the corrected cone point resistance, u is the measured pore pressure, and a is the area ratio for a given piezocone. The undrained shear strength is given by

$$s_u = (q_T - \sigma_{v0})/N_{kT} \tag{7.9}$$

For normally consolidated soils, N_{kT} increases linearly with the plasticity index and may be taken as 13 ± 2 for $I_p = 0$ and 18.5 ± 2 for $I_p = 50$ (Aas, 1986).

Recent cone designs incorporate circuitry to allow measurements of soil resistivity as the cone is pushed into the ground. These measurements are useful in detection contamination plumes in groundwater.

7.4.3 Pressuremeter Test

A pressuremeter is a cylindrical expandable probe designed to apply uniform pressure on the sides of a borehole by means of a flexible membrane (see Figure 7.9). The pressuremeter setup consists of a probe with an inflatable rubber membrane, a pressure–volume control unit, and connecting rigid plastic tubes. In the Menard-type pressuremeter (MPM), the probe is lowered into a slightly oversized prebored hole. The MPM is used for firm soils and

Figure 7.9
Pressuremeter setup
(courtesy Roctest)

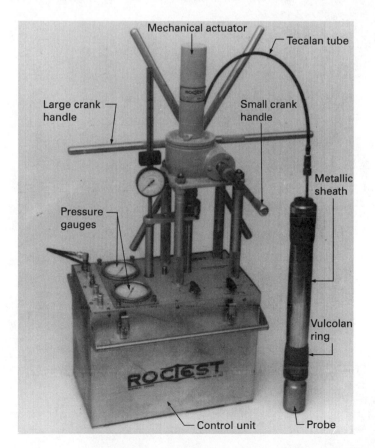

has been used extensively in testing municipal waste. The self-boring pressure-meter (SBP) bores itself into the ground and is generally used for testing soft soils. Figure 7.10 shows the raw data from a pressuremeter sounding in municipal refuse.

The pressuremeter modulus E_m is determined as

$$E_m = 2(1 + v)(v_0 + v_m)(\Delta p/\Delta v) \tag{7.10}$$

where E_m is the pressuremeter modulus, v_0 is the initial volume of the measuring cell, v_m is the volume change at a pressure corresponding to the mean pressure in the pseudoelastic range, $\Delta p/\Delta v$ is the slope of the pressure–volume curve in the linear range, and v is Poisson's ratio. Using the data in Figure 7.10, we can make the following calculations.

Elastic zone begins at pressure 117.81 kPa; volume = 302.91 cc
Elastic zone ends at pressure 285.32 kPa; volume = 517.84 cc
Average volume, v_m = 410.38 cc
Δp = 167.51 kPa
Δv = 214.93 cc
v = 0.33 (assumption)
v_0 = 750 cc (known)
E_m = 2.66 $(v_0 + v_m)$ (167.51/214.93) = 2432.4 kPa (24.8 tsf)

Figure 7.10
Pressuremeter
results in refuse

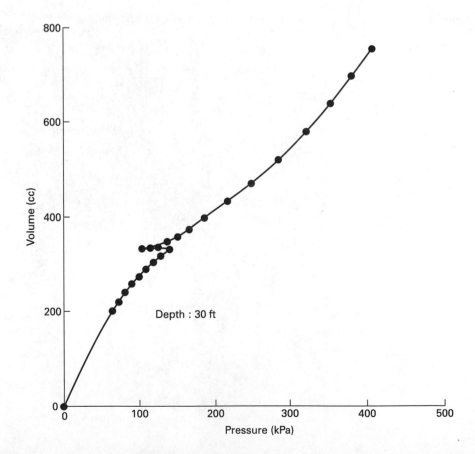

The slope of the unload curve may be considered to represent a modulus at small strain and thus close to moduli obtained from seismic methods (Mair and Wood, 1987). The data may also be used to estimate shear strength parameters (Briaud, 1992) based on the analysis of expansion of a cylindrical cavity.

7.4.4 *Field Load Tests*

For materials such as municipal refuse, a reasonable way to assess compressibility is to conduct a load test to stress deep layers of the refuse. The general approach encompasses the following steps.

  1. Determine the thickness, composition, and condition of refuse using investigative techniques such as borings, pressuremeter, etc.

  2. Select areas where types of waste appear to be different either in composition or strength.

  3. Install settlement plates (Figure 7.11). The subgrade for the plates is

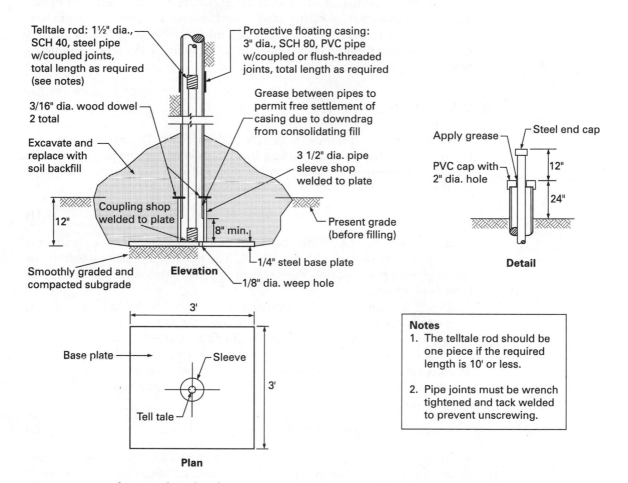

Figure 7.11   Settlement plate details

prepared by excavating to a level platform and filling depressions before a plate is installed. Take initial elevation readings.

4. Initially place a mound of soil carefully around the plate and riser for protection from construction equipment. Also place load fill carefully to avoid damage.

5. The width of the loaded area should be sufficient to stress the column of waste. For cap construction application, the width should be at least equal to the thickness of the waste.

6. Monitor elevations and report data as settlement versus fill elevation.

7.5   GEOPHYSICAL METHODS

Geophysics encompasses a variety of methods used as indirect investigative techniques. Such methods can be grouped as (a) electrical methods, (b) electromagnetic methods, (c) seismic methods, and (d) nuclear methods. Such methods are used to investigate contamination plumes, depth to groundwater, depth to rock, lithology, and discontinuities; locate buried drums; and obtain in situ measurements of elastic moduli and densities. Although geophysical methods are no substitute for direct site investigation such as drilling and sampling, indirect methods can be valuable in limiting the scope and delineating areas of direct investigations.

7.5.1 *Electrical Methods*

7.5.1.1 Resistivity   *Electrical resistivity* (ER) is primarily used for mapping contamination plumes and detecting leaks from lined lagoons but can also be used for mapping groundwater table and saltwater intrusions. The basic theory of the method is Ohm's law, which is expressed as

$$V = IR \tag{7.11}$$

where V is the potential of the electric circuit (volts), I is the current (amperes), and R is the measured resistance (ohms). The measured resistivity will be the *apparent resistivity* because the resistivity of the subsurface material varies from point to point.

Figure 7.12 shows the field installation utilizing the Schlumberger array. Two electrodes at M and N are needed to measure potential V. The distance a is kept fixed, and s varies. The ratio a/s should be small (between 3 and 30). In the limit as a approaches zero, the quantity V/a approaches the potential gradient at the midpoint of the array. As s is increased, the measurement will be affected by deeper geologic materials. The apparent resistivity (ohm-meter or ohm-feet) is (COE, 1979)

$$\rho_{app} = (\pi/I)s^2(V/a) \qquad \text{for} \qquad s \gg a \tag{7.12}$$

where ρ_{app} is the apparent resistivity (ohm-meters), $2s$ is the distance between electrodes A and B (meters), and a is the distance between electrodes M and N (meters).

Figure 7.12
Surface resistivity electrode configuration (after COE, 1979)

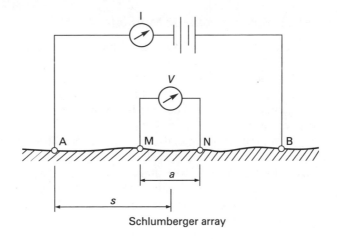

Schlumberger array

Apparent resistivity at small s reflects the true resistivity of the upper layers of geologic material whereas at large s it reflects the lower layers. The electrode spacing is often empirically assumed to be the effective depth of investigation. Electrode spacings are increased until the relationship between apparent resistivity and spacing becomes asymptotic. In the Wenner array (used in the ASTM Standard Method G-57-78), all the four electrodes are moved between successive observations. The Schlumberger array is less time consuming.

The minerals of most geologic materials are nonconductive. Electric current flows almost entirely through pore fluid. Resistivity is therefore sensitive to saturation and the type of pore fluid. Pure water (without salts) is nonconductive; it becomes conductive with the addition of salts. Conductivity is therefore proportional to salinity. For example, groundwater in overburden has a typical resistivity of 100 ohm-meter whereas seawater has a typical resistivity of 0.2 ohm-meter. This is the reason resistivity surveys are effective in mapping groundwater table and contamination plumes in areas where the background resistivity can be contrasted from that of the plume. Table 7.8 provides typical resistivity values for earth materials (COE, 1979).

Table 7.8
Typical electrical resistivities of earth materials

Material	Resistivity (ohm-meters)
Clay	1–20
Sand, wet to moist	20–200
Shale	1–500
Porous limestone	100–1,000
Dense limestone	1,000–1,000,000
Metamorphic rocks	50–1,000,000
Igneous rocks	100–1,000,000

Source: COE, 1979

7.5.2 *Electromagnetic Methods*

The electromagnetic (EM) method measures a composite conductivity (1/resistivity) in milliohms/meter of the subsurface reflecting the combined effect of the thickness of the geologic materials, their depth and their specific conductivity (Wait, 1982). A specific conductivity cannot be assigned to a particular geologic material. The results are influenced more by shallow layers than deeper layers.

Contaminant plumes, drums, and voids will result in conductivity contrasts (anomalies) that indirectly help mapping such features. EM surveys are generally inexpensive but have limitations with respect to penetration of electromagnetic waves and sensitivity. Both natural anomalies, such as faults or a change in lithology, and cultural features, such as power lines, could present problems in EM survey interpretation.

7.5.2.1 Ground Penetrating Radar (GPR)   The basic measurement in GPR is the travel time and amplitude of a reflected signal microwave (Uriksen, 1982). A radar antenna (source) emits an electromagnetic (EM) pulse that is directed into the ground (Figure 7.13). Contrasts in the conductivities of the subsurface materials cause waves to be reflected to the ground and picked up by an antenna.

The distance of wave penetration partially depends on the conductivity of the subsurface material. For example, a buried drum is a good conductor and reflects the entire EM wave. Also, wet soils are better conductors than dry soils, so the EM wave penetration in wet soil is limited because EM waves get reflected. Therefore, underlying drums or cavities will not be detected if the surface soils are wet.

The frequency of the emitted pulse also affects the distance of wave penetration. High frequency waves have shorter penetration (but better resolution) than do lower frequency waves. Under suitable field conditions (where EM waves can penetrate the subsurface), the technique is effective in detecting anomalies generated by drums as well as water table, cavities, and other geologic structures.

7.5.2.2 Magnetic Methods   A magnetometer (Heiland, 1963) registers the earth's magnetic field. Anomalous magnetic field strengths may infer the presence of ferrous metal objects such as pipes, drums, or metal-containing waste. The depth of the anomaly may be inferred as well. Because the technique is highly sensitive to all metal, the presence of any metallic object influences the data.

7.5.3 *Seismic Methods*

The basic theory of all seismic methods is that the velocity of seismic waves in subsurface material is a function of the elastic properties of the materials. Two types of waves are of interest in most techniques: A compressional or body wave (P wave) and a shear (secondary) wave (S wave).

Figure 7.13
Ground penetrating
radar operation
(courtesy Jim Werle)

Ground penetrating radar

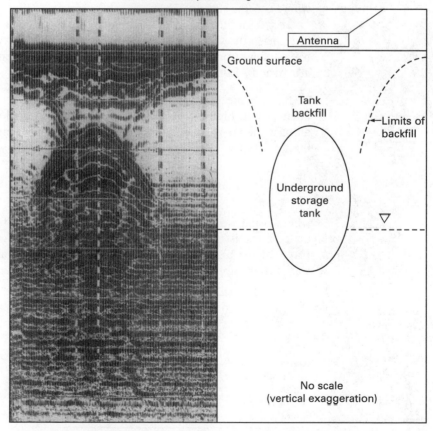

Actual profile printout   Interpretation of profile

If we place a plate on the surface of the ground and strike it with a sledge hammer, particles of the subsurface will vibrate forward and backward along the line of wave advance from the point of impact. This type of wave is a compressional (P) wave. A point away from the impact will feel the impact after a finite time that is equivalent to the distance divided by the velocity of the compressional wave propagation, v_p, expressed as

$$v_p = \sqrt{\frac{E(1-v)}{\rho(1+v)(1-2v)}} \tag{7.13}$$

where E is Young's modulus, v is Poissons's ratio, and ρ is the mass density of the material.

The second type of wave is the shear or secondary (S) wave where particles vibrate in a direction transverse to the direction of wave propagation. Its propagation velocity, v_s, is expressed as

$$v_s = \sqrt{\frac{G}{\rho}} \tag{7.14}$$

where G is the shear modulus.

It is clear from the above relationships that the compressional wave is faster than the shear wave. In addition, since fluids have no shear resistance, S waves cannot propagate through water. Some representative P wave velocities are given in Table 7.9.

7.5.3.1 Seismic Refraction The principal technique of seismic refraction is illustrated in Figure 7.14. Seismic waves are generated at the surface and travel along a radial path until they get refracted when encountering a layer with higher velocity. The law of refraction follows Snell's law:

$$\frac{\sin i}{\sin r_1} = \frac{v_1}{v_2} \tag{7.15}$$

where i and r are measured perpendicular to the interface. The critical incidence angle is that for which angle r is 90°. At this angle the refracted rays travel at the higher velocity and get refracted back and generate new (head) waves in the upper material. If the angle of incidence is greater than the critical

Table 7.9
Representative values of P wave velocities

Air (sea level, 20°C)	1127 ft/s
Fresh water (20°C)	4850 ft/s
Dry sand	1500–3000 ft/s
Clay	3000–6000 ft/s
Saturated loose sand	5000 ft/s
Weathered rock	5000–10,000 ft/s
Shale	12,000 ft/s
Hard granite or limestone	22,000 ft/s
Steel	20,000 ft/s

Source: COE, 1979

Figure 7.14
(a) Refraction of
ray paths across
a boundary between
two elastic media;
(b) schematic of seismic
refraction survey
(after COE, 1979)

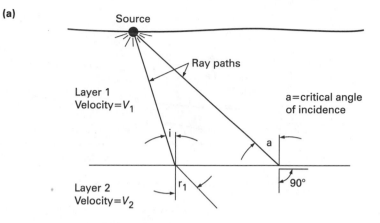

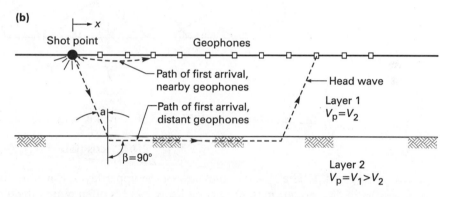

angle, the energy is reflected back into the lower (upper) velocity layer. At some critical distance, the head waves will reach the geophones faster than the reflected waves. Based on this, the velocities of a multilayer system are derived from the interpretation of the distance from the source–arrival-time relationship.

This technique is applicable where the velocities increase with depth. Sloping boundaries present difficulty in interpretation.

7.5.3.2 Seismic Reflection   When a wave is generated at the surface, some waves (direct waves) are propagated and picked up by geophones close to the source (Figure 7.15). These are the first waves to arrive. The time of arrival is computed as

$$t_{A1} = R_g / V_{p1} \tag{7.16}$$

where R_g is the distance of the geophone from the source and V_{p1} is the P wave velocity of the surface layer.

Other waves impinge on the interface of two layers and get reflected back to the upper layer. At geophones close to the source, these waves arrive after

Figure 7.15
Wave front and arrival
time diagrams (after
COE, 1979)

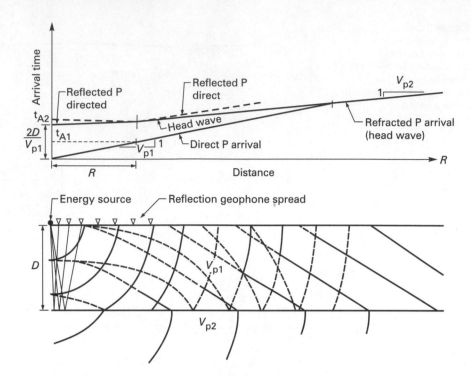

the direct waves and the time of arrival is computed as

$$t_{A2} = 2D/V_{p1} \cos(\tan^{-1} R_g/2D) \qquad (7.17)$$

where D is the thickness of the upper layer. For a multilayer system, the interpretation is more complicated and is often quite difficult.

Seismic refraction and reflection in geotechnical engineering has been used primarily for identification at top-of-rock or groundwater levels. For dynamic response analyses, the shear wave (S) velocity is of a primary interest. The S wave is a function of the shear modulus, which in turn is related to effective stress, shear strength, and density. Reliable determination of the shear wave velocity can be made using the crosshole shear wave velocity measurements.

7.5.4 *Borehole Geophysics*

7.5.4.1 Crosshole Seismic Testing In the crosshole test, seismic waves are generated in a borehole at a known depth and arrival time of seismic signals is measured at geophones placed in another hole (see Figure 7.16). The test is covered under ASTM D4428/D4428M-91 (Standard Test Methods for Crosshole Seismic Testing). In geotechnical applications, the shear wave velocity is of primary interest. A borehole hammer can be used to generate shear waves by exerting shear force along the side of the borehole. The mechanism receiving the strike of the hammer is wedged against the borehole side using a plate that is inflated by air or water. A sample oscillograph record is shown in Figure 7.16. As shown in the figure, P waves are first to arrive followed by the

Figure 7.16
Cross-hole seismic
testing.
(a) Basic concept;
(b) sample oscillograph
record (COE, 1979)

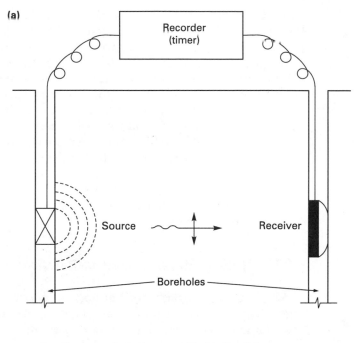

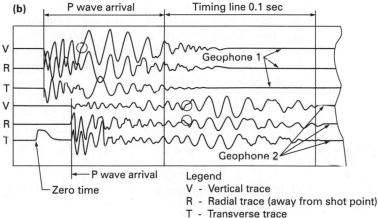

higher-amplitude S waves. With proper installation and expert interpretation, the test is considered reliable. The primary use is property determination.

7.5.4.2 Electrical Measurements   The two techniques of geotechnical interest are the spontaneous potential (or self-potential) log and borehole resistivity log. Data from both techniques provide electrical properties (that can be used for strata correlation from one borehole to the other), lithological characteristics, and pore fluid identification.

   The self-potential (SP) log is a record of variation of naturally occurring electric potential between an electrode at a particular depth in fluid-filled borehole and an electrode at the surface. Electrical resistivity is determined by passing a current between an electrode in the borehole and an electrode at the surface (Figure 7.17). According to a 1979 (COE) manual the following conclusions may be made (COE, 1979). A high resistance with negative SP means clean porous sand with fresh water. A high resistance with zero SP means a carbonate or crystalline lithology or a dry sand. A high resistance with positive SP means relatively dry silt or clay or shale. A low resistance with negative SP means saturated sand with water containing many dissolved solids. A low resistance with zero SP means saturated clayey sand or partially saturated clay or shale. A low resistance with positive SP means silty clay or shale with low porosity or fractures in carbonate or crystalline rocks. Anomalies in resistivity measured along the depth of the borehole may also be indicative of a contamination plume. Borehole electrical measurements are increasingly used in environmental investigations.

7.5.5 *Nuclear Measurements in Borehole*

7.5.5.1 Nuclear Logging   Nuclear logging in a borehole can provide a continuous logging of bulk density, water content, clay content, and porosity. The two basic methods used are the natural gamma log and the gamma-gamma log.

The natural gamma log measures the amount of natural gamma radiation emitted from the stratum being logged. The primary source of natural gamma radiation is the potassium isotope K^{40} in clay minerals. Natural gamma provides a means for identifying clay layers and strata correlation from borehole to borehole.

Figure 7.17
Generalized circuitry for the combination of SP and single-point resistance logs (COE, 1979)

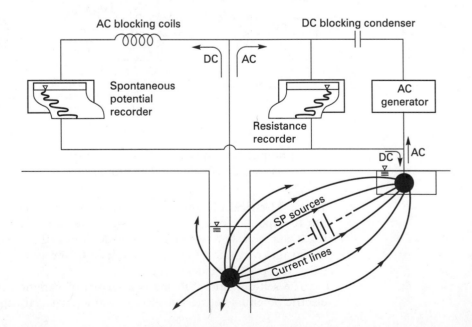

The basic operation for the gamma-gamma log is shown in Figure 7.18. A sonde with an isotopic source of gamma radiation and a detector shielded from the source is lowered in the hole. The emitted gamma photons that pass through the stratum get scattered (Compton scattering), resulting in loss of energy. Scattering is proportional to the electron density in the logged material, which is also a function of the bulk density and composition.

7.5.5.2 Neutron Water Detector A neutron mass is nearly equal in mass and diameter to that of a hydrogen atom found commonly in water. The devise is similar in concept to the gamma-gamma device (Figure 7.18). A source emits fast neutrons that get slowed down by collision with hydrogen atoms. The amount of neutron radiation received by the detector is calibrated against the amount of hydrogen (and therefore water) present. Low counts mean higher

Figure 7.18
Schematic of the borehole gamma-gamma density tool (COE, 1979)

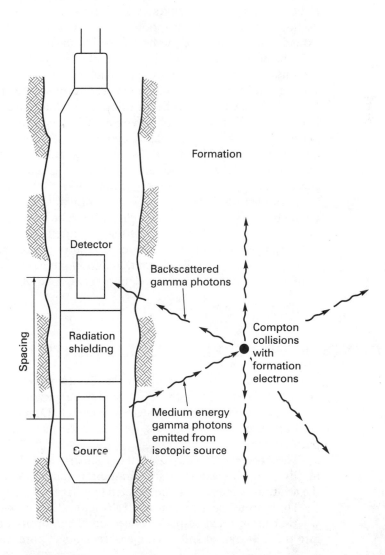

water content as more neutrons are slowed down. In addition to the hydrogen present in the soil and water, hydrocarbons adsorbed to clay minerals affect the count and thus the reliability of water content (that is present in pore space) measurements.

Nuclear methods similar in operation to those above are used to measure density and water content of compacted fills and clay liners.

7.5.6 *Support Measurements*

In conjunction with the above field geophysical measurements, caliper logging of the borehole, borehole fluid resistivity, and fluid temperature are typically measured to aid in interpreting and correcting geophysical data. The caliper log is a record of the changes in borehole size. Borehole diameter logs provide useful information such as lithology and fracture frequency. A probe with arms extended to the sides of the hole is pulled up the hole. Very small changes in hole diameter or wall condition are converted to electric signals, monitored at the surface, and recorded as a strip chart.

Fluid resistivity or conductivity (the reciprocal of resistivity) is measured through a downhole probe housing two plates (electrodes) with space in between to allow passage of borehole fluid. A constant, controlled AC current is transmitted between the plates, and changes in electrical resistance are transmitted uphole. Fluid resistivity is related to (and thus can be used to evaluate) water quality. The data are also used in interpreting other geophysical measurements.

Fluid temperature logs provide continuous fluid temperature recording throughout the depth of the borehole. A probe with one or more sensors connected to surface electronic instruments monitors and records temperature data. Temperature logs are used for several purposes, such as correcting other geophysical data, identifying contamination plumes or aquifers intersecting the borehole, and determining correlation between boreholes.

7.6   GROUNDWATER LEVEL MEASUREMENTS

Groundwater level measurements usually are made using three common categories of construction: the open standpipe piezometer, the porous element piezometer, and the air-actuated piezometer. Other types that may also be used are electrical piezometers, oil pneumatic piezometers, and water pressure piezometers.

The open standpipe piezometer is illustrated in Figure 7.19. It is used primarily through reasonably free-draining materials. Where several strata exist, the stratum of interest can be isolated (see Figure 7.22, p. 177).

The porous element piezometer (Figure 7.20) is used in fine-grained soils where a more rapid response is needed. The porous tip is driven or inserted into the fine-grained soil when measurements of pore water pressure is desired.

Figure 7.19
Open standpipe
piezometer. Test
sections may be
perforated with slots
or drilled holes
(reproduced from
NAVFAC, 1982)

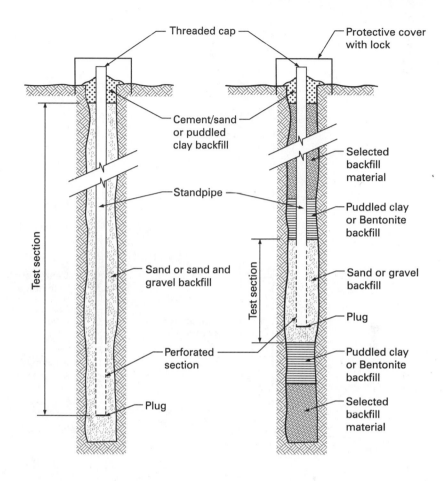

A metallic tip is less susceptible to damage than a ceramic tip. If size of the pores in the porous tip is small—that is, it has a high air entry value—then only water (and no air) will enter the tip in the unsaturated zones. This allows accurate measurement of pore water pressure without interference by air.

In the air-actuated piezometer (Figure 7.21), a ceramic tip and two air tubes are connected with a flexible diaphragm between them. Compressed air or nitrogen is introduced at increasing pressure through one of the air tubes. The applied air pressure is resisted by the pore water pressure acting on the other side of the diaphragm. When the air pressure is equal to the water pressure, the membrane is pushed away, allowing the air to flow up through the other tube to an indicator where the air bubbles become visible. The pneumatic device is sensitive to pressure changes as low as 1 cm.

Figure 7.20
Porous element
piezometer
(reproduced from
NAVFAC, 1982)

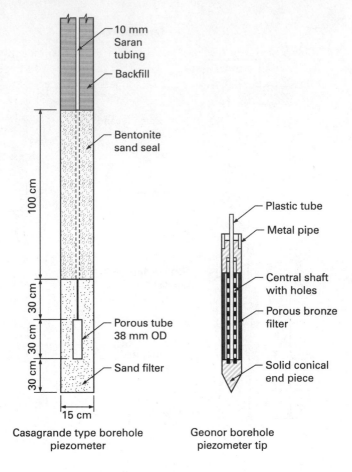

10 mm
Saran
tubing

Backfill

Bentonite
sand seal

100 cm

30 cm

30 cm

30 cm

30 cm

Porous tube
38 mm OD

Sand filter

15 cm

Plastic tube

Metal pipe

Central shaft
with holes

Porous bronze
filter

Solid conical
end piece

Casagrande type borehole
piezometer

Geonor borehole
piezometer tip

Figure 7.21
Air-actuated piezometer

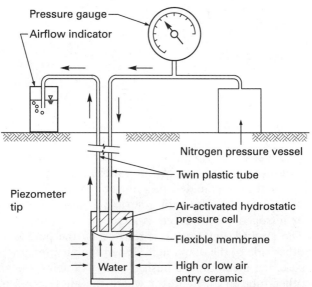

Pressure gauge

Airflow indicator

Nitrogen pressure vessel

Twin plastic tube

Piezometer
tip

Air-activated hydrostatic
pressure cell

Flexible membrane

Water

High or low air
entry ceramic

7.7 FIELD PERMEABILITY TESTS

The purpose of the field permeability tests is to measure the hydraulic conductivity of in situ materials. The test is usually conducted in a test boring or in a monitoring well. The test is influenced by the position of the water level, type of material, depth of test zone, hydraulic conductivity of the test zone, and heterogeneity and anisotropy of the test zone. The tests are grouped into three broad categories: (a) the constant head test, (b) the variable (falling or rising) head test, and (c) the pump test.

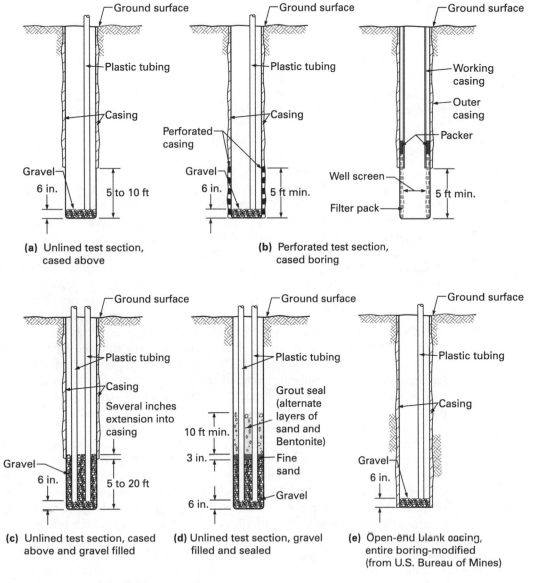

(a) Unlined test section, cased above

(b) Perforated test section, cased boring

(c) Unlined test section, cased above and gravel filled

(d) Unlined test section, gravel filled and sealed

(e) Open-end blank casing, entire boring-modified (from U.S. Bureau of Mines)

Figure 7.22 Test zone isolation methods

The test zone can be isolated in several ways, as illustrated in Figure 7.22. Where flow rates are very high or very low, the falling head test is more suitable. The falling head also is more applicable in unsaturated zones. In reasonably pervious soils below the water table, a rising head test is recommended.

Since the flow surface is very limited for the isolation methods of Figure 7.22, the test results are sensitive to the conditions of the borehole (the hole should be clean) and the drilling methods. Both the use of circulating drilling fluid and the driving of casing during advancement of the borehole negatively impact the results of the tests. In stable formations, the use of augering or air rotary drilling techniques is usually preferable if permeability tests are to be performed.

Figures 7.23 and 7.24 can be used for interpreting the variable head tests (NAVFAC, 1982). Because of the several factors impacting the results in borehole tests, the hydraulic conductivity is frequently determined from the laboratory tests or estimated based on the index properties (Chapter 6). Estimates may also be made using an empirical relation between effective particle diameter D_{10} (mm) and the hydraulic conductivity, k (cm/sec) (Hazen, 1982):

$$k = C(D_{10})^2 \qquad (7.18)$$

where $C = 100$ (varies between 40 and 150).

A more reliable assessment of the hydraulic conductivity can be obtained from pumping tests. However, pumping tests are much more expensive to perform and require much greater time to complete. An excellent treatment of this subject is presented by Driscoll (1986).

In municipal refuse, the corrosive nature of the materials as well as the probable high temperatures inside the landfill impact the selection of materials for the well. Figure 7.25 (p. 181) shows the cross section of a test well installed through a sanitary landfill (Oweis et al., 1990). High gas flow rate from the observation wells, which were screened partly in the vadose zone, along with high fluid conductivity and persistent foaming, caused considerable difficulties with the accurate measurement of fluid levels. This was resolved by inserting 3/4-in. diameter high-temperature CPVC pipe into the bottom of the well screens prior to the commencement of the pumping test. This allowed measurements to be taken within the CPVC pipes using M-scopes with conductivity gain controls. In addition, pressure transducers were attached to the tip of the CPVC pipes for automatic recording of head changes.

Special well caps were constructed to allow access for the monitoring equipment and to prevent free gas discharge at the surface. This made working in the area tolerable and prevented any possible explosion hazard. In addition, sealing the well head reduced the amount of bubbling and foaming in the well, which allowed reliable water level measurements. A threaded nipple in the cap permitted gas pressures to be periodically measured in each of the observation wells using a water manometer. Maximum measured gas pressure heads in three observation wells were 1.92 ft, 1.53 ft, and 0.48 ft.

The temperature was measured by lowering a thermometer in the observation wells; temperature ranged from 140°F to 150°F. Water temperature

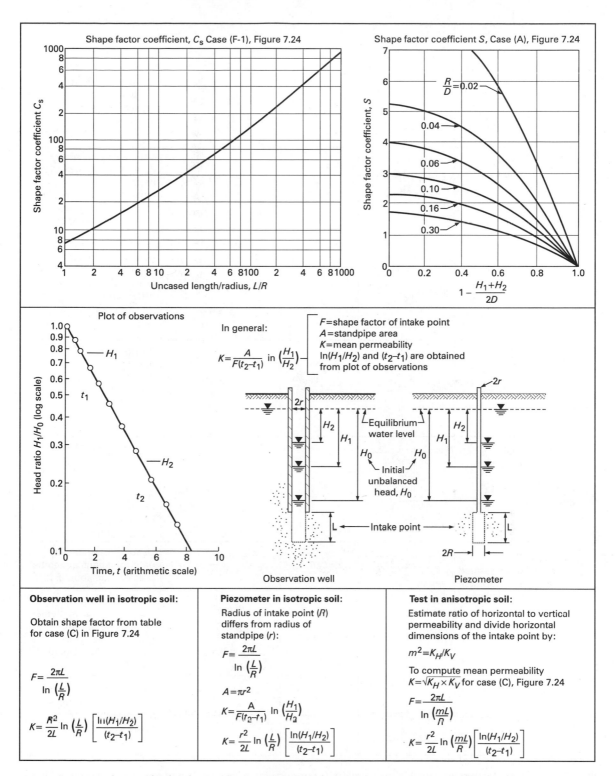

Figure 7.23 Analysis of hydraulic conductivity by variable head tests (reproduced from NAVFAC, 1982)

	Condition	Diagram	Shape factor, F	Permeability, K, by variable head test	Applicability
Observation well or piezometer in saturated isotropic stratum of infinite depth	(A) Uncased hole	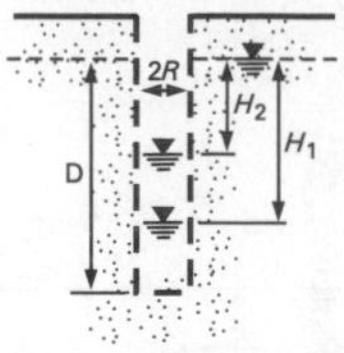	$F = 16\pi DSR$	(For observation well of constant cross section) $$K = \frac{R}{16DS} \times \frac{(H_2 - H_1)}{(t_2 - t_1)}$$ for $\frac{D}{R} < 50$	Simplest method for permeability determination. Not applicable in stratified soils.
	(B) Cased hole, soil flush with bottom	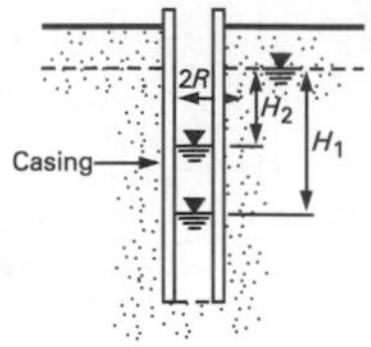	$F = \frac{11R}{2}$	$$K = \frac{2\pi R}{11(t_2 - t_1)} \ln\left(\frac{H_1}{H_2}\right)$$ for $6" \le D \le 60"$	Used for permeability determination at shallow depths below the water table. May yield unreliable results in falling head test with silting of bottom of hole
	(C) Cased hole, uncased or perforated extension of length L	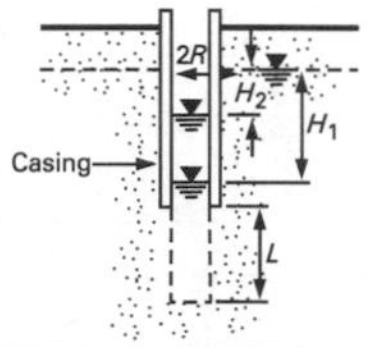	$F = \frac{2\pi L}{\ln\left(\frac{L}{R}\right)}$	$$K = \frac{R^2}{2L(t_2 - t_1)} \ln\left(\frac{L}{R}\right) \ln\left(\frac{H_1}{H_2}\right)$$ for $\frac{L}{R} > 8$	Used for permeability determinations at greater depths below water table.
	(D) Cased hole, column of soil inside casing to height L	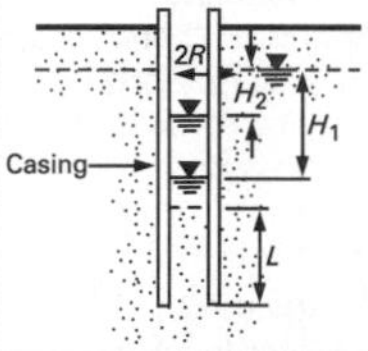	$F = \frac{11\pi R^2}{2\pi R + 11L}$	$$K = \frac{2\pi R + 11L}{11(t_2 - t_1)} \ln\left(\frac{H_1}{H_2}\right)$$	Principal use is for permeability in vertical direction in anisotropic soils.
Observation well or piezometer in aquifer with impervious upper layer	(E) Cased hole, opening flush with upper boundary of aquifer of infinite depth	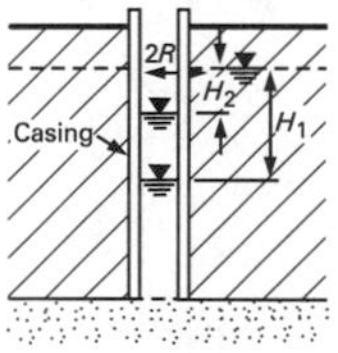	$F = 4R$	$$K = \frac{\pi R}{4(t_2 - t_1)} \ln\left(\frac{H_1}{H_2}\right)$$	Used for permeability determination when surface impervious layer is relatively thin. May yield unreliable results in falling head test with silting of bottom of hole.
	(F) Cased hole, uncased or perforated extension into aquifer of infinite thickness: (1) $\frac{L_1}{T} \leq 0.2$ (2) $0.2 < \frac{L_2}{T} < 0.85$ (3) $\frac{L_3}{T} = 1.00$ Note: R_0 equals effective radius to source at constant head.		(1) $F = C_s R$ (2) $F = \frac{2\pi L_2}{\ln(L_2/R)}$ (3) $F = \frac{2\pi L_3}{\ln\left(\frac{R_0}{R}\right)}$	(1) $$K = \frac{\pi R}{C_s(t_2 - t_1)} \ln\left(\frac{H_1}{H_2}\right)$$ (2) $$K = \frac{R^2 \ln\left(\frac{L_2}{R}\right)}{2L_2(t_2 - t_1)} \ln\left(\frac{H_1}{H_2}\right)$$ for $\frac{L_2}{R} > 8$ (3) $$K = \frac{R^2 \ln\frac{R_0}{R}}{2L_3(t_2 - t_1)} \ln\left(\frac{H_1}{H_2}\right)$$	(1) Used for permeability determinations at depths greater than about 5 ft. (2) Used for permeability determinations at greater depths and for fine-grained soils using porous intake point of piezometer. (3) Assume value of $R_0/R = 200$ for estimates unless observation wells are made to determine actual value of R_0.

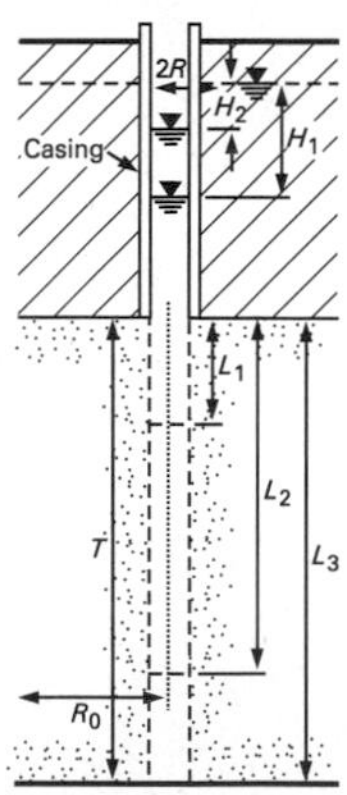

Figure 7.24 Shape factors for calculation of hydraulic conductivity from variable head tests (reproduced from NAVFAC, 1982)

Figure 7.25
A test well in refuse

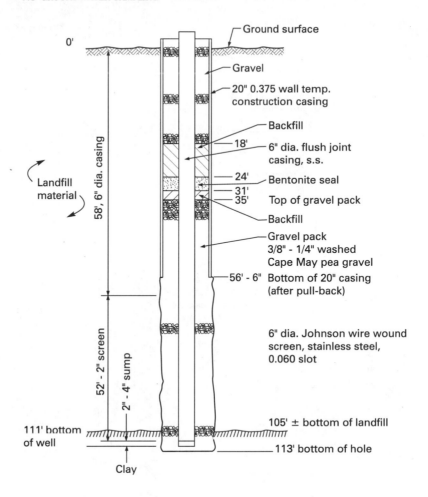

Ground surface

0'

Gravel

20" 0.375 wall temp. construction casing

Backfill

18'

6" dia. flush joint casing, s.s.

24' — Bentonite seal

31'

35' — Top of gravel pack

Backfill

Landfill material

58', 6" dia. casing

Gravel pack 3/8" - 1/4" washed Cape May pea gravel

56' - 6" Bottom of 20" casing (after pull-back)

6" dia. Johnson wire wound screen, stainless steel, 0.060 slot

52' - 2" screen

2" - 4" sump

105' ± bottom of landfill

111' bottom of well

113' bottom of hole

Clay

during the pumping tests was measured at 132°F. Use of a flow meter to measure discharge rates during pumping was aborted due to plugging by fibers during the early portion of the test. Flow calculations were subsequently based on measured discharge temporarily diverted into drums.

7.8 GROUNDWATER SAMPLING

7.8.1 Monitoring Wells

Monitoring wells are installed to sample and assess the quality of the groundwater for environmental purposes. Figure 7.26 (EPA, 1992) shows a typical construction of a monitoring well. When this construction is used in rock, the hole is larger and the rock is not screened. In the construction shown, the screen is sized to retain 90% of the filter. If the natural formation is coarse-grained soil, no filter pack is necessary. The well should be developed until the

Figure 7.26
Cross section of a
typical monitoring well
(EPA, 1992)

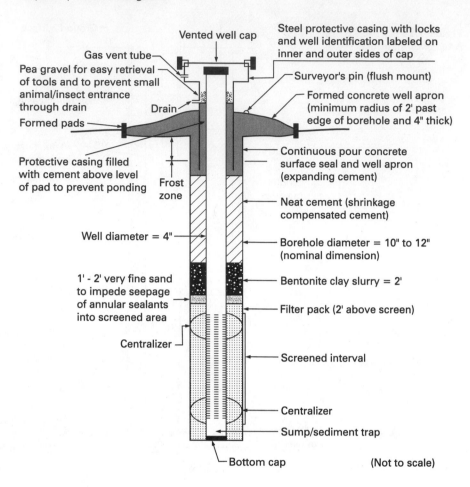

water pumped has low turbidity (e.g., <5 NTUS). Indicator parameters (temperature, conductivity) are recorded before each sampling round. Evacuation of three to six well volumes is recommended before sampling. Well-defined protocols should be followed for field measurements, sampling, sample preservation, and sample custody. More guidance is given by EPA (1992).

7.8.2 *Lysimeters*

Pressure–vacuum lysimeters are used to obtain soil moisture samples in unsaturated soil. In unsaturated soils, water is held by soil–water tension. The basic theory of a lysimeter is to apply vacuum (suction) in a borehole to induce a gradient and flow into a cup. Figure 7.27 (EPA, 1987) shows a typical construction of a lysimeter. The application of vacuum may remove some volatile organic from the water. After the water is collected, the accumulated water is forced to the surface by applying positive inert gas pressure.

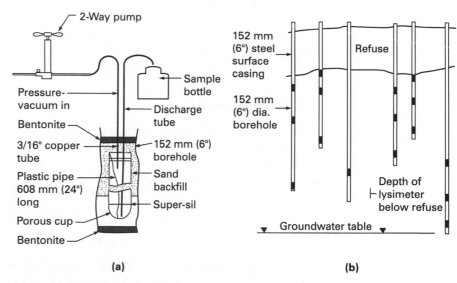

Figure 7.27
Pressure–vacuum
lysimeter installation
(EPA, 1987).
(a) Cross section of a
typical pressure–vacuum
lysimeter installation;
(b) cross section of a
lysimeter network

7.9 FIELD MONITORING

Field observations are made either to provide information on the behavior of the structure during construction and loading or to determine whether there are any indications of failure under service loads. Some of the measurements made in field monitoring are vertical displacements, lateral displacements, rotation, pore water pressure, suction, loads, contact pressure, etc. (The methods for monitoring the pore water pressure are described in previous sections.)

If disposal facilities are built on marginal sites, soil movements and excess pore water pressure may need to be monitored to ensure stability during land filling phases. To monitor movements, reliable benchmarks are established that are sufficiently removed from the site so that they are not affected by the construction activities or by the subsequent stress changes.

7.9.1 Displacements

Surface monuments can be used for monitoring horizontal and vertical displacements (Figure 7.28). Settlements are obtained from level readings and the lateral displacements by surveying. Where the area to be monitored is loaded by a fill, settlement platforms are installed as described in Section 7.4.4. The top of the pipe is monitored for vertical displacement. As the height of the fill increases, the pipe is extended.

Figure 7.28
Surface monuments

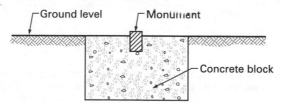

7.9.2 Pore Pressure Measurements

Properly installed piezometers (see Figures 7.19–7.21) are used for pore water pressure measurements. This information is used in assessing the stability of a structure such as a landfill (see Chapter 13).

7.9.3 Tensiometers

A tensiometer is a device for measuring the pressure head in the soil. In unsaturated soil, this pressure is less than atmospheric because the fluid is under tension. Tensiometer installation can be used effectively to evaluate test sections for measuring field hydraulic conductivity of clay liners (Chapter 10).

The basic theory is illustrated in Figure 7.29 (ASTM Standard Guide D 3404-91). In general, the value of the tension and pressure are related as follows:

$$T_f = P_{atm} - P_f \tag{7.19}$$

where T_f is the tension of an elemental volume of fluid (M/LT^2), P_{atm} is the absolute pressure of the standard atmosphere, and P_f is the absolute pressure of the same elemental volume of fluid (M/LT^2). If we consider a porous cup inside the soils, and the walls of the cup are saturated and hydraulically connected to the soil water, the pressure in the porous cup is expressed as

$$P_w = P_g + T_w \tag{7.20}$$

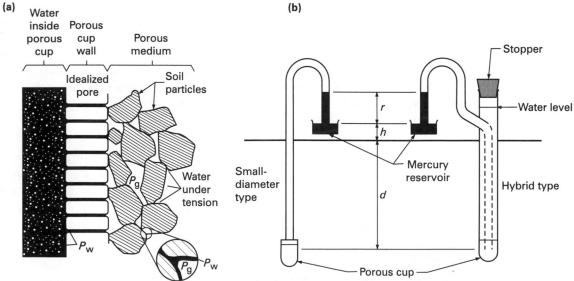

P_w - Absolute pressure of soil water
P_g - Absolute pressure of soil gas

Figure 7.29 Tensiometer concept and measurements. (a) Enlarged cross section of porous cup–porous medium interface; (b) monometer type measurement. (Copyright ASTM. Reprinted with permission.)

where P_w is the absolute pressure of an elemental volume of soil water (M/LT^2), P_g is the absolute pressure of the surrounding soil gas (M/LT^2), and T_w is the tension of the same elemental volume of soil water (M/LT^2).

The construction of a tensiometer is illustrated in Figure 7.29. A porous cup is attached to an airtight, water-filled tube connected to a mercury reservoir, vacuum gauge, or pressure transducer. The saturated porous cup material provides the link between the soil water and bulk water inside the instrument. In the case of a mercury manometer, we can apply Eq. 7.19 and, assuming the gas is under atmospheric pressure, express the soil water tension as

$$T_w = P_{atm} - P_w = (\rho_{Hg} - \rho_{H_2O})r - \rho_{H_2O}(h + d) \tag{7.21}$$

Various units are used to express tension or pressure in soil water. The "bar" unit is $100.0 \text{ kPa} = 0.9869 \text{ atm} = 1020 \text{ cm}$ of water at $4°C = 1020$ g/cm^2 in a standard gravitational field. The "pF" unit is $\log_{10}$ cm of water tension (dimensionless).

Soil tension is a function of the degree of saturation and soil type. Soil water tension may also be measured in the laboratory (Walsh et al., 1993) using the filter paper method and may be correlated to degree of saturation and soil type, as illustrated in Figure 7.30 (Walsh et al., 1993).

7.9.4 Inclinometers

Soil inclinometers are excellent devices for measuring lateral deformations to evaluate the stability of slopes and retaining structures. The key elements to an

Figure 7.30
Relationship between soil suction, degree of saturation, and percent fines (reproduced from Walsh et al., 1993, by permission)

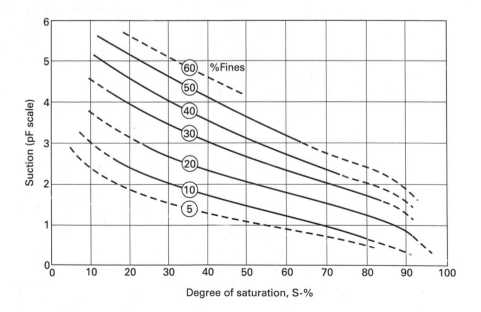

Figure 7.31
Soil inclinometer

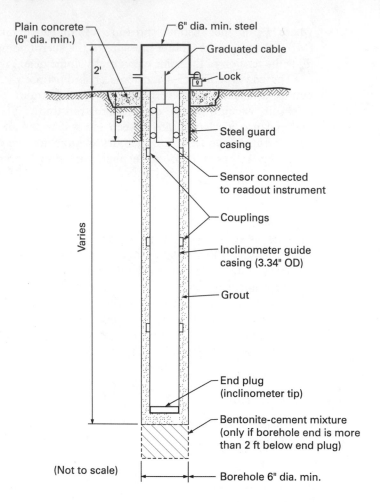

Plain concrete (6" dia. min.)
6" dia. min. steel
Graduated cable
Lock
2'
5'
Steel guard casing
Sensor connected to readout instrument
Couplings
Inclinometer guide casing (3.34" OD)
Varies
Grout
End plug (inclinometer tip)
Bentonite-cement mixture (only if borehole end is more than 2 ft below end plug)
(Not to scale)
Borehole 6" dia. min.

inclinometer system are a probe, a guide casing, a cable, and a readout unit (Figure 7.31).

The probe is mounted on guide wheels that travel up and down in the guide casing. The electrical cable is used to transmit the signals from the sensing element to a readout unit at the surface. In addition, the cable has graduated marks on it to control the depth at which the sensing element is operating. The inclinometer measures the average inclination of the casing between the upper and lower sets of wheels. By multiplying the distance between the two sets of wheels by the sine of the angle of inclination of the probe, one can obtain the displacement of the upper set of wheels with respect to the lower set. Starting at the bottom of the casing and summing these displacements up to the top, one can compute the total displacement of the top of the casing with respect to the bottom. Repeating such measurements periodically provides data on the depth, magnitude, direction, and rate of casing deflection. The casing should extend either into a firm stratum not expected to deform significantly or to well below any critical failure surface.

7.10   SUMMARY

In any investigation program where refuse or potentially toxic waste is expected, proper environmental safety measures should be followed. The appropriate level of protection (A–D) should be pursued, with Level A being applicable to the worst conditions. The proper selection of protective tools and clothing should be determined by a qualified individual. Investigation of large sites consists of a desk study, preliminary investigations, and detailed investigations. Applicable regulations and guidelines are provided. For hazardous wastes special conditions apply regarding cleaning of the equipment, record keeping, storage of the samples, etc.

In situ measurements are made using standard penetration tests, cone penetration tests, pressuremeter tests, load tests, geophysical methods, and field permeability tests. Field monitoring provides information on the behavior of structures under construction and those in service. Many types of equipment are available for measuring vertical and lateral displacement, pore pressure, etc.

NOTATIONS

a	area ratio for the given piezocone	q_c	cone penetration resistance (F/L^2)	a	distance between electrodes M and N (L)
C	constant for permeability computation in sand	q_T	corrected cone penetration resistance for a jointless cone (F/L^2)	D	thickness of the upper layer (L)
C_n	correction factor			E_m	pressure meter modulus (F/L^2)
D_{10}	effective particle diameter (L)	s_u	undrained shear strength (F/L^2)		
D_{50}	particle size (L) at which 50% by weight of the soil is fine	u	measured pore water pressure (F/L^2)	N_{55}	SPT blow count corresponding to an energy ratio of 55%
D_r	relative density	α	empirical parameter	P_{atm}	absolute pressure of the standard atmosphere (F/L^2)
E	Young's modulus (F/L^2)	σ_v'	effective overburden stress (F/L^2)	P_f	absolute pressure of the same elemental volume of fluid (F/L^2)
G	Shear modulus (F/L^2)	σ_v	total vertical stress (F/L^2)		
I_p	plasticity index	ϕ	angle of friction	P_g	absolute pressure of surrounding soil gas (F/L^2)
k	hydraulic conductivity (L/T)	ρ	mass density of the material (M/L^3)		
N	standard penetration resistance (Blows/30 cm)	v	Poisson's ratio	P_w	absolute pressure of an elemental volume of soil water (F/L^2)
N'	corrected N for overburden	ρ_{app}	apparent resistivity (ohm-meter)		
N_k	cone factor			R_g	distance of the geophone from the source (L)
N_{kT}	cone correction factor for soil plasticity	$2s$	distance between electrodes (L)		

T_f tension of an elemental volume of fluid (F/L^2)

T_w tension of the elemental volume of soil water (F/L^2)

v_0 initial volume of the measuring cell

V potential of electric circuit (volt)

I current (ampere)

R resistance (ohms)

v_m volume change at a pressure corresponding to the mean

pressure in the pseudoelastic range

V_p P wave velocity (L/T)

V_{p1} P wave velocity of the surface layer (L/T)

V_s shear wave velocity (L/T)

CHAPTER 8

Site Selection

8.1 INTRODUCTION

There are several issues that impact the site selection for a solid waste land disposal facility. In broad terms, the three major issues are environmental, economic, and political. The geotechnical and hydrogeological parameters fall within the environmental category. The political factor is heavily impacted by public attitude. For a siting study to achieve public acceptance, citizen groups should participate in identifying the siting criteria and their relative importance. The ultimate goal is to select a site that will provide the greatest protection to the environment in case the technology designed to afford protection fails. With this goal in mind, some states are identifying areas where waste facilities can be located safely. The outcome of a siting study depends on the relative importance given to different criteria. Once a site has been accepted, construction of the landfill should begin immediately; waiting often makes the site unacceptable due to various socioeconomic and regulatory reasons.

8.2 THE SITING PROCESS

A flow diagram showing the general siting process is shown in Figure 8.1. A large amount of information must be handled systematically in several stages. The purpose of each stage is to progressively narrow the scope of the study from a wide geographic area to one or more sites for detailed investigation and analysis. The factors that may be considered in assessing the suitability of a site are arranged alphabetically in Table 8.1. Some of the factors listed are applicable to the regional selection process, some to site-specific or local conditions, and some to both.

8.2.1 Graphical Procedure

One procedure to narrow the scope of a study is to use overlay maps prepared for the entire geographic area in which the proposed landfill is to be located (Noble, 1992). In each overlay a different criterion (from Table 8.1) is shaded. These overlays are then superposed on each other yielding a composite site suitability map, as shown in Figure 8.2. The areas that show clear through the superposed overlays have the least number of restrictions. One of the shortcomings of such sieving or screening is that the importance given to each of the mapped criterion is the same (Robinson, 1986).

Figure 8.1
Siting process

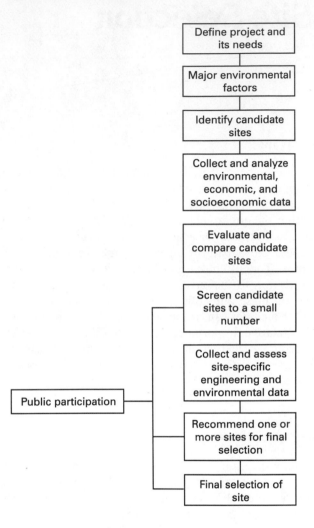

Figure 8.2
Overlays for site screening

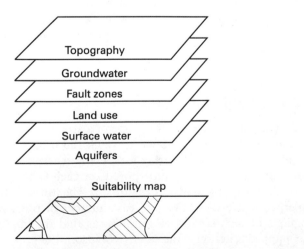

Table 8.1 Factors to consider in site assessment

Economic	Socioeconomic	Environmental/geotechnical
Access to highways and available highway capacity	Archaeological and historical sites	Aesthetic impact
Compatibility with existing solid waste management systems	Cultural patterns	Agriculture preservation areas
Cost of police, fire, and road maintenance	Dedicated land	Air quality, gas compositions, and particulate matter
Development, operation, and maintenance cost	Economic and community resources	Areas with high groundwater level
Distance to waste generation locations	Emergency response	Climate and atmospheric conditions
Economic effects on community	Land use and zoning	Distance from water supply wells
Effect on property value	Noise impact	Fault areas
Flexibility	Proximity to school and residences	Flood plains and wetlands
Highly productive agricultural areas	Public safety and health	Forest, wilderness, and scenic areas
Land development	Sensitive receptors	Geology

8.2.2 *Numerical Procedure*

Numerical ranking of sites can also be used to aid in site selection. Such rankings can be obtained by assigning weight to a criterion (or subcriterion) and giving it a suitability rating based on its significance. The ranking score C is computed as

$$C = \sum_{i=1}^{n} W_i \cdot S_i$$

where W is the weight, S is the suitability rating, i is the criterion number, and n is the number of criteria. Various siting criteria and their significance with respect to environmental performance are summarized in Table 8.2. Using this summary, weights are assigned to various criteria and subcriteria. For example, a weight of 5 may be assigned to those criteria considered most important and a weight of 1 may be assigned to criteria considered least important. A site that is considered good with respect to a specific criterion could be given a suitability rating of 3, moderate a rating of 2, and fair a rating of 1. Of course, other options for numerical ranking could be formulated. The words *good, moderate, fair*, etc. are somewhat subjective and so must have a logical basis.

For example, a subsoil permeability of less than 10^{-7} cm/sec is considered good since this value is specified for a soil liner by many regulatory agencies. Sites with a high cation exchange capacity (CEC) also can be considered good. A CEC greater than 25 mg/100 grams (milliequivalents per 100 grams) could have a rating of 3 and a CEC less than 15 a rating of 1. Soils with a pH greater than 6.5 are positively effective in the removal of metal cations. Thus,

Table 8.2   Definition and significance of siting criteria

Criteria	Subcriteria	Definition	Significance
Soils	Permeability	Soil property that governs the rate at which water moves through it	Subsoil permeability impact, release of pollutants to groundwater. Lower subsoil permeability is preferable for siting.
	pH	Indication of acidity and alkalinity (pH 7 = neutral)	Characterizes tendency of soil for sorption of heavy metals. Greater pH is preferable for siting.
	Cation exchange capacity (CEC)	Capacity of soil to exchange cations expressed as a sum for all exchangeable cations	Indicates the ability of soil to attenuate some contaminants, particularly heavy metals. Higher CEC is preferable for siting.
	Surfacial soils	Unconsolidated materials at Earth's surface	Affects degree of attenuation and the need for liners. Surfacial soils with lower permeability are preferable for siting.
Geology	Bedrock and outcropping		Carbonate rocks are susceptible to solution; fractured rock facilitates pollution migration. Sites with more overburden are preferable.
	Continuity and mass permeability	Related to open discontinuities, solution channels	Controls the potential for migration of contaminants.
	Faults	Mapped planes or zones of rock fracture along which displacement has occurred	Impacts the stability of a facility and potential release of pollutants.
	Seismic impact	Related to the peak rock acceleration expected at the site	Impacts the stability of a facility and potential release of pollutants.
Groundwater	Aquifer/well yield	Relates to a geological formation or group of formations that is capable of yielding useable quantities of groundwater to wells or springs	Sites with high aquifer capabilities may be off-limits for some facilities.

Table 8.2 *Continued*

Criteria	Subcriteria	Definition	Significance
Groundwater (*cont.*)	Wellhead penetration zone	Related to the capture zone of a pumped well	Capture zones may be off-limits for landfill.
	Aquifer use	Use of aquifer could be actually within a specified distance from facility, potential, or sole source (sole or principal supply of drinking water to a large percentage of a populated area)	Impacts the water supply. Aquifers with low actual or potential use are preferable for siting purposes. Sole-source aquifers are considered very significant even if yields are low.
	Groundwater quality	The natural quality of groundwater as measured against drinking water standard	Areas with poor natural groundwater quality represent more suitable locations, all else being equal.
	Groundwater flow system	Refers to the occurrence and movement of groundwater with regard to direction and velocity	Sites where direction of groundwater flow is away from use, or where flow is upward, or where water is deep, are preferable, all else being equal.
	Seasonal high groundwater level	The maximum level to which groundwater is expected to rise	Unsaturated zones act as a barrier (no direct mixing) between base of facility and groundwater. Most regulations specify a minimum of 1.5 m (5 ft)
Monitoring aspects		Refers to RCRA requirements for groundwater monitoring	Sites that are easier to monitor (e.g., presence of a layer of sand and gravel) or have a known discharge body (e.g., a lake) are preferable, all else being equal.
Cover	Cover material	Refers to earth material (available on-site) used for daily waste cover	Sites with an abundance of workable and relatively impervious soils and preferable, all else being equal.

Table 8.2 *Continued*

Criteria	Subcriteria	Definition	Significance
Slope		Deviation of the land surface from the horizontal measured as the average topographic relief for the site	Impacts release of contaminants, site development, and operation. Sites with gentler slopes are preferable, all else being equal.
Surface and groundwater hydrology	Proximity to streams/lakes	Refers to overland proximity and protected uses of the nearest lake or stream	Impacts opportunity for runoff and contaminants polluting lakes/streams.
	Proximity to wells/aquifer	See Aquifer under Groundwater on page 192.	Impacts groundwater resources. Sites closer than 800 m (2500 ft) to a high-yield well (70 gal/min) may be excluded.
	Proximity to flood-prone areas	Land areas inundated by flood of specified frequency (usually 100 yr)	Impacts transport of hazardous waste.
	Proximity to recharge areas	Refers to lands draining to existing or planned storage reservoirs	Impacts drinking water supply.
Topography	Slope erodability	Migration of soil particles by surface water or other natural phenomena	Potential of soil erosion impacts facility construction and operation.
	Runon and runoff	Runoff refers to rainwater or leachate that drains overland away from facility. Runon refers to drainage overland onto any part of the facility.	Sites with little need for control of run on from upland and slow runoff are preferable. Runon is usually controlled by berms, stream diversion, etc. Runoff control is impacted by velocity of water traversing the site.

subsoils with a pH greater than 6.5 are considered good for siting whereas those with values between 5 and 6.5 are moderate and those with values less than 5 are fair. Sites where the groundwater quality is poor are considered good sites. Poor quality may be defined in terms of total dissolved solids (TDS) greater than 10,000 mg/l. Fair sites may be defined as those where the TDSs are less than 500 mg/l. In seismically active areas, the distance to a causative fault with the potential for seismic instability or expected rock acceleration with a given recurrence interval (say 250 years) could be used for rating.

Example 8.1

Partial subsurface data at two possible landfill sites are given in Table 8.3. Determine which of the two sites is more suitable for locating the landfill.

Solution: To compare the two sites we assign weights to various criteria as shown in Table 8.4. The assumed suitability number of various criteria are listed in Table 8.5. Summaries of the ranking for sites A and B are given in Tables 8.6 and 8.7, respectively. As shown in Tables 8.6 and 8.7, site A scores higher than site B and is thus more suitable for locating the landfill.

Although Example 8.1 illustrates the general methodology, it is important to realize that for actual siting most of the criteria listed in Table 8.1 must be evaluated. Both the weighting and suitability ratings have to be determined. Although an engineer may have the technical know-how and experience to select appropriate weights and suitability numbers, he or she must yield to the fact that the public must have a greater say in selecting these numbers if siting process is to come to fruition.

Table 8.3 Site data

Site	Aquifer hydraulic conductivity (m/s)	Type of soil	Depth of soil (m)	Water table depth (m)	Slope (%)	Annual recharge (mm)
A	10^{-5}	Silty sand	12	10	5	10
B	10^{-3}	Silty clay	3	2	2	300

Table 8.4 Criteria and assigned weights

Criteria/subcriteria	Weight
Aquifer hydraulic conductivity	4
Type of soil	3
Depth of soil	2
Water table depth	5
Slope	1
Annual recharge	4

Table 8.5
Suitability numbers

Criteria/subcriteria	Description	Suitability rating
Aquifer hydraulic conductivity (m/s)	Less than 10^{-8}	10
	10^{-8} to 10^{-7}	8
	10^{-6} to 10^{-5}	6
	10^{-4} to 10^{-3}	3
	Greater than 10^{-3}	1
Type of soil	Low plasticity clay	10
	Silty clay	9
	Sandy clay	7
	Glacial till	5
	Silty sand	3
	Sand and gravel	1
Depth of soil (m)	Greater than 30	10
	26 to 30	9
	21 to 25	8
	16 to 20	6
	11 to 15	5
	6 to 10	3
	2 to 5	2
	1 or less	1
Water table depth (m)	Greater than 30	10
	26 to 30	9
	21 to 25	8
	16 to 20	6
	11 to 15	5
	6 to 10	3
	2 to 5	2
	1 or less	1
Slope (%)	0 to 1	10
	1 to 2	9
	2 to 4	7
	4 to 6	5
	6 to 12	3
	12 to 15	2
	Greater than 15	1
Annual recharge (mm)	0 to 50	10
	51 to 100	8
	101 to 200	6
	201 to 400	4
	401 to 600	2
	Greater than 600	1

Table 8.6
Ranking of site A

(1) Criteria/subcriteria	(2) Weight from Table 8.4	(3) Suitability number from Table 8.5	(4) Ranking (col. 2 × col. 3)
Aquifer hydraulic conductivity	4	8	32
Type of soil	3	3	9
Depth of soil	2	5	10
Water table depth	5	3	15
Slope	1	5	5
Annual recharge	4	10	40
Total			111

Table 8.7
Ranking of site B

(1) Criteria/subcriteria	(2) Weight from Table 8.4	(3) Suitability number from Table 8.5	(4) Ranking (col. 2 × col. 3)
Aquifer hydraulic conductivity	4	6	24
Type of soil	3	9	27
Depth of soil	2	2	4
Water table depth	5	2	10
Slope	1	7	7
Annual recharge	4	4	16
Total			88

8.3 SUSSEX COUNTY

The siting methodology for a landfill in Sussex County is illustrated in Table 8.8. The numbers in parentheses, which were established with the participation of citizen groups and not entirely by professionals, indicate importance or weight given to each category. As the result of citizen-group participation, slopes were given more weight (41) than the permeability (29.6) of the subsoils. However, the siting criteria in Table 8.8 is consistent with that adopted by the New Jersey Department of Environmental Protection.

In Level 1, *exclusionary criteria* (*fatal flaws*) are applied. The exclusionary criteria normally encompass regulatory exclusions (if any) and may include inland or coastal wetlands, distance from water supply wells, agricultural preservation areas, flood plains, archaeological and historical sites, dedicated lands, habitats for threatened or endangered species, and developed land. Overlay maps can be used in screening the sites based on exclusionary criteria. In Sussex County, four exclusionary criteria (fatal flaws) were applied in Level 1, which narrowed the possible sites to 91 and eliminated approximately 50% of the county.

Level 2 screening involved application of further exclusionary criteria and ranking the remaining sites based on the geotechnical/hydrogeological criteria shown in Table 8.8. Exclusions could include areas with high seasonal water tables, areas with very steep slopes, critical recharge areas, heavily fractured rocks, rocks with solution channels, active faults, etc.

Table 8.8 Siting methodology in Sussex County

Level of screening	Description	Criteria[1]	Subcriteria
1	Initial screening—apply broad exclusionary criteria		
2	Screening—apply geotechnical/hydrogeo-logical exclusionary criteria; rank sites based on geotechnical/hydrogeological criteria	Soils (74)	Permeability (29.6), pH (7.4), cation exchange capacity (14.8), surfacial soils (22.2)
		Geology (74)	Bedrock type (20.6), bedrock continuity (16.4), faults (37)
		Groundwater (99)	Aquifer yield (26.4), aquifer use (26.4), groundwater quality (9.9), seasonal water table (23.1), groundwater flow system (13.2)
3	Rank sites based on further hydrogeological and environmental criteria	Monitoring aspects (28) Cover material (34), slope (41) Land use (57), aesthetic impact (54), site access (43), sensitive areas (59), utilities (45), soils (83), hazards (40), surface and groundwater hydrology (94), geology (85), topography (70) (slope, erodability runon/runoff), flora and fauna (42), air quality (59), noise (54), odors (65)	
4	Rank further based on eight additional economic and socioeconomic criteria; select a site or sites for detailed specific investigation and testing		

[1] The numbers in parentheses indicate the weight attached to the category.

After the number of possible sites were reduced in the Level 2 process, further screening, based on hydrogeological and other criteria, was performed at Level 3. After this level of screening, the geotechnical/hydrogeological suitability of the top three or four sites are usually within a few score points of each other. Therefore, detailed, site-specific studies are made at this level.

Level 4 screening is usually based on economic and socioeconomic issues only. Environmental assessment of the proposed landfill site must be made carefully and mitigation of any adverse effects of the siting should be included in the assessment. For example, a higher suitability number may be assigned to a given criterion if the mitigation potential for its adverse effect is high. The advantage of this process is its ability to advance sites with the most hydrogeological attributes for consideration.

8.4 DATA SOURCES IN THE UNITED STATES

There are several data sources available in the United States to assist in siting processes. U.S. Geological Survey (USGS) data available include geological index maps, professional papers, water supply papers, topographic maps, and other data. U.S. Department of Agriculture (USDA) Soil Conservation Services (SCS) data include surveys of surface soils described in agricultural terms. Data such as pH, CEC, seasonal water levels, etc. are provided for the counties surveyed.

Other federal agencies through which data may be obtained are Bureau of Sport; Fisheries & Wildlife; Bureau of Mines; Bureau of Indian Affairs; National Oceanic and Atmospheric Administration; Federal Aviation Administration; etc. Many state geological surveys also provide publications with excellent detailed geological maps covering local areas.

Aerial photography is available in 9-in. forms with overlap for stereoscopic viewing. Scales could range from 1:12,000 to 1:80,000 photos used for topographic–geological mapping. Drainage patterns are available from USGS or SCS. In addition, imagery obtained by satellite with scales varying from 1:1,000,000 to 1:250,000 are available from Earth Research Observation System Date Center (EROS). Skylab imagery from orbits 770 miles above the Earth and aerial photography produced by National Aeronautics and Space Agency (NASA) are also available from EROS.

8.5 SUMMARY

Major issues in site selection are environmental, economic, and political, with the political factor being impacted heavily by public opinion. Although the criteria considered important by the public may not necessarily be the soundest environmentally, they are the most crucial for getting acceptance or rejection.

In the several stages of the siting process, either graphical or numerical (or both) procedures may be used to narrow the scope of the siting study and eventually locate a suitable site. To assist in this process, many government and private agencies provide various types of data.

PROBLEMS

8.1 What are the similarities and differences between regional site selection and local site selection processes?

8.2 What is a fatal flaw? How are fatal flaws determined?

8.3 Give some examples of exclusionary criteria and your assessment as to why it is appropriate to consider them as such.

8.4 Why is it important that environmental assessment and mitigation be considered as an integral part of the site selection process?

8.5 Who do you consider as most qualified for assigning weight and suitability rating to various criteria and subcriteria? Should people other than professionals be involved in this process? If so, who and why?

8.6 What weights would you assign to the following criteria in the siting process for a landfill? Give your reasoning for each assignment.

(**a**) Existing land use and zoning

(**b**) Access roads

(**c**) Distance to waste generation locations

(**d**) Noise impact on residential area

(**e**) Visual impact

CHAPTER 9

Ground Modification and Compaction

The conditions at a given site may not always provide proper support for the foundation. However, it may be possible to improve the properties of the subsurface materials for better support. This is done by surface compaction, deep dynamic compaction, preloading or precompression, drainage, reinforcing, use of additives, grouting, and replacement of the poor materials.

9.1 COMPACTION

Compaction of soils or waste materials increases their density and strength, and reduces their permeability. The loose materials are placed in lifts varying in thickness from a few inches to a few feet (with a maximum of about 3 ft in the case of garbage). For design and control of construction, both laboratory and field tests are conducted. For heterogeneous materials such as those found in a municipal landfill, reliable laboratory tests are difficult to design and conduct. In general, field tests — although expensive and time consuming — yield more reliable data.

9.1.1 *Coarse-grained Soils*

For coarse-grained soils with up to 12% fines, the extent of compaction is measured in terms of its relative density:

$$D_r = \frac{e_{max} - e}{e_{max} - e_{min}} \tag{9.1}$$

where e_{max} is the void ratio of the soil in its loosest state, e_{min} is the void ratio of the soil in its densest state, and e is the actual void ratio. In terms of unit weight, the relative density is given by

$$D_r = \frac{\gamma_{d\text{-}max}(\gamma_d - \gamma_{d\text{-}min})}{\gamma_d(\gamma_{d\text{-}max} - \gamma_{d\text{-}min})} \tag{9.2}$$

where $\gamma_{d\text{-}max}$ is the dry unit weight of the soil in its densest state, $\gamma_{d\text{-}min}$ is the dry unit weight of the soil in its loosest state, and γ_d is the actual dry unit weight of the soil. For details of the testing method, see ASTM D 2049.

9.1.2 Mixed Soils

In the standard compaction test (ASTM D 698) for a mixed soil, the soil is compacted in a mold with a volume $\frac{1}{30}$ ft^3 in 3 layers with each layer receiving 25 blows of a 5.5 lb hammer falling 12 in. This procedure produces energy of 12,400 ft-lb/ft^3 of soil (600 kN-m/m^3). This method is preferred for control of compaction for clay liners. An alternate method (ASTM D 1557) uses 5 layers, 25 blows, a 10 lb hammer, and an 18 in. drop. This procedure produces an energy of 56,000 ft-lb/ft^3 of soil (2700 kN-m/m^3).

The degree of compaction is affected by type of soil, energy input, and water content. As shown in Figure 9.1, for different energy input the typical plot of dry density and water content is a paraboloid. Other shapes of curves are also reported (Winterkorn and Fang, 1975). The water content corresponding to the maximum dry density is known as the *optimum moisture content*.

Figure 9.1
Effect of compactive effort on compaction parameters

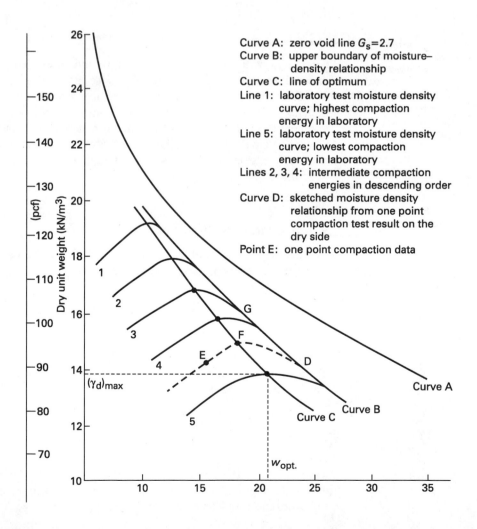

Curve A: zero void line G_s=2.7
Curve B: upper boundary of moisture–density relationship
Curve C: line of optimum
Line 1: laboratory test moisture density curve; highest compaction energy in laboratory
Line 5: laboratory test moisture density curve; lowest compaction energy in laboratory
Lines 2, 3, 4: intermediate compaction energies in descending order
Curve D: sketched moisture density relationship from one point compaction test result on the dry side
Point E: one point compaction data

Both the optimum moisture content and the maximum dry density depend on the energy input. The 100% saturation or zero-air-void curve (Figure 9.1) represents the upper limit of the soil density at any given moisture content and serves as a check on the test data.

In comparing field and laboratory data, the dry density relative compaction (RC) (or percent compaction) is specified as

$$RC = \frac{\rho_{d(\text{field})}}{\rho_{d(\text{max})}} \times 100 \tag{9.3}$$

An increase in energy input increases the maximum dry density but decreases the optimum moisture content. The relative increase in dry density is greater for soils of higher plasticity (Monahan, 1994). Thus poor control in compaction operation during the construction of a liner, which by design must contain materials of high plasticity, may result in considerable variation in its permeability.

The compaction parameters and their effects on strength, permeability, and shrinkage are discussed in Chapter 10. For a description of compaction equipment, see NAVFAC DM 7 (1982).

9.1.3 *Compaction of Waste Products*

Compaction has proven to be an effective way to reduce the volume of waste, increase the life of disposal sites, and improve the engineering properties of waste. Compaction curves for some mineral and industrial wastes are given in Figure 9.2; however, for a given material there would be a range of values rather than a single curve. The curves shown for fly ash indicate a variation in the maximum dry unit weight and optimum moisture content of more than 30% and 100%, respectively.

9.1.3.1 Coal and Incineration Ashes For fly ash, the maximum dry unit weight ranges from 74.4 lb/ft^3 (11.7 kN/m^3) to 127.5 lb/ft^3 (20 kN/m^3) and the optimum moisture content from 9% to 50%. The average dry unit weight and optimum moisture content for Class F fly ash are about 82.8 lb/ft^3 (13 kN/m^3) and 25%, respectively. The corresponding values for Class C fly ash are 94.2 lb/ft^3 (14.8 kN/m^3) and 20%, respectively.

In compaction tests on fresh fly ash, new samples must be used for each test as the recompacted fly ash behaves quite differently. The dry unit weight increases as the specific gravity of the fly ash solids increase. Presence of unburned carbon reduces its specific gravity, unit weight, and strength. Usually the higher unit weights are associated with lower optimum moisture content. Some compaction characteristics of fly ash are given in Table 9.1.

Kelly et al. (1991) described extensive field compaction tests for fly ash from western Pennsylvania. Three types of compactors were used: a dual drum asphalt roller, a single smooth drum roller, and a padfoot roller. All produced RC higher than 95% of standard Proctor maximum dry unit weight.

Figure 9.3
Percent compaction for seven passes of three different rollers (after Kelly et al., 1991)

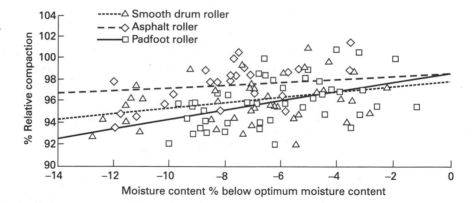

2500 vpm for the entire moisture content range of 0.4% to 13.4% below the OMC. As shown in Figure 9.3, relative compaction achieved in the field with the three types of rollers depended much less on moisture content than when specimens were prepared in the laboratory. Based on this data it was concluded that 95% of the standard Proctor maximum dry density could be accomplished with different types of equipment (including a dozer tractor alone) without a tight control on moisture content.

For fresh fly ash, compaction must be done immediately after placement because of the rapid commencement of pozzolanic action. Borex has been used as retardant to mitigate the difficulty associated with rapid hydration (Manz and Manz, 1985).

Considerable scatter in the field compaction data has been reported (Leonards and Bailey, 1982). Typical curves for laboratory compacted bottom ash samples are shown in Figure 9.4. For comparison, a typical compaction curve for clean coarse-grained soils (clean sand and sandy gravel) is depicted in

Figure 9.4
Standard compaction data for some bottom ash samples (redrawn after Huang and Lowell, 1990)

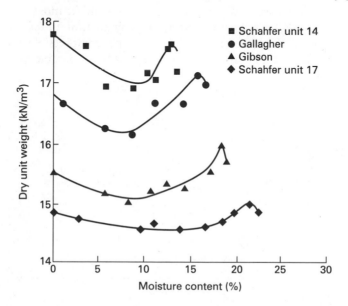

Figure 9.5. Note the similarity between the response of bottom ash and clean coarse-grained soils. The unit weights are the highest when these materials are either dry or fully saturated. For such materials densest packing or highest dry density is obtained by vibratory compaction. Maximum and minimum relative densities for boiler slag, which behaves in the same way, are in the range of 91–110 lb/ft³ (13.2–16 kN/m³) and 71–88 lb/ft³ (10.3–12.8 kN/m³), respectively (DiGiolia et al., 1977). For incinerator residue the maximum dry unit weight ranges between 82.8 lb/ft³ (13 kN/m³) to 108.2 lb/ft³ (17 kN/m³) and the optimum water content is 15% to 26%.

9.1.3.2 Coal Mine Waste   The dry unit weight of mine rock and coarse coal refuse from coal mining operations may range from 12.5 kN/m³ to 21 kN/m³ and the optimum moisture content between 6% to 14%. Because of the varied nature of the coarse refuse materials, considerable variations will exist from test to test. Handling and disposal and the mechanical action of compaction equipment will cause considerable particle breakdown. While selecting strength parameters during the design phase, the effect of both physical breakdown and chemical weathering must be considered. Some compaction characteristics of coal waste are given in Table 9.2. The type and size of compaction equipment and the thickness of lift had considerable influence on the strength and permeability of compacted coal refuse (Saxena et al., 1984). Smooth drum, vibratory, and vibratory sheepsfoot rollers were all found to be effective, with the smooth drum roller performing the best. Materials compacted with the smooth drum roller had the highest shear strength, the lowest measured field permeability, and a relatively impermeable surface. The recommended lift thickness is 0.3 m or less, and the compaction moisture content is near the optimum value.

As shown in Figure 9.2, coal waste shows a clearly defined maximum dry density and optimum moisture content. At most disposal sites, the moisture content is not controlled during compaction. In the absence of proper compaction, the materials become soft and compressible upon saturation. Furthermore, coal mine waste contains a considerable proportion of combustible

Figure 9.5
Typical laboratory compaction data for clean, coarse-grained soils (after Foster, 1962)

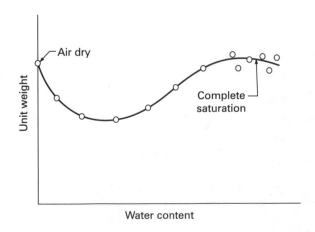

Table 9.2
Compaction of coal
waste

Source	Optimum moisture %	Max. dry unit weight kN/m^3	Specific gravity of solids, G_s
Coarse waste			
Appalachian	21–9	14.5–18.9	1.75–2.5
UK	16–4	14.8–20.4	1.8–2.7
Fine waste			
Appalachian	20–18	10.2–11.8	1.4–1.66

Source: Chen et al., 1976; Holubec, 1976; Saxena et al., 1984; Usmen, 1986

materials. To reduce the danger of rapid ignition and sustained combustion, and to obtain higher strength and lower permeability, proper management of compaction is essential. Even from an environmental and economical view-point it makes sense to control compaction because the water that seeps through the uncompacted waste is acidic and requires costly treatment.

Selected mine stone waste has been used in Great Britain to construct liners and covers at disposal sites for municipal, commercial, and inert wastes. A hydraulic conductivity of 10^{-5} cm/s is considered acceptable at such sites (Rainbow and Nutting, 1986).

9.1.3.3 Flue Gas Desulfurization Sludge   The maximum dry unit weight of FDG ranges from 70 lb/ft^3 (11 kN/m^3) to 100 lb/ft^3 (16 kN/m^3) and the optimum water content from 45% to 15%. The sulfite sludges are at the lower end of the density scale and the higher end of the optimum water content. The reverse is true of gypsum or sulfate sludge.

Because of the high water content of processed sludge, the field dry density is less than the laboratory values, ranging from 5 kN/m^3 (31.8 lb/ft^3) to 10 kN/m^3 (63.7 lb/ft^3). At these water contents the application of energy greater than 20% to 40% of the standard compaction value is nonproductive. Laboratory tests show an increase in optimum moisture content and a decrease in dry density if dry fly ash, lime, or soil are added to the sludge. In the field this advantage has not been realized fully because of the difficulty of obtaining a homogeneous mix (Krizek et al., 1987; Ullrich and Hagerty, 1987).

9.1.3.4 Municipal Landfill   To minimize settlement-related problems in a municipal landfill, articles such as tires, washing machines, etc., must be removed. The sorted materials are more easily and uniformly compacted so the risk of larger sudden and long-term settlements is reduced. It is usually necessary to place a few inches to a few feet of thick granular fill on the refuse before compaction, especially if the conditions are wet, to make it possible to operate the equipment. A thicker blanket of fill serves as a rigid mat for the placement of the foundation and reduces the amount of differential settlement.

Volume reductions of 2% to 17% have been reported from compaction. Most of the reduction occurs in five passes (Shoemaker, 1972). Excess pore pressures that develop in saturated landfill materials from the weight of the roller make it difficult to achieve any degree of compaction.

Several other variables—some controllable and some not—affect the compactibility of refuse. The controllable parameters (Surprenant and Lemke, 1994) are the condition of the refuse cell base, the compactor wheel tip design and condition, composition of refuse, compactor operator experience, cover material, working-face slope angle, lift thickness, number of compactor passes, and type of machine (full-pass versus open center). The uncontrollable parameters are the physical characteristics of refuse ambient temperature and humidity. In a detailed test study, Surprenant and Lemke (1994) arrived at the following conclusions regarding the effect of these various parameters:

1. Up to 14% increase of unit weight is achieved if a 100,000 lb class compactor is used compared to a 70,000 lb one. The maximum density achieved was 1600 lb/yd^3 using a 1-ft lift.

2. Up to 25% increase in density is achieved if a 1-ft lift is used compared to a 2.5-ft lift.

3. Most of the compaction is achieved in four passes. With 2.5-ft lifts, a 9% increase in unit weight was achieved in four additional passes. Using a 1-ft lift, the increase was only 3%.

The tests were conducted by placing the municipal solid waste on a flat base with a well-compacted cover soil. Refuse was placed and tests conducted by pushing uphill while maintaining a slope of 5 H : 1 V. No cover material was used.

9.1.4 Quality Control

The number of verification testings needed for compacted soils can be assessed based on probability techniques (Spigolon and Kelly, 1984; Harrop-Williams, 1987). The tests include laboratory and field density tests, laboratory hydraulic conductivity tests on "undisturbed samples," field permeability tests, and certain index properties tests. A key to these procedures is the estimation of the coefficient of variation for the property under study.

Suppose, for example, that the number of field density tests for each lift of linear material placed daily is to be determined. The size of the lift (the block) is chosen such that the testing does not impede construction, and the locations to be tested are selected at random. The average number of tests is determined based on (Spigolon and Kelly, 1984)

$$n_{\mathrm{u}} = \left(\frac{tv'}{S_{\mathrm{e}}}\right)^2 \tag{9.4}$$

where n_{u} is number of units in the sample, t is the probability factor, S_{e} is the allowable sampling error of the expected mean, and v' the coefficient of variation (standard deviation ÷ mean) of the known or estimated value of the block. The probability factor t is obtained from t-distribution tables based on a confidence level and sample size (a tabulation for large samples (size > 30) is given in Table 9.3).

A block is an isolated quantity of the same composition and produced by essentially the same process. Its size is much greater than a sample (covered by

Table 9.3
t-factors for large samples

t-factor	PROBABILITY OF EXCEEDING *E* SIGNIFICANCE	
	One-sided tail	Two-sided tail
3	0.0013	0.0026
2.575	0.005	0.01
2.32	0.010	0.020
2.0	0.023	0.046
1.96	0.025	0.050
1.645	0.05	0.10

Eq. 9.4) taken from it. In practice, the coefficient of variation, v', is established based on experience with similar installation of the same materials. If v' is not known, a small sample can be used to estimate its value; then the value of n_u can be determined from Eq. 9.4. If the sample used was less than n_u, then additional samples are called for.

A range of values of v' for various tests are given in Table 9.4. Most of the spread in the values of v' is due to different methods of testing, differences in testing procedures, deviation from the specified testing procedure, different testing apparatus, different operators, etc. For better results, strict control of these parameters cannot be overemphasized (Lumb, 1974).

Example 9.1

Consider an area block 100 ft × 200 ft is to be compacted. Determine the number of tests **(a)** for compaction control if v' for relative compaction is 3% and it is desired that the error (S_e) of the average of the values tested is not to

Table 9.4
Coefficient of variation for soil tests

Test type (%)	Range of *v'* values (%)	Suggested
Angle of friction, sand, fly ash	5–10	10
Coefficient of consolidation	25–100	50
Maximum dry unit weight		
Soils	1–7	5
Fly ash	10–17	15
Liquid limit	2–48	10
Optimum moisture content	11–40	
Clay soils		20
Sandy gravelly soils, fly ashes		40
Plastic limit	9–29	30
Plasticity index	7–79	
Clay soils		30
Slurry wall		
Slurry density	1–7	5
Backfill moisture content	17–20	20
Backfill, percent fines	15–17	20
Relative density	5–25	10
Standard penetration test	27–85	30
Unconfined compressive strength	6–100	40

Source: Lee et al., 1983; Bergstrom et al., 1987; McLaren and DiGioia, 1987

vary by more than 2% of the mean for the block for 99% confidence (a probability of 1 in 100 that the error will exceed S_e), (b) for hydraulic conductivity control if v' for hydraulic conductivity is 50% and assuming S_e to be 15% of the mean and for 95% confidence (a probability of 5 in 100 that S_e will exceed 15%).

Solution:
(a) $n_u = (2.32 \times 3/2)^2 = 12$.
(b) $n_u = (1.645 \times 50/15)^2 = 30$.

These assessments are sensitive to the value estimated for the coefficient of variation and other statistical parameters. The number of tests for quality control is usually regulated or based on experience with successful installations. The New Jersey Department of Environmental Protection requires hydraulic conductivity testing on "undisturbed core samples" of the clay liner from locations at 200 ft grid. A field infiltration test using a double-ring infiltrometer (see Chapter 10) is required for each 10 acres. Field density and moisture content are required at 50 ft grid. Clay stockpiles are to be tested (one per 16,000 yd^3) for a particular site for particle size distribution, plasticity, and hydraulic conductivity. These requirements, especially for field density, are somewhat excessive but perhaps can be used to establish the statistical basis for less stringent requirements at later stages in the project.

Further details on compaction equipment, quality control, and quality assurance may be found in EPA (1986) and Hilf (1975).

9.2 DYNAMIC DEEP COMPACTION

The basic concepts of dynamic compaction (also known as dynamic consolidation, impact densification, or heavy tamping) as it is used today were presented by Ménard and Broise (1975). The method consists of dropping heavy blocks of steel, concrete, or thick shells filled with concrete or sand weighing 5 to 20 tons from heights up to 30 m (100 ft). Drop weights as high as 180 tons have been used. The imprints of these blocks may be square, circular, or octagonal. The blocks are lifted either by cranes with a load capacity of up to 25 tons (23 tonne) and 100 ft (30 m) drop or by tripods, which have a higher load capacity and a greater drop height. Tripods permit a free drop, which is more efficient than a crane drop. The applied energy in most instances is between 60 and 300 ft-kip/ft^2 (100 and 400 tonne-m/m^2).

The materials treated by dynamic compaction exhibit a higher bearing capacity and lower postconstruction settlement. It is most suited for loose coarse-grained soils, rubble fills, and nonhazardous landfills with low percentage of rubble. Materials below the water table cannot be treated very efficiently with this method.

9.2.1 Drop Weight Treatment

The area to be treated is divided into grid patterns with each grid point receiving several blows in a given pass. Several passes may be necessary to obtain the desired results. Field studies are conducted for grid spacings, energy level of blows, number of blows per pass, number of passes, etc. To evaluate the extent of ground improvements, measurements are taken for drop weight imprints, heave, average settlement, changes in soil properties, pore water pressure, and ground vibration.

The initial grid spacing is generally greater than the thickness of the layer to be improved and the energy per blow is high enough that the compaction energy will influence the entire depth of the stratum. The high energy drops may form imprints 1 to 2 m (3 to 6 ft) deep, with the greatest crater depth occurring in the first six drops. The energy input per unit area determines the total amount of settlement. At the end of each pass the imprints are filled and the ground is leveled using the material heaved around them. The amount of compression at the end of each pass is determined from topographic maps prepared after each pass. The final, low-energy pass is called *ironing*; for this pass, a weight with square footprint is used. If the top 0–1 m (0–3 ft) of the materials still are not compacted properly after the ironing pass, standard compaction equipment must be used for densification. At sites where the upper materials are very soft, a working mat 1–2 m (3–6 ft) thick consisting of granular materials is added to facilitate treatment (Charles et al., 1981).

9.2.2 Depth of Influence

The maximum depth to which a soil is influenced, D_{max} (m) is estimated by

$$D_{max} = I(WH)^{0.5} \tag{9.5}$$

where I is the influence factor, W is the weight of the dropping block in metric tonne, and H the drop height in meters. Ménard and Broise (1975) used 1 for I. Other suggested values are 0.5 (Leonards et al., 1980), 0.65 to 0.80 (Lukas, 1980), and 0.3 to 0.8 (Mayne et al., 1984). An I value of 0.5 appears to be most widely accepted at this time.

Factors that influence I are soil type, initial soil density, depth to groundwater table, area of falling block, grid spacing, number and sequence of drops, number of passes, time elapsed between successive passes, method of determining ground improvement, and definition of what constitutes improvement. The ground improvement is determined by comparing the values before and after the treatment using cone penetration resistance, standard penetration resistance, vane shear strength, pressure meter modulus and limiting pressure, etc. In heterogeneous materials such as municipal landfills, large-scale plate load tests may be most appropriate (Baker, 1982).

9.2.3 Construction Vibrations

The impact from the weight causes transient vibrations in the ground. The guideline for estimating possible damage to the adjacent structures is based on peak particle velocity V_p, which depends on the site conditions and the scaled energy factor $(WH)^{0.5}/d$, where d is the horizontal distance from the impact

Figure 9.6
Scaled energy factor versus particle velocity: ■, most data; □, upper limit; ◇, rubble; △, wet sand; ◆, decomposed garbage; ▲, loose sand (data from: ■, □, Mayne et al., 1984; ◇, △, ◆, Lukas, 1980; ▲, Leonards et al., 1980)

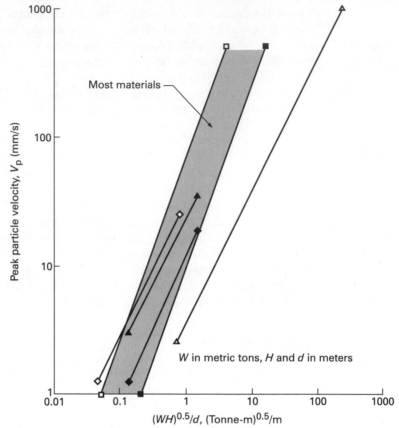

point. Figure 9.6 shows the relationship between particle velocity and scaled energy and may be used as a guide for determining the safe energy limits. Most data from sand, rubble, and silty soils lie in a narrow range (Mayne et al., 1984). A particle velocity of 50 mm/sec (2 in./sec) is considered the threshold of damage to residential structures (Wiss, 1981).

9.2.4 Dynamic Compaction of Municipal Landfill

Case histories of stabilization by dynamic compaction and useful guidance on design and construction are given by Lukas (1980). Lukas (1992) and Lukas and Seiler (1994) evaluated other case histories including some of those cited above. Table 9.5 (Lucas, 1992) shows the densification (as a percent reduction in thickness) produced by dynamic compaction (D.C.) compared to stabilization by other means such as surcharging (S), recompaction (R) in lifts, or surface compaction (S.C.). The Arkansas site responded most favorably to dynamic compaction and was treated early after closure (about 3 years). The Indiana site was densified about 13 years after closure.

As previously discussed, dynamic compaction has been shown to be effective in producing large settlements compared to other stabilization

Table 9.5 Immediate ground compression

Site	Method	Compression (%)	Service loading (psf)	Elastic compression (%)	Secondary compression index
Location, thickness in ft, age (in years) after closure					
C (Arkansas, 16–39, 3)	D.C.	20–25	360–1800	1.7	0.014–0.023
D (Indiana, 26–30, 11–13)	D.C.	10	1320–2400	2.7–3.4	0.001
New York, 10–20, M	S.C.	10	240–360	3–4.5	0.03
Connecticut, 42, 0	R	19.4	720	3	0.018–0.04
California, 20, M	S	3	Cut 10 feet		0.013
New Jersey, 30, M	S	5–7	$\Delta p/p_\sigma < 1$		0.013
		11–14	$\Delta p/p_\sigma > 1.4$		

D.C. = dynamic compaction; S.C. = surface compaction with 30- to 540-ton roller; R = recompaction; S = surcharge; p_σ = existing stress; O = old (20–25 years); M = medium age.
Source: After Lukas, 1992

methods. Lukas (1994) presented further evidence that the settlements are higher than those produced by primary compression of untreated refuse under load (see Table 9.5). Obviously, this depends on the magnitude of the imposed load but is generally true for road embankments and light building loads. Conservatively, it can be assumed that more than 70% of the primary settlement can be eliminated. To accomplish this, the energy per unit volume of refuse imparted by typical dynamic compaction is comparable to the energy imparted in the standard Proctor test or less than 1/2 the modified effort (see Section 9.1.2). Following treatment by dynamic compaction, the refuse will compress under imposed load as an elastic material and settlement may be calculated using data from a pressuremeter after densification.

For site D in Figure 9.7, the predynamic compaction pressure meter modulus E_r values ranged from 10 to 50 tsf. After dynamic compaction, the range was 70 to 110 tsf. At site F (rubble fill), the postdynamic compaction pressuremeter E_r value ranged from 130 to 265 tsf. Predynamic compaction pressure meter modulus E_r value for site G (mine spoil) varied from 100 to 150 tsf.

Based on these case histories, it appears that dynamic compaction would essentially eliminate the primary compression under loading. The immediate elastic settlements may be calculated using an elastic modulus for refuse from the pressure meter, plate load tests, or other means (see Chapter 4). The long-term secondary settlement (creep) is still a problem for young landfills developed shortly after closure (sites C, D, and E in Figure 9.7). It appears, however, that dynamic compaction may reduce the rate of creep by as much

Figure 9.7
Variation in average
normalized settlement
with time (Lukas, 1992)

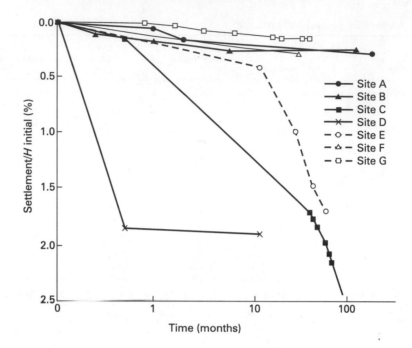

as 50% compared to the same site without stabilization with dynamic compaction.

Charles et al. (1981) used dynamic compaction on a municipal landfill that was 15 years old and up to 6 m (20 ft) thick. A 15-tonne weight with a base area of 4 m² was dropped ten times at each location from heights up to 20 m (66 ft). The primary grid spacing was 5 m (16 ft). There were one or two additional tamping stages at grid points offset from the original grid with the maximum energy input of 2600 kNm/m². High excess pore pressures were observed in some instances of dynamic compaction. The average settlement was 0.5 m (20 in.). Under identical embankment loads, the immediate settlement of the untreated refuse was three times larger than the treated refuse. After about five years, the long-term settlement of the treated section was about 35 mm (1.4 in.).

Welsh (1983) cites a roadway site with 20 ft (6 m) to 40 ft (12 m) of refuse with 3 ft (1 m) cover. Stabilization was accomplished in three passes by using 18 ton weight, 92 ft drop, 10 drops per location, and at 15 ft (4.5 m) centers. Settlement of the treated area after seven days was 1/20 of untreated areas.

Ménard (1984) cites a case for a warehouse designed with floor loads of 400 lb/ft² and spread footing with 3 kips/ft². Refuse 20 to 55 ft thick was stabilized with three to six passes and with applied energy of 36 to 75 ton-ft/ft². The pressure meter modulus E increased by a factor of 5 near the top and 2.3 at a depth of 21 feet. Load tests after treatment showed settlement of 0.4 to 0.5 in. under 3 kips/ft² stress. For a railroad site where dynamic compaction was used to homogenize the fill, up to six passes were necessary. Drop heights of 30 to 60 feet were used with weights of 9 to 15 tons.

Because of the substantial densification of refuse, the dynamic compaction would be expected to reduce the rate of decomposition and secondary compression due to the reduction of the surface area. Proper tamping could perhaps eliminate or conservatively reduce the primary settlement by 70% and reduce the secondary settlement by 50%. The dynamic compaction process introduces high excess pore water pressure. To accelerate consolidation, vertical drains are used in conjunction with dynamic compaction, especially in the zone of footings supporting the columns. In employing densification by dynamic compaction, the possibility of contaminants migrating to the surface during densification must be evaluated and environmental impact assessed.

The use of stabilized refuse as a foundation material is generally limited to light structures or highways and where thickness of the refuse is 30 ft (9 m) or less.

Example 9.2

A 20-ft-thick refuse is to be stabilized by dynamic compaction using a 15-ton tamper with a drop of 20 m.
(a) Determine the adequacy of this process.
(b) Assuming a target energy of 1.5 times standard effort, determine the spacing of impact points and the number of passes.

Solution:
(a) Use $I = 0.5$ and Eq. 9.5.

$$D_{max} = 0.5(15 \times 20)^{0.5} = 8.7 \text{ m or } 28.4 \text{ ft} > 20 \text{ ft}$$

The depth of influence is more than adequate.
(b) Assume 5 m by 5 m spacing and 10 tamps.

$$\text{Energy} = \frac{15 \times 1000 \times 9.81 \times 20 \times 10}{5 \times 5 \times 6.1 \times 1000} = 192 \text{ kN-m/m}^3$$

The number of passes for 1.5 times standard effort $= 1.5 \times 600/192 = 3.1$. Four passes will suffice.

9.3 PRELOADING

Precompression is an effective and economical means of improving weak and compressible materials. It is commonly done through the use of earth fill or water as the surcharge. Before placing the earth fill on the compressible material, a drainage layer is placed on top of it, as shown in Figure 9.8. In water loading, the area is surrounded by an embankment and the drainage blanket is covered with an impermeable membrane, which is then filled with water to provide the surcharge load. Under the surcharge load the underlying weak soils decrease in volume and gain in strength. Various instruments are used to monitor pore water pressure, settlement, and stability of the loaded system.

Figure 9.8
Preloading.

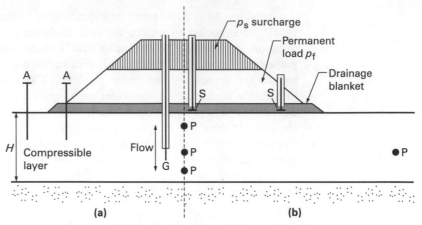

P = Piezometer; S = Settlement plates; A = Alignment stakes; G = Settlement probes

Surcharging beyond the design load can further reduce postconstruction settlement. Subsequently, when the actual loads are applied the resulting settlements are considerably smaller. When refuse is stabilized by preloading, the primary compression will be mostly eliminated. Upon subsequent load application the immediate elastic settlement and long-term creep could still develop. As illustrated in Table 9.5, preloading is less effective than dynamic compaction in stabilization of refuse. For other methods of precompression, see Holtz and Wagner (1975).

9.3.1 *Factors Affecting Settlement Rate*

In a soil not fully saturated, surcharging causes rapid settlement due to compression and dissolution of gases. In a municipal landfill, the rate of decomposition will affect the settlement rate. For example, during dry or cold seasons, bacterial activity and chemical decomposition decrease, thus reducing the settlement rate. On the other hand, during wet or warm seasons, biological activity and chemical decomposition accelerate and increase the settlement rate.

9.3.2 *Construction Control in Preloading*

A preloading job requires proper design, construction supervision, and field monitoring. Prior to placement of the surcharge, a layer of clean sand ($<3\%$ passing No. 200 sieve) material is placed on the site. The drainage blanket may contain a perforated pipe to assist drainage of the sand. Since the materials being stabilized are of low shear strength, surcharge load is applied at a controlled rate to allow the materials to consolidate and possibly gain strength. The initial load increment is determined based on the field and laboratory test data for undrained conditions. The subsequent load increments are determined from consolidated undrained conditions. The strength data may be obtained from a vane shear test, cone penetration test, or pressure meter test. An appropriate safety factor should be used against shear failure. The stability analysis is based on an appropriate method such as a circular failure surface, sliding wedge, or others (see Chapter 13). If the stability conditions cannot be satisfied, it may be necessary to provide berms, which would require additional fill material.

9.3.3 Instrumentation

Field instruments such as those shown in Figure 9.8 are required in many cases to monitor settlements, rate of settlement, and lateral movement. Surface settlement plates and deeper settlement probes are installed so that both the surface and subsurface settlements can be determined. The groundwater profile and excess pore water pressure distribution are obtained from piezometers placed at selected depths (see Chapter 7 for details of instrumentation). Lateral movement and heave can be detected by using alignment stakes at varying distances from the toe of the embankment. Inclinometers at the edge of the surcharge can provide warnings about potential stability problems. Settlement observations generally provide more reliable data than piezometers; piezometers have not yielded useful information for landfills (Sheuns and Khera, 1980). Instrument readings are taken more frequently (weekly) initially and then less frequently (monthly). (For a case history where inclinometers were used see Chapter 5.)

9.3.4 Primary Compression of Municipal Waste

The settlement from primary compression may be computed using Eq. 4.17. Various parameters for municipal landfill materials that were given in Table 4.4 may be used for preliminary estimates. Figure 9.9 shows a load test result for a site in Long Island, New York. The load test was conducted by installing settlement plates and filling to a height of 8 ft to 10 ft. The extent of fill was designed to produce a stress increase of at least 5% at the bottom of the waste. At this location (B-2), the refuse thickness was about 85 ft and the width of the embankment was about 125 ft. Pressuremeter test data and modulus are shown in Figure 9.10.

The refuse at the B-2 location contained sand, silt, clay, newspapers, rubber, glass, and metals. No odors were evident and the age of landfill was estimated to be 18 years after it became inactive. At another location (B-6), the variability of the refuse properties is illustrated by back-calculating the compressibility parameters; average values were assumed.

9.3.5 Secondary Compression for Municipal Waste

The secondary compression of a landfill continues after the primary compression. The long-term settlement appears to be linear on a log–time scale (as described in Section 4.6) and can be determined by Eq. 4.33.

Figure 9.9
Loading and settlement
curves for fill load
test at location B-2
of Long Island landfill.
Reproduced from Kabir
et al., 1994, by
permission.

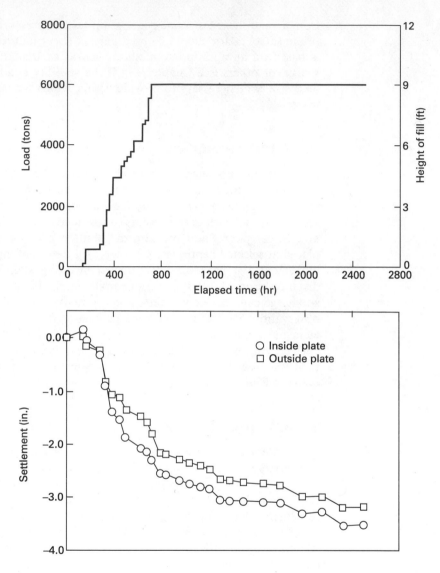

Figure 9.9
Loading and settlement curves for fill load test at location B-2 of Long Island landfill. Reproduced from Kabir et al., 1994, by permission.

9.3.6 *Surcharging of Municipal Waste Materials*

Experience with preloading of municipal waste has shown that the primary compression is completed within a month or two, depending on the thickness of the landfill. To reduce the amount of postconstruction settlement, surcharge is left in place beyond the time needed for primary compression under this load. A considerable amount of secondary compression occurs during this period (t_1), as represented by *cd* in Figure 9.11.

If a structure is to be built on the treated landfill, then the surcharge load and the load corresponding to the permanent load must be removed. When the load is removed, rebound occurs along *de* at a slope of *RR*. The construc-

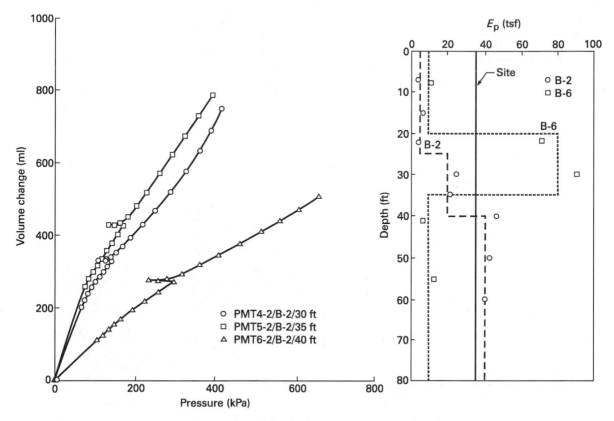

Figure 9.10 Pressuremeter test data and modulus for location B-2. Reproduced from Kabir et al., 1994, by permission.

tion of the structure imposes the permanent load. The settlement of the structure takes place along *ed* and can be computed at any time *t* from the following equations:

$$\delta = H \cdot RR \cdot \log \frac{\sigma_o' + p_f}{\sigma_o'} + C_\alpha' \cdot H \cdot \log \frac{t}{t_2} \qquad \text{for } t > t_2 \qquad (9.6)$$

$$\delta = H \cdot RR \cdot \log \frac{\sigma_o' + p_f}{\sigma_o'} \qquad \text{for } t \leq t_2 \qquad (9.7)$$

The value of t_2 is determined from

$$\log \frac{t_2}{t_1} = \frac{CR - RR}{C_\alpha'} \cdot \log \frac{\sigma_o' + p_f + p_s}{\sigma_o' + p_f} \qquad (9.8)$$

where *CR* is the compression ratio and *RR* is the recompression ratio.

Figure 9.11
Surcharge design for
reduction of secondary
settlement (reproduced
from Bjerrum, 1972, by
permission)

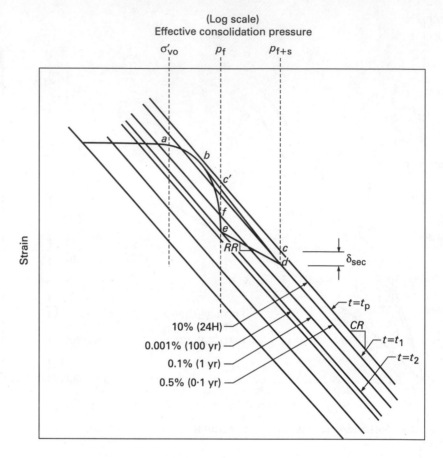

Example 9.3

A 20-ft-thick layer of refuse is to be stabilized to support a warehouse having
an average load of 500 lb/ft^2. For this refuse, $CR = 0.3$, $RR = 0.05$, $C_\alpha = 0.05$,
and unit weight $= 55$ lb/ft^3. The site is to be graded, which requires adding 3 ft
of sand with a unit weight of 120 lb/ft^3. Estimate the settlement after 10 years
if the preload of 500 lb/ft^2 and a surcharge of 500 lb/ft^2 are kept for a period of
1 year prior to construction. Assume t_{100} of refuse to be 1 month.

Solution:

$$\sigma_o' = 3 \times 120 + 10 \times 55 = 910$$

From Eq. 9.8

$$t_2/t_1 = 4.56$$

or

$$t_2 = 4.56 \text{ yr}$$

From Eq. 9.7

$$\delta = 20 \cdot 0.05 \cdot \log\left(\frac{910 + 500}{910}\right) + 0.05 \cdot 20 \cdot \log\left(\frac{10}{4.56}\right)$$

$$= 0.53 \text{ ft}$$

9.3.6.1 Expressway in New Jersey An experimental embankment was constructed for an expressway in the late 1970s in northern New Jersey. The landfill materials within the project area were in various degrees of decomposition with thin layers of soil cover. Information from various sources indicated that the landfill was between 5 and 15 years old.

The depth of landfill ranged from 16 to 28 ft (4.9 to 8.5 m). The underlying materials consisted of an average of 4 ft of gray-black clayey organic silt and about 3 ft of brown peat (locally referred to as the New Jersey Meadowmat), followed by 20 to 25 ft (6 to 7.5 m) of well-graded dense sand and gravel with less than 10% silt. The soils below this layer were varved clays.

The final pavement elevation of the eastbound roadway was established below the top of the landfill with the excavation extending to a point 5 ft (1.5 m) below the subgrade level. This required a removal of 5 to 21 ft (1.5 to 6.4 m) of the refuse. A section of the eastbound roadway is shown in Figure 9.12. The loading sequence, settlement observations, and the soil profile for one station of the eastbound roadway are shown in Figure 9.13.

The materials excavated from the eastbound roadway were cleared of large objects, placed on the westbound roadway in lifts about 2 ft (0.6 m) thick, and compacted using 3 to 5 passes of a 35-ton sheepsfoot compactor. Data for the westbound roadway is presented in Figure 9.14. The surcharge load was increased between 400 and 500 days, at which time the settlement rate increased. This trend is more clearly evident starting at about 80 days, where the height of the added surcharge load was considerably greater. After 400 days the load was increased again and the rate of settlement showed a corresponding increase.

Figure 9.12
Section of eastbound roadway (reproduced from Oweis and Khera, 1986, by permission)

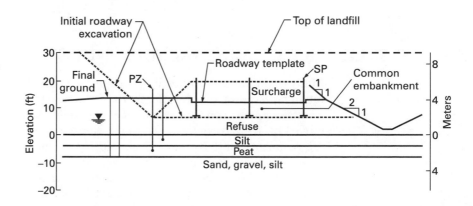

Figure 9.13
Load-settlement-time
response of eastbound
roadway (reproduced
from Oweis and Khera,
1986, by permission)

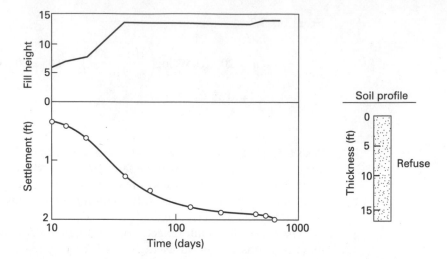

Figure 9.14
Load-settlement-time
response of westbound
roadway on organic
soil and refuse
(reproduced from
Oweis and Khera, 1986,
by permission)

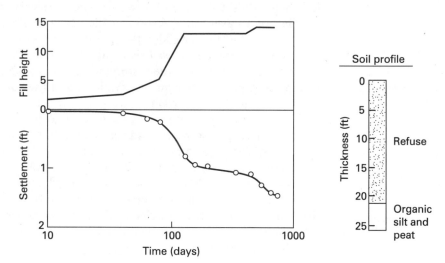

9.4   SETTLEMENT OF MINERAL AND INDUSTRIAL WASTE

Many of the wastes composed of fine-grained materials are disposed of behind
retaining dams or dikes in the form of slurries. In the land disposal of these
wastes one must consider the limited availability of land area and the waste's
impact on the environment. To make efficient use of the land, the volume of
the sediments must be reduced and their strength increased. The techniques
used consist of desiccation; drainage; consolidation under self-weight; surcharging; stabilization with chemicals, lime, cement, or fly ash; or mechanical
methods that reduce the water content before disposal. Vertical drains also

have been successfully used to promote drainage and accelerate precompression. For details see Hansbo (1979) and Van Zanten (1986).

9.5 OTHER STABILIZING TECHNIQUES

Moulton et al. (1976) described a large-scale test where a landfill $40 \times 40 \times 18$ ft ($12 \times 12 \times 5.5$ m) deep was grouted under gravity with fly ash at 100% moisture content. Because of the lack of positive confinement, the grouting did not result in any compaction. Initially, the rate of settlement for the grouted section was high but in two years it dropped to half that of the ungrouted section. Thus some beneficial effects were indicated.

9.6 CONSTRUCTION ON LANDFILLS

Construction on or in refuse poses several problems not usually encountered on other sites and many arguments against such construction have been presented over the years (Eliassen, 1947). The major problems are settlement, odor control, and gas control; others relate to the durability of construction materials in the contaminated soil environment. Eliassen (1947) cites methods for housing construction on landfills where the objectives were gas and odor control. These early methods for gas control are similar in concept to the current passive techniques. The gas problem was handled by leaving a 12- to 24-in. (305- to 610-mm) air space between the floor and the top of a 12-in. (305-mm) cinder layer overlying the landfill cap. The air space was enclosed with wood siding which was vented through screened openings on all four sides. A gastight paper membrane covered with aluminum foil was secured under the wood flooring. (Details on gas management are given in Chapter 14.)

Eliassen (1947) also cites an allowable bearing pressure for footings of 2000 lb/ft^2 (100 kPa) that sanitary landfills may carry without appreciable localized settlements. Without stabilization, large total and differential settlements may preclude the support of buildings and roadways. Even with stabilization, only light structures (1 to 2 stories) may be supported on treated refuse because the ongoing long-term settlement due to decomposition cannot be totally eliminated by stabilization. Restriction or prevention of moisture entering the landfill can substantially reduce the rate of settlement from decomposition and can be accomplished by paving or proper placement of a geomembrane over the landfill.

For light structures, support of refuse may be impractical for refuse thicknesses more than about 20 ft (6 m) because of the difficulty in stabilization and subsequent settlement. For heavier structures, deep foundations would be necessary. Removal or excavation of refuse is usually not an attractive alternative because of the inherent environmental and health problems.

In addition to the conventional soil mechanics parameters, other information—such as the nature of the contaminants, soil resistivity, pH, ion contents, and presence of bacteria—should also be determined during the foundation design. Although little is known about the effect of contaminated soils on construction materials, such information may have some value in selecting appropriate materials. For example, ions of sulfate, chloride, and calcium affect the durability of steel. The structural elements of foundation — such as shallow footings, piles, and piers — and underground building components — such as vapor barriers, pipes, and drains — are vulnerable in a waste environment. The exact rate of deterioration is difficult to predict as our current knowledge is limited to uncontaminated soil environments.

9.6.1 *Piles in Refuse*

Sometimes shallow foundations are deemed infeasible because of expected large settlements, the impossibility of removing the refuse, or other factors. In these cases, pile foundations become necessary. The factors controlling pile design are drivability, downdrag, and corrosion potential.

Driving piles through refuse is usually difficult. Techniques such as pre-augering or predrilling may facilitate the process. The use of heavy pile sections coupled with protective tips and a large hammer with more than 32,000 ft-lb of delivered energy also may be effective. The use of vibrating hammers is unlikely to be effective because of obstructions. Despite all this, foundation designs need to include pile deviations of 5 ft or more from planned locations to accommodate for the relocation that may be necessary to avoid unexpected obstructions.

All pile types (steel, concrete, and timber) are subject to deterioration and their corrosion potential should be considered in the selection of pile types. The use of concrete-filled steel pipe piles may be an option. The heavy steel pipe allows hard driving and provides protection of the concrete; the steel contribution to the structural support is ignored. Dense concrete with acid resistance and low water–cement ratio is usually recommended. Another option is steel piles (without concrete). If steel piles are used, corrosion should be allowed for by using a much thicker section than required for structural support. A third option is creosote-treated timber piles, although protection of the creosote or the protective chemical during driving and the need to pre-auger for each timber pile make this an expensive option. A summary of the existing technology of pile protection in a corrosive environment is presented by Gauffreau (1987).

Very few data are available for estimating the downdrag on piles due to settlement of refuse. The limited observations suggest that the downdrag force may be about 10% of the weight of overlying refuse. For example, if the depth of refuse is 50 ft and the unit weight is 40 lb/ft^3 (6.3 kN/m^3), then the downdrag stress is 200 lb/ft^2 (9.6 kPa).

Data are also lacking on the lateral pressure from refuse. It is logical to consider that the refuse is somewhat reinforced by various objects as evidenced by the near-vertical slopes in refuse over 25 ft. The active lateral earth pressure coefficient, K_a, is expected to be less than that of loose sand (which is about 1/3) and a value of 0.2 appears reasonable for unsoaked refuse.

9.7 SUMMARY

The methods for the reduction of waste volume and improving its load-carrying capacity include surface compaction, dynamic compaction, and pre-loading. Surface compaction is used for municipal waste and some industrial waste where the water content of the waste can be controlled, such as natural soils, blended soils, fly ash, and coal waste. If the compaction is for a soil liner then the design and construction require extensive supervision, quality assurance, and quality control. The number of tests needed can be determined based on the coefficient of variation.

In dynamic compaction, heavy blocks are dropped in a grid pattern, resulting in a decrease in the volume of the treated materials, an increase in the load-bearing capacity, and a reduction in postconstruction settlements. Dynamic compaction is most suited for loose coarse-grained soils, rubble fills, and nonhazardous landfills (especially municipal landfills where it has found extensive use). Materials below the water table cannot be treated very efficiently.

Preloading with surcharge provides an effective means of treating very soft soils such as dredged waste, municipal waste, etc. This method requires more time than the others but can be applied to deep, very soft materials.

NOTATIONS

C_α'	secondary compression ratio with respect to strain	$\gamma_{d\text{-max}}$	dry unit weight of the soil in the densest state (F/L^3)	q_c	cone penetration resistance (F/L^2)
C_c	compression index	$\gamma_{d\text{-min}}$	dry unit weight of the soil in the loosest state (F/L^3)	RC	relative compaction (%)
CR	compression ratio			RR	recompression ratio
C_t	secondary compression index with respect to void ratio	γ_d	dry unit weight of the soil (F/L^3)	σ_o'	effective overburden stress (F/L^2)
S_e	allowable sampling error of the expected mean	H	drop height of weight (L)	t	probability factors from the t-distribution tables
e	void ratio	I	influence factor	t_1	time at which settlement value is desired (T)
$e_{\max}$	void ratio of the soil in the loosest state	k	permeability (L/T)		
		N	standard penetration resistance	t_p	completion time for primary consolidation (T)
$e_{\min}$	void ratio of the soil in the densest state	n_u	number of units in the sample	u_e	excess pore pressure (F/L^2)
e_0	initial void ratio	p_f	final or permanent load (F/L^2)	v'	coefficient of variation
E_r	refuse modulus from pressuremeter (F/L^2)	p_s	surcharge load (F/L^2)	W	weight of the dropping block (F)
				w_n	natural water content (%)

PROBLEMS

9.1 An old landfill with 25 ft of refuse has been abandoned for nearly 30 years. Develop a dynamic compaction program to densify the refuse. The available cranes can lift and drop up to 25 tons of weight a distance of up to 135 ft. Design the program to result in compaction energy equivalent to 1.5 standard Proctor.

9.2 Pressuremeter tests after the densification of the landfill in Problem 9.1 indicated an average modulus of 80 t/ft^2. Determine the elastic settlement and estimate the long-term settlement caused by the structural load after 50 years. Assume an average slab area load of 200 psf, a concentrated column load of 50 tons, and a design bearing pressure of 1 t/ft^2.

9.3 The designer for the structure in Problem 9.2 specified "light" dynamic compaction for the parking areas. Determine the compaction requirement assuming 0.8 standard Proctor energy.

9.4 A portion of the landfill in Problem 9.1 is to be stabilized by preloading to support a light industrial building imposing an average uniform load of 250 psf. To achieve proper grading, a layer of sand 2 ft thick will cover the building site. The specification requires the sand fill to have an average unit weight of 115 pcf. The total thickness of sand placed was 10 ft and it was kept for 2 years in area A and 3 years in area B. Assuming $CR = 0.3$, $RR = 0.05$, and $C_\alpha' = 0.02$, determine the total settlements after 30 years following construction in areas A and B, respectively. Consider a unit weight of 50 pcf for the refuse and no water table within the refuse.

9.5 A proposed roadway is to be located on a landfill that was closed 10 years ago. The landfill is 18 ft deep and a surcharge of 12 ft with a unit weight of 105 pcf was placed on it. Assuming appropriate consolidation parameters, determine the settlement of the landfill due to the applied surcharge.

10 Liners

10.1 INTRODUCTION

Liners are broadly classified into four major groups: soil (earthen) liners, geomembranes (also called flexible membrane liners), geosynthetic clay liners, and vertical cutoff walls. The purpose of all liners is to restrict fluid migration from the facility into the environment (soil, groundwater, or air). The fluid could be leachate in a landfill, waste in a lined pond, or gas in a landfill. Caps designed to restrict rainwater intrusion into landfill or to restrict gas emissions to the atmosphere are constructed in a manner similar to liners. We will discuss caps separately (in Chapter 12) because of other parameters (e.g., surface erosion) that pertain to them.

Soil liners are usually made of natural inorganic clays or clayey soils to achieve low in situ hydraulic conductivity. Soils classified as CH, CL, and SC (see Chapter 2) are typical clay liners. If natural clay or clayey soils are not available, commercially available high swelling clay (bentonite) can be mixed with silty sands or sands to produce a usable soil liner.

A geomembrane, if intact, is virtually impermeable to liquids; however, gases and vapors of liquids can permeate a geomembrane on a molecular level by chemical diffusion (Haxo, 1990). A geomembrane consists of one or more sheets made by a synthetic process utilizing resins such as polyethylene, polyvinyl chloride, and other ingredients. Different geomembranes have different structural properties and resistance to chemicals.

A geosynthetic clay liner (GCL) typically consists of a relatively thin (4 to 6 mm) clay (typically bentonite) layer contained by two (upper and lower) geotextiles or bonded (by adhesive) to a geomembrane.

Vertical cutoff walls are rarely used as a part of a liner structure for a new landfill. However, in older landfills with no liners or base leachate collection system, a vertical cutoff wall may be installed to restrict leachate (or gas) release to the environment. The wall is constructed of either soil bentonite, cement bentonite, plastic concrete, grouted sheet piles, or some similar material; it is toed in several feet into a layer with low hydraulic conductivity.

Modern designs use composite liners. A composite liner (Figure 10.1) is a geomembrane placed directly above and in intimate and uniform contact with a soil liner or a geosynthetic soil liner. The federal regulations for Subtitle D facilities require a composite liner with a minimum thickness of 2 ft (61 cm) of a compacted soil liner having a hydraulic conductivity of no more than 10^{-7} cm/s. The minimum thickness of the geomembrane is usually 30 mils (0.03 in.); if the geomembrane is high density polyethylene (HDPE), the minimum must be 60 mils (0.06 in.). The leachate collection and removal system is designed to limit leachate buildup to less than 1 ft (30 cm) above the liner.

Figure 10.1
Common liner designs

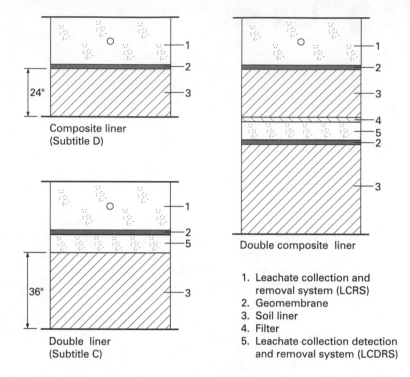

Composite liner
(Subtitle D)

Double liner
(Subtitle C)

Double composite liner

1. Leachate collection and
   removal system (LCRS)
2. Geomembrane
3. Soil liner
4. Filter
5. Leachate collection detection
   and removal system (LCDRS)

Subtitle C regulations require a 3-ft (91-cm) soil liner. The leakage detection system under Subtitle C regulations can be constructed of sand with either a hydraulic conductivity of 0.01 cm/s or more and 12 in. (30.5 cm) or more in thickness or a synthetic drainage layer (geonet or a geocomposite) with a transmissivity of 3×10^{-5} m²/s or more. A minimum bottom slope of 1% is stipulated. No minimum requirements on the geomembranes are given but thicknesses of 60 to 100 mils (1.5 to 2.5 mm) have typically been used.

Some state regulations require double (or dual) composite liners for sanitary landfills (see Figure 10.1). The minimum thickness of the soil component and the geomembrane is usually dictated by the regulations. Typically, a minimum 2-ft-thick component of soil liner is specified together with a 60-mil geomembrane.

A leachate collection system overlies the upper component of the liner system. The system usually consists of material with high hydraulic conductivity (0.01–1.0 cm/s) with perforated pipes for collection of waste fluid. Materials such as coarse sands and gravel are preferable because of lower clogging potential by sedimentation from landfill leachate and biological growth (as discussed in Chapter 11).

A leak detection system (see Figure 10.1) separates the upper and lower composites in a double composite liner. In some designs geonets are used. A geonet is made by a synthetic process (typically using polyethylene resin) and consists of parallel sets of ribs overlying and connected to parallel ribs at various angles. The openings (apertures) created are typically 0.3 in. × 0.3 in. (7.6 mm × 7.6 mm), which allows planer flow (transmissivity) through the geonet. A transmissivity of no less than 5×10^{-4} m^2/s is usually specified. Transmissivity is the product of thickness and hydraulic conductivity. The transmissivity of a 1-ft (30.5 cm) soil layer with 0.1 cm/s in hydraulic conductivity is 3×10^{-4} m^2/s.

The transition between a soil liner and a drainage layer often requires a filter to preclude erosion (piping or migration) of soil particles from the liner or base into the drainage layer. In many applications, a geotextile filter is bonded to the geonet to form what is generically called a *geocomposite*. A geotextile is a permeable synthetic material composed of textiles (fabrics) manufactured by various techniques (see Chapter 11).

10.2   SUBGRADE PREPARATION

In a typical liner construction, the foundation layer over which a liner component is supported is generically termed *subgrade*. In the case of the bottom layer of the liner structure, the area receiving the liner is excavated or filled to meet design grades. In a composite liner, the clay component on which the geomembrane is founded also is termed subgrade.

As a part of the subsurface investigation, the natural undisturbed soil receiving the landfill must be proven to be stable enough to support the landfill with an adequate safety factor against foundation failure and tolerable settlements under the loads imposed by the facility. If stability is not proven, the site is excavated to a stable geologic unit or the weak stratum is stabilized (e.g., by dynamic compaction or other means).

For a site to be acceptable, regulations require that the seasonal highest water table level be at least 5 ft (1.5 m) below the base of the liner structure. Thus, groundwater is not supposed to be a problem. Under some situations, a drainage system (thick stone layer and piping) has been used to suppress the water table and satisfy the regulatory requirements. However, in some glacial deposits perched water is encountered. In such cases, the water must be drained before the exposed subgrade is compacted. Typically, a portion of the water-bearing base soil is excavated, a filter is placed in contact with the base soil, and then a drain stone layer and a pipe to collect and convey the water off-site are added.

In temperate climate with moderate to heavy rainfall, surface runoff greatly affects the subgrade preparation for a landfill cell. The site should always be graded to effect rapid surface drainage. In addition, perimeter construction swales may be essential to keep the site firm enough to allow mobility of construction equipment.

Subgrades with a high percentage of fines (passing a No. 200 U.S. standard sieve) are difficult to compact in wet weather because of their sensitivity to moisture. If the soil becomes completely or nearly saturated, the compaction causes excess pore water pressure in the soil. One solution in this case is a very slow compaction as to allow the soil to drain. Other solutions would be to compact the soil in dry weather or to aerate the soil and allow it to dry prior to compaction.

Figure 10.2
Tie-in of new soil liner
to existing soil liner
(EPA, 1993)

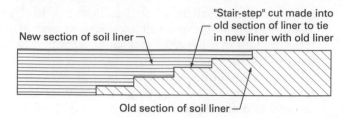

In the case of a clay liner as a subgrade, the liner must be smoothly graded to avoid gaps between the liner and the geomembrane on top. Where a new section of liner is to be tied in to an old section, the old section should be cut in a stepped fashion over a distance of 3 to 6 m (EPA, 1993), as illustrated in Figure 10.2. The surface of each step in the old liner must be scarified to maximize bonding between the old and new sections.

The moisture density (compaction) curve serves as a good guide for subgrade preparation. The compaction moisture content should be close to the specified moisture content. If the subgrade is not part of the liner and the hydraulic conductivity is not a design issue, the field moisture content is not usually specified although compacting dry-of-optimum may lead to settlement of some compacted fills when wetted. In the case of a clay liner subgrade, both density and moisture content are specified.

During the subgrade compaction, any soft spots or areas should be removed and replaced by compacted soil. Any weaving of the subgrade under the compactor is an indication of high moisture content and poor compaction. Any obvious cracks is an indication that the moisture content is too low. In this case, the surface should be excavated, wetted, and recompacted.

10.3 COMPACTED SOIL LINERS

10.3.1 Material Selection

The primary factors controlling the performance of compacted soil liners are the hydraulic conductivity, strength, and potential for shrinkage with moisture content as an index. These three components are tied directly or indirectly to compaction. To achieve a low hydraulic conductivity, the U.S. Bureau of Reclamation (1974) recommends a minimum plasticity index of 10 or preferably 12. Limit on the plasticity index is 25 and on the liquid limit is 45. The limitation on the liquid limit is significant in limiting the swell potential.

Clay liners are usually subject to drying during construction, and cracking due to shrinkage is a potential problem unless the liner is protected during construction. Table 10.1 (Lutton et al., 1979) provides guidance on the potential for volume change in terms of the moisture content. Clays with high shrinkage limits are preferable to those with lower shrinkage limits. In liner construction, the clay is compacted wet-of-optimum and the field moisture

Table 10.1
Guidance for selection
of liner materials

Likelihood of volume change with change in moisture	PLASTICITY INDEX		
	PI in arid regions	PI in humid regions	Shrinkage limit (%)
Little	0–15	0–30	12 or more
Little to moderate	15–30	30–50	10–12
Moderate to severe	30 or more	50 or more	10 and less

content is typically higher than the shrinkage and plastic limits. In actual liner compaction, the closer the moisture content to the shrinkage limit the lower the cracking due to shrinkage. Table 10.2 (U.S. Bureau of Reclamation, 1974) provides an index of volume expansion when a dry clay lining becomes wet. With the criteria on stability and hydraulic conductivity satisfied, clay with low plasticity index and high shrinkage limit is preferable. Kleppe and Olson (1985) and Daniel and Koerner (EPA, 1993) provide more extensive guidance on shrinkage.

The source of the material (the borrow area) is a factor in determining the suitability of a particular clay for liner construction. If the material at the source or as delivered has a high moisture content (e.g., 5% above the target moisture content), it is usually difficult to condition by normal aeration. Normal aeration involves turning the soil with a harrow or plow (disking) or turning the soil with a dozer or a grader. On the other hand, if the material is too dry, it is difficult to break up the chunks of clay (clods) and achieve proper compaction. (This is one reason for limiting the plasticity index to about 25 — highly plastic clay is tougher and more difficult to break and condition.) In some cases pulverization of the clay using special equipment may be necessary to achieve proper compaction.

Once a borrow area is identified, the clay is tested to assess its suitability. Table 10.3 (EPA, 1993) provides guidance on the type and frequency of tests used for testing borrow areas. In the case of soil bentonite liners, the bentonite is tested for liquid limit (ASTM D 4318), grain size (ASTM D 422) and free swell (no ASTM standard). The recommended testing frequency is one per

Table 10.2
Relation of soil index
properties and
probable volume
changes for highly
plastic soils

Colloid content (% minus 0.001 mm)	Plasticity index	Shrinkage limit (%)	Estimated expansion (% total volume from dry to saturated condition under 1 psi pressure)
>28	>35	<11	>30 (very high)
20–31	25–41	7–12	20–30 (high)
13–23	15–28	10–16	10–20 (medium)
<15	<18	>15	<10 (low)

Table 10.3
Guidance for minimum quality control tests on borrow areas

| Parameter | ASTM test method | FREQUENCY* | |
		Test/yd³	Test/m³
Water content	D 2216	1 test per 2600	1 test per 2000
Atterberg limits	D 4318	1 test per 6500	1 test per 5000
Percent fines	D 1140	1 test per 6500	1 test per 5000
Percent gravel	D 422	1 test per 6500	1 test per 5000
Compaction curve	D 698 or D 1557	1 test per 6500	1 test per 5000
Hydraulic conductivity		1 test per 13,000	1 test per 10,000

* A change of material type as visually determined by an experienced soil engineer may increase the frequency of the testing.
Source: After EPA, 1993

truckload. Table 10.4 provides guidance on the type and frequency of tests on clay as it is spread in a lift.

10.3.2 *Compaction Parameters*

Figure 9.1 shows the parameters of the compaction–moisture density relationship. Moisture–density relationships for various compaction energies are shown as curves 1, 2, and 3. Curves 4 and 5 are for the same compaction energy but the materials (although from the same source and the same clay pit) are different. If a single moisture–density data point E is plotted, then a moisture–density curve D can be sketched and a point F can be defined as the approximate maximum dry density and optimum moisture content for the material tested (using the same compaction energy that was used to establish curves 4 and 5). A necessary requirement for this one-point compaction method is that point E be on the dry side of the optimum and several curves previously be established to cover the variability of the material from the same source. A complete moisture–density curve may take several days to develop. The one-point compaction is usually accomplished in less than a day (or even few hours) depending on the initial moisture content and the time required to cure the sample.

Table 10.4
Recommended minimum quality control tests on loose liner lift before compaction

| Parameter | ASTM test method | FREQUENCY* | |
		Test/yd³	Test/m³
Percent fines	D 1140	1 per 1000	1 per 800
Percent gravel	D 422	1 per 1000	1 per 800
Atterberg limits	D 4318	1 per 1000	1 per 800
Percent bentonite	None**	1 per 1000	1 per 800
Compaction curve specification		1 per 5200	1 per 4000

* Test frequency should be increased for testing suspected material. Except for compaction curve, at least one test should be performed each day the soil is placed.
** The methylene blue test (Alther, 1983) may be used. This test is applicable for soil–bentonite liners only.

The acceptance criteria may be in terms of percent compaction. If the measured dry field density is 95 pcf and the established maximum dry density is 115 pcf, then the percent compaction is $(95/115) \times 100$, or 83%, which is less than the specified 90%. A relatively homogeneous material could have a variation in the maximum dry density of ± 5 pcf and in the optimum moisture content of $\pm 3\%$ (EPA, 1993). In many construction projects, the maximum dry density for the material being compacted may be less than the control reference density. An easy way to resolve this problem is one-point compaction. As explained later, possible errors in field measurements of density and moisture content also must be considered.

For all the compaction curves, two values of moisture contents are defined for a given dry density (except the maximum density, where a unique w_{opt} is defined for a given compaction energy and given soil). As explained later, the hydraulic conductivity depends on the molding moisture content and the density. For this reason, the dry density alone cannot be used as an index for hydraulic conductivity. At a high moisture content, the dry density is not reflective of the compaction energy; for example, point G is common to curves 1, 2, and 3 and curve B is the upper boundary for moisture–density laboratory tests.

Curve C (line of optimums) is useful for compaction control (EPA, 1993). If field values plot above this line, then a proper field compaction energy is delivered. If only 80 or 90% of the field values are above this line, then the compaction effort in the field and/or the reliability of measurements may be questionable.

10.3.3 Effects on Hydraulic Conductivity

The hydraulic conductivity of a liner in the field depends on several factors that have been identified as important through many years of research and field experiences. These factors are index properties, compaction energy, compaction water content and density, and climate.

10.3.3.1 Index Properties Other factors being equal, the hydraulic conductivity decreases with increasing plasticity index. Thus CH soils are less permeable than CL or SC soils. Because of variability in field conditions and soil types, the hydraulic conductivity for a given plasticity index may vary by as much as a factor of 50 (based on data collected by Benson et al. (1994)).

Benson et al. (1994) correlated the fine content and found, as expected, a reduction of hydraulic conductivity with increasing fine content as well as a reduction in hydraulic conductivity with increasing liquid limit and increasing initial saturation. The scatter in this data is also large but it appears that a minimum of 50% fines results in hydraulic conductivity of less than 10^{-6} cm/s. Obviously, less than 50% fines may be appropriate depending on the plasticity and other factors.

To protect the geomembrane from puncture as it presses against the soil liner, liner material should have limited or preferably no gravel (fine or coarse) or large particles. The U.S. Bureau of Reclamation (1974) allows clayey gravel

(GC) for earth lining. EPA (1993) emphasizes the importance of uniform mixing but limits the gravel content to no more than 50% by weight, based on hydraulic conductivity considerations. This means that the soil classification is on the border of gravelly clay–clayey gravel. Gravel size should be limited to the fine range.

Large clods in liner material cannot be broken by construction equipment, especially if the soil is highly plastic. Even if the moisture content is at optimum, it may be very difficult to break the clods for proper hydration and compaction in highly plastic soils. Larger clods produce liners with higher hydraulic conductivity, and the effect is more pronounced for thin liners and if clods are not uniformly distributed. The specification may require that a clod pass a 3 in. or even a 2 in. sieve to ensure that sizes are below this. Suggested maximum clod size in liner material varies from one in. to the full lift thickness (EPA, 1988). The smaller size restriction (1 to 2 in.) applies to areas compacted by hand-operated tampers. The less restrictive clod size (up to 1/2 lift thickness) applied to areas compacted by rollers. Removal of obviously over-sized particles can be done in the field either by hand labor or use of appropriate equipment.

10.3.3.2 Compaction Energy   The parameters controlling the energy per unit volume of soil are essentially the same in the field as in the laboratory. The larger the weight of the compactor, the larger the delivered energy per unit volume. Thinner lifts result in more energy delivered per unit volume of soil. More passes of the compactor deliver more energy, which affects compaction up to a limited number of passes (typically 10 passes for lifts 12 in. or less).

The hydraulic conductivity decreases with higher compaction energy. The type of compactor used plays a significant role; sheepsfoot rollers (generic name for padfoot, tamping foot, and sheepsfoot rollers) have been statistically shown to produce lower hydraulic conductivities than rubber tire compactors (Benson et al., 1994). The kneading action of a sheepsfoot causes greater soil shearing and lesser permeability.

Loose contact or poor bonding between lifts in a soil liner could result in a substantial increase in the hydraulic conductivity, especially if the loose bond planes are interconnected through hydraulic defects such as vertical or sub-vertical cracks, loose planes, etc. A good specification usually requires the lower lift in place to be scarified with a disc before placing the new lift.

10.3.3.3 Compaction Water Content and Density   The effect of compaction moisture content on the hydraulic conductivity is illustrated in Figure 10.3. An increase in moisture content leads to a decrease in hydraulic conductivity, due to the larger degree of dispersion in a soil structure with higher moisture content. The kneading action of a plunger-type compactor — such as a sheepsfoot roller — results in more soil shearing and dispersion, which causes lower hydraulic conductivity. Low-density soil at high moisture content may achieve the conductivity requirements but is likely to fail the strength and shrinkage requirements.

Figure 10.3
Compaction curve and
effect on permeability
(reproduced from
Mitchell, 1976, by
permission of John
Wiley & Sons, Inc.)

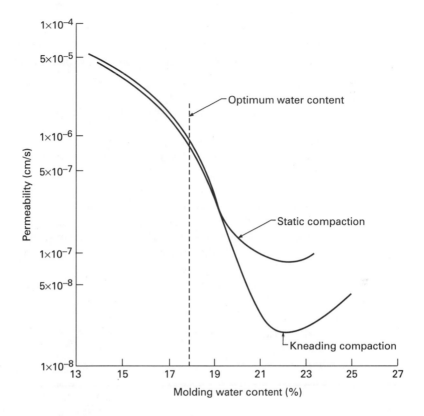

10.3.3.4 Climate   Shrinkage cracks are the direct result of the drying of the
clay liner being constructed. A shrinkage crack effectively reduces the thickness of the uncracked portion of the liner. A fully penetrating crack increases
the hydraulic conductivity of the liner by several orders of magnitude, depending on the width and number of cracks (see Chapter 12). Cracked portions of
the liner should be excavated and/or wetted and then recompacted. On some
projects, it may be prudent to protect the liner during installation to minimize
shrinkage.

Kleppe and Olson (1985) performed tests on three soils and concluded that
volumetric shrinkage of less than 5% led to relatively minor cracking. Severe
cracks were generally associated with volumetric strains above 10%. Shrinkage strains increased essentially linearly with compaction moisture content
and were independent of the dry density. Addition of sand to the clay reduced
drying shrinkage. Cracking became more severe when the samples were
soaked then dried.

Frozen soil must not be used in liner construction. Because freezing and
thawing cause cracks that increase the hydraulic conductivity, it is recommended that liner construction be performed in nonfreezing temperatures or
that the liner be protected from freezing during construction. Kim and Daniel
(1992) found that the permeability laboratory samples compacted wet-of-

optimum (which is the recommended practice) increased by two orders of magnitude after five freeze–thaw cycles. If the liner inadvertently becomes frozen, the frozen portion should be conditioned or removed and recompacted.

10.3.4 *Compaction Effect on Shear Strength*

The shear strength of compacted clay depends on the density as well as the moisture content. In well-built liners, the compaction moisture content should be a few percentage points above optimum. The strength on the wet side is maximum at the maximum dry density and optimum moisture content; it then decreases with increasing water content along curve B of Figure 9.1. The relationship between the peak shear strength (at maximum dry density and optimum water content) and the shear strength along curve B for standard compaction at higher moisture content may be approximated (Leroueil et al., 1992) as

$$C_w/C_{w_{opt}} = \exp[-5.8(w - w_{opt})/PI] \qquad (10.1)$$

Where C_w is the undrained shear strength of compacted sample in the standard Proctor test at moisture content w greater or equal to w_{opt}, w is the water content, w_{opt} is the optimum water content in the standard Proctor test, and $C_{w_{opt}}$ is the undrained shear strength at $w = w_{opt}$.

    The value of $C_{w_{opt}}$ depends on the maximum dry density and soil type. If such value is determined for the control testing used for the project, the strength at higher moisture may be approximated from Eq. 10.1. Eq. 10.1 also may be used to anticipate mobility difficulty at higher moisture contents during compaction (Leroueil et al., 1992). If the compactor exerts a pressure of 6000 psf and the bearing capacity is 5 times shear strength, the minimum field shear strength required to avoid rutting is 1200 psf. If the peak shear strength at $w_{opt} = 15\%$ is 4000 psf and $PI = 20$, then rutting would be expected at or beyond a field moisture content of about 19%.

10.3.5 *Field Control*

The increase or decrease in compaction moisture content produces opposite trends with respect to the hydraulic conductivity and to strength and shrinkage. Strength is dictated by stability requirements. The field moisture content should not be too high above the shrinkage limit. Figure 10.4 (Daniel and Wu, 1993) shows the acceptable zone on the moisture–density curve for strength, shrinkage, and hydraulic conductivity requirements.

    The acceptable zone based on strength criteria defines the minimum dry density and range of moisture content that will produce compacted soil with acceptable strength. The zone for shrinkage is defined in terms of the upper limit of the moisture content. The upper boundary of all limits is the zero-air-void curve A in Figure 9.1.

Figure 10.4
Acceptable zone based on low hydraulic conductivity, low desiccation-induced shrinkage, and high unconfined compressive strength (Daniel and Wu, 1993, by permission)

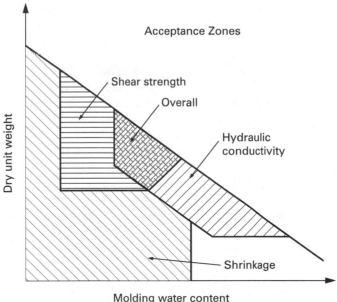

The acceptable zone for hydraulic conductivity may be determined in at least two procedures. The first procedure, adopted in many commercial laboratories, is to compact the soil to a given density using the specified procedure (standard or modified or 15 blows). The hydraulic conductivity is determined and an envelope drawn to define the zone where moisture–density combinations produce acceptable hydraulic conductivities. Typically, 8 to 10 hydraulic conductivity tests are required.

A more fundamental approach (Daniel and Wu, 1993) accounts for possible compaction efforts in the field with the modified test (ASTM D 1557) representing an upper limit on the compaction energy that can be produced in the field. This approach involves the following steps recommended by Daniel and Koerner (EPA, 1993) and Daniel and Wu (1993):

1. Prepare compacted samples using the modified, standard, and reduced (15 blows) compaction procedures to produce compaction curves as shown in Figure 9.1. In most applications, the modified procedure is not applicable because conventional compaction equipment is used. In this case, the standard and reduced compaction procedures are sufficient.

2. Permeate the samples (e.g., ASTM D 5084, U.S. Corps of Engineers EM1110-2-1906, APP.V11, permeability test with back-pressure) and plot the hydraulic conductivity results on the density–moisture content plot for successful tests. A confining stress no more than 5 psi (35 kPa) is recommended.

3. Perform unconfined compression tests (e.g., ASTM D 2166 or U.S. Corps of Engineers EM1110-2-1906, Appendix XI). Test specimens 1.4 to 2.8 in. in diameter with height-to-diameter ratio of not less than 2 : 1 may be trimmed from compacted samples. For strength equal to or higher than the specified value, plot the results on the density-moisture content plot.

4. From drying shrinkage tests, determine the range of moisture contents for which cracking is not severe (Daniel and Wu, 1993). The threshold must be specific for the soil and the project. Plot the test results on the density–moisture content plot.

5. Superimpose the plots in steps 2 through 4. The zone where all the permissible ranges overlap (Figure 10.4) is the acceptable zone to be used for field control. The line connecting the points of optimum for different compaction efforts can be used approximately to define the acceptable moisture–density zone for acceptable permeability.

In many cases, the incoming clay is wet and limitations on moisture content to limit shrinkage cracks may be impractical. The liner may be covered with topsoil or a membrane and topsoil to limit shrinkage. If the liner is founded on a very dry subgrade, it may be necessary to separate the liner from the subgrade by installing a geomembrane (Daniel and Wu, 1993). However, the weight of the waste in landfills tends to help close the drying shrinkage cracks and thus the long-term effect may not be significant (especially as compared to caps or lagoon liners where the overburden stress is relatively small).

A key requirement for the success of an acceptable zone on moisture and density determined as described above is that measurements of these quantities in the field be taken in a reliable manner. Both the nuclear density (ASTM D 2922) and the nuclear moisture gauges (ASTM D 3017) are popular because of the quick results that reduce delay in construction. We will not discuss the nuclear method in detail but a brief background (EPA, 1988; EPA, 1993; Manufacturer's literature) could help reduce the misuse of such equipment.

Nuclear density gauges consist of a source of gamma rays and one or more detectors. Nuclear moisture gauges consist of a source of fast neutrons (americium-beryllium) and a slow neutron detector. Current nuclear gauges contain both density- and moisture-measuring capabilities (Figure 10.5). The gauges for density are used in either a backscattering or transmission mode. In the transmission mode, the source is lowed into a preformed hole in the soil. In the backscattering mode, gamma rays are directed downward into the soil. In both cases, the total unit weight of the soil is a function of the intensity of radiation detected at the surface. The probability of scattering (Compton scattering) is a function of the electron density of the medium.

Measurement of density by backscattering is not recommended for liners as it is heavily influenced by the density in the upper 1 to 2 in. (25 mm to 50 mm). The direct transmission technique is more applicable for liners. To overcome the problem of the chemical composition of the soil, the gauge may be calibrated by measurements of field densities using the sand cone method (ASTM D 1556), the drive cylinder method (ASTM D 2937), or the rubber balloon method (ASTM D 2167).

A typical requirement is a density test at grid points 100 ft (30.5 m) apart for every lift. Daniel and Koerner (EPA, 1993) recommend that for every 20 tests the rapid results be calibrated against results from the sand cone or rubber balloon methods. A third method suggested for calibration is to recover undisturbed samples (ASTM D 1587) and determine the unit weight.

Figure 10.5
Schematic diagram of nuclear water content–density device (EPA, 1993)

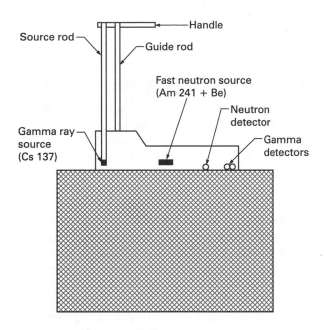

This method is suitable for liners with no large particles (equal to or larger than 1/6th of sample diameter).

The principle of operation for the moisture gauge is that hydrogen in water, having the same nuclear mass as a neutron, is effective in slowing down or reducing the energy of the emitted fast neutron. The response of the moisture gauge is biased toward the moisture content near the detector. Hydrogen-containing material (other than water) has a moderating influence on the neutrons and affects the results. To eliminate the influence of soil composition, the gauge should be calibrated by direct measurement of water content using ASTM D 2216 (direct oven-drying method).

Another potential problem with the nuclear moisture measurements is that the presence of trenches and objects within a short distance (less than 3 ft) from the instrument can cause an overestimate of the moisture content. Imperfect seating of the instrument on the soil surface also causes errors.

The relatively quick (rapid) procedures in addition to the nuclear method (ASTM D 3017) are

ASTM D 4643 Determination of water (moisture) content of soil by the microwave oven method

ASTM D 4944 Determination of water (moisture) content of soil by the calcium carbide gas pressure tester method

ASTM D 4959 Determination of water (moisture) content of soil by the direct heating method

None of these rapid methods (including the nuclear method) should replace the standard ASTM D 2216 (direct oven-drying method). Daniel and Koerner (EPA, 1993) recommend that every tenth sample tested with a rapid method be also tested as per ASTM D 2216. They also recommend a minimum rapid testing frequency of 5/acre/lift (13/hec/lift), which is nearly the same as specified by many state regulations.

10.3.6 *Soil Bentonite Liners*

In some cases, suitable clay may not be available for liners. If this is the case, blended soil—which is usually granular soil mixed with bentonite—may be used. Bentonite is primarily composed of the smectite group of minerals. Such minerals are characterized by large cation-exchange capacity, high swelling potential, and low hydraulic conductivity. Sodium bentonite (characterized by sodium as the external cation adsorbed on the surface of the mineral) is more widely used than calcium bentonite (characterized by calcium as the external cation). Sodium bentonite has larger swelling capacity. In order to achieve a low hydraulic conductivity using calcium bentonite, three times as much bentonite may be needed. However, calcium bentonite may be more resistant to chemicals. Calcium bentonite exhibits higher shear strength (Gleason et al., 1997).

When mixed with other soil and compacted, bentonite yields a blended soil with an acceptable hydraulic conductivity (usually less than 10^{-7} cm/s). Silty sands or soil with a large percentage of fines require less bentonite than do coarser soils. When bentonite is mixed with permeable soil and water in relatively small amounts (1 to 8% of the soil's dry weight), it decreases the permeability because of this expansion property.

The recommended (one per truckload) quality control tests (EPA, 1993) for bentonite are a liquid limit test (ASTM D 4318) and a free-swell test (which measures the amount of swell when bentonite is exposed to water in a glass beaker). High-quality sodium bentonite has a liquid limit in excess of 500% and free swell in excess of 18 cc. Calcium bentonite free swell is less than 6 cc.

The methylene blue test suggested by Alther (1983) may be used to determine the bentonite content. A water absorption test (filter press test) may be used to determine the bentonite content if bentonite clay is the only clay mineral in the mix. Granualor or powdered sodium bentonite may be specified. Gleason et al. (1997) reported that powdered bentonite exhibited lower hydraulic conductivity when permeated with strong calcium chloride solution. Consequently an additional quality control test is the grain size of dry bentonite (ASTM D 422).

Other quality control tests are usually specified and will typically include hydraulic conductivity (ASTM D 5084) at one test per acre per lift. For mix design, one compaction test (ASTM D 698 or ASTM D 1557) is recommended for 5000 m^3 of material (EPA, 1993).

10.3.7 *Field Hydraulic Conductivity*

Some regulations insist on a field permeability test without actually specifying what the test should be. Two common tests are discussed here.

10.3.7.1 Sealed Double-ring Infiltrometers   ASTM D 5093 describes a test procedure using the sealed double-ring infiltrometer (SDRI), as diagrammed in Figure 10.6. The SDRI (ASTM D 5093) consists of an inner and an outer ring that are preferably square but also can be circular. The inner ring has a minimum diameter (or width) of 24 in. (610 mm) and the outer ring consists of 4 aluminum sheets that are each 12 ft wide × 3 ft high × 0.08 in. thick (3.6 m × 610 mm × 2 mm). A minimum distance of 2 ft (610 mm) is maintained between the inner and outer ring. The inner ring has a depth of 4 to 6 in. and is centered within the outer ring, which is open. The inner ring has a top and is shorter than the outer ring.

Measurement of flow is made by connecting a flexible bag filled with a known weight of water to a port in the inner ring. Air is flushed out through a second port as the inner ring is filled. As water infiltrates the liner, the water level in the inner ring drops. After an interval of time, the lost water in the inner ring is replaced with an equal amount drawn from the plastic bag. The bag is removed and weighed. The weight loss converted to volume is the volume of water infiltrating the liner during the known interval of time. The result is a plot of infiltration rate versus time. Because the inner ring is sealed, no correction for evaporation is needed. The outer ring is embedded at least 45 cm and the inner ring 10 cm; this forces one-dimensional flow for 45 cm, which is more than 2/3 of a typical liner thickness (60 cm).

Calculating the hydraulic conductivity requires some assumptions. If we assume a fully soaked liner, atmospheric pressure at the base of the liner, and one-dimensional infiltration, then the gradient is defined as in Figure 10.6. The conductivity in this case is apparently overestimated if the thickness of the wetted front is less than the thickness of the liner. The specific discharge in this case is equal to the rate of flow (measured volume of flow divided by the time over which it occurs) divided by the area of the cylinder. Shackelford et al. (1994) investigated the capillary barrier effect when unsaturated flow occurs through fine-grained soil overlying a coarse-grain soil. At the time the wetting front reaches the fine-grained/coarse-grained interface, continuity requires equal capillary head and flux on both sides of the interface. Because of the difference in the wetting curves, flux continuity requires that water be reflected upward into the fine-grained soil. Under such a situation, interpretation of the SDRI as per Figure 10.6 may yield too low (and hence unconservative) values of hydraulic conductivity. Complete saturation may require several years depending on the initial degree of saturation.

The thickness of the wetted front can be assessed based on data from tensiometers. Chen and Yamamoto (1989) reported an SDRI test on montmorillonitic clay (CH) where they measured suction by installing tensiometer-type probes between the two rings at different depths. The measured average suction below the wetted front was 620 cm compared to a wetted front thickness of 43 cm and a water depth of 46 cm. Reimbold (1988) on the other hand found suction to be close to zero for clay liners. Most probably the real situation would depend on the initial saturation and the compaction moisture content. Clay compacted well beyond optimum (a very common occurrence) will exhibit low suction whereas drier clay near or below optimum compaction will exhibit high suction.

Figure 10.6
Sealed double-ring
infiltrometer.

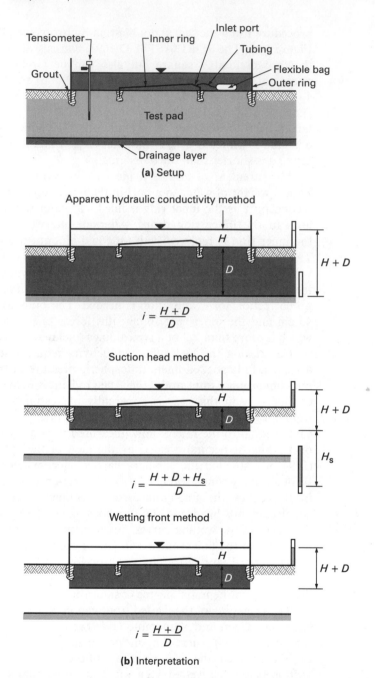

(a) Setup

Apparent hydraulic conductivity method

$$i = \frac{H + D}{D}$$

Suction head method

$$i = \frac{H + D + H_s}{D}$$

Wetting front method

$$i = \frac{H + D}{D}$$

(b) Interpretation

10.3.7.2 Test Pad   A test pad is constructed using the proposed soil and construction procedures. The dimensions of the pad may be 50 ft × 100 ft or sometimes less depending on the compactor to be used. The liner is built at least to the same thickness as called for in the specifications. A free-draining material as well as drainage pipes are recommended as a foundation for the liner. This will allow water infiltrating through the liner (if any) during the double-ring infiltrometer test to be drained.

10.3.7.3 Two-stage Borehole Permeability Tests A two-stage borehole permeability test (Daniel, 1989; Boutwell, 1992; Trautwein and Boutwell, 1994) is accepted by some regulatory agencies for measuring the hydraulic conductivity of clay liners in the field. A hole is drilled into the liner and a casing is installed and sealed. A conventional falling head test is then performed and a steady-state value of stage 1 vertical hydraulic conductivity is determined. The true vertical permeability is a function of a parameter m that relates the vertical-to-horizontal permeability. In the second stage of the test, the hole is made deeper, another falling head test is performed, and a value of permeability is determined in terms of the parameter m characterizing the horizontal conductivity. Because both vertical permeabilities are equivalent, the two expressions from both tests are solved for m to give the vertical and horizontal permeabilities.

The test setup is shown in Figure 10.7. The hole diameter is 5 cm larger than the casing. The stage 2 extension is a minimum of 1.5 times the casing diameter. The length of casing is a minimum of 2.5 diameters.

The test is simple: It involves installing a 4-in. (10-cm) casing in an oversized hole with a diameter of about 15 cm. The annulus is then grouted using bentonite pellets or other appropriate material to effect a seal. Bentonite pellets are placed in 2- to 3-cm layers and are tamped while the casing is held in place. A measuring standpipe (1/2 or 3/4 in. inside diameter) (1 to 2 cm) is

Figure 10.7
Two-stage field permeability test (after Boutwell, 1992)

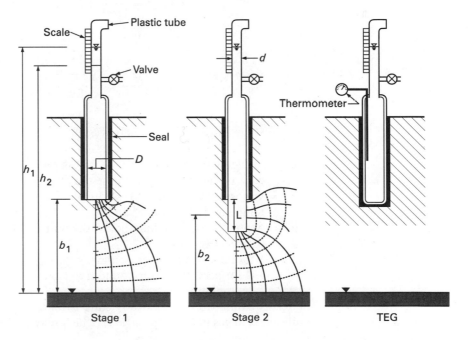

attached to the casing. To avoid hydraulic fracturing, the water height above the ground surface should be limited to no more than 1.5 times the lengths of the casing below the ground surface. A dummy test setup (temperature effect gauge or TEG) is installed with the bottom of the casing capped. Any change of water level in the standpipe of the TEG will be due to temperature fluctuations.

If the measured head after time t_2 is h_2, then the corrected head h'_2 is expressed as

$$h'_2 = h_2 - c \tag{10.2}$$

where c is the increase in TEG (dummy setup) standpipe water level during the time period t_1 to t_2, h_2 is the measured head at time t_2, and h'_2 is the corrected head at time t_2. Corrections to the calculated permeability due to viscosity changes are made using Eq. 6.4.

The plane from which the head is measured is the water table. If the water table is too deep, a limitation of $20D$ on the dimension below the bottom of the casing is imposed (see Figure 10.7). If the water table is less than $20D$, the actual dimension is used. D is usually the inside diameter of the casing unless water seeps into the casing during sealing; in this case the outside diameter is used. For the second-stage test, the inside diameter is used.

As water level drops with time, the following calculations are made:

$$k_{v1} = (G1_m)(F) \tag{10.3}$$

$$(G1_m) = \frac{\pi d^2}{11mD}\left(1 + a\,\frac{D}{4mb_1}\right) \tag{10.4}$$

$$(F) = R_c\,\frac{\ln\dfrac{h_1}{h_2}}{t_2 - t_1} \tag{10.5}$$

where
  k_{v1} = stage 1 hydraulic conductivity
  d = inside diameter (ID) of standpipe
  D = inside diameter of the casing unless the wetted surface extends to the edge of the seal, in which case the outside diameter is used
  R_c = viscosity factor (ASTM D 5084) of water at 20°C for the average temperature between t_1 and t_2
  b_1 = depth of tested medium below the bottom of the casing
  a = 0 for b_1 = infinity
       1 for impervious lower boundary
       −1 for pervious lower boundary
  m^2 = k_h/k_v (unknown)

The test is continued until a steady state is reached, which is indicated by no significant changes in k_1 (k_v for $m = 1$) (which may take few days or weeks).

After stage 1 is complete, stage 2 is initiated. The hole is advanced beneath the bottom of casing and the test is repeated. The conductivity k_{v2} is calculated as

$$k_{v2} = (F)(G2_m) \qquad (10.6)$$

where

$$G2_m = \frac{d^2}{16Lm^2(1 - 0.5623\ e^{-1.566L/D})} (2 \ln A + a \ln B + p \ln C) \qquad (10.7)$$

where

$$A = \frac{mL}{D + 2T} + \left[1 + \left(\frac{mL}{D + 2T}\right)^2\right]^{1/2} \qquad (10.8)$$

$$B = \frac{\dfrac{4mb_2}{D} + \dfrac{mL}{D} + \left[1 + \left(4\dfrac{mb_2}{D} + \dfrac{mL}{D}\right)^2\right]^{1/2}}{\dfrac{4mb_2}{D} - \dfrac{mL}{D} + \left[1 + \left(\dfrac{4mb_2}{D} - \dfrac{mL}{D}\right)^2\right]^{1/2}} \qquad (10.9)$$

$$C = \frac{\dfrac{mL}{D} + \left[1 + \left(\dfrac{mL}{D}\right)^2\right]^{1/2}}{A} \qquad (10.10)$$

where
  k_{v2} = stage 2 hydraulic conductivity
  L = length of stage 2 extension
  T = thickness of smear zone
  p = ratio of k_h to conductivity of smear zone
  b_2 = $b_1 - (L/2)$

and

$$m = (k_h/k_v)^{0.5} \qquad (10.11)$$

If the soil is isotropic ($m = 1$), the limiting values k_1 (k stage 1) and k_2 (k stage 2) are calculated from Eq. 10.3 and Eq. 10.6 for $m = 1$. If the tested soil is homogeneous and anisotropic, then k_v will not equal k_h but k_v should be the same whether calculated from Eq. 10.3 or Eq. 10.6.

To solve these equations, a value of m is assumed and the k values are calculated by trial and error. If a k calculated based on $m = 1$ is less than a specified k value, then we know that the real k would be less, so lengthy test times and interpretation would not be necessary for regulatory purposes.

For $m = 1$, the limiting values for permeabilities are

$$k_1 = F(G1_1) \qquad (10.12)$$

$$(G1)_1 = \pi \frac{d^2}{11D} \left(1 + a \frac{D}{4b_1} \right) \qquad (10.13)$$

$$k_2 = (G2)_1 F \qquad (10.14)$$

$$(G2)_1 = \frac{d^2}{16L(1 - 0.5623 \, e^{-1.566L/D})} (2 \ln A' + a \ln B') \qquad (10.15)$$

$$A' = \frac{L}{D} + \left[1 + \left(\frac{L}{D} \right)^2 \right]^{1/2} \qquad (10.16)$$

$$B' = \frac{4 \dfrac{b_2}{D} + \dfrac{L}{D} + \left[1 + \left(4 \dfrac{b_2}{D} + \dfrac{L}{D} \right)^2 \right]^{1/2}}{4 \dfrac{b_2}{D} - \dfrac{L}{D} + \left[1 + \left(4 \dfrac{b_2}{D} - \dfrac{L}{D} \right)^2 \right]^{1/2}} \qquad (10.17)$$

From Eq. 10.12 and Eq. 10.14, it follows that

$$k_{v1} = FG1_m \qquad k_1 = FG1_1 \qquad k_1 = k_{v1} G1_1 / G1_m$$
$$k_{v2} = FG2_m \qquad k_2 = FG2_2 \qquad k_2 = k_{v2} G2_1 / G2_m$$
$$k_{v1} = k_{v2}$$

and

$$\frac{k_2}{k_1} = \frac{\dfrac{G1_m}{G1_1}}{\dfrac{G2_m}{G2_1}} \qquad (10.18)$$

where $p = k_h/k_s$, k_s is the hydraulic conductivity of the smear zone, and T is the thickness of the smear zone (0.25 in. = 0.6 cm).

For $k_2/k_1 < 1.1$, Boutwell (1992) recommends the values of p shown in Table 10.5.

Table 10.5
Recommended values of p

k_2/k_1	p
>1.1	1
0.9–1.1	1.2
0.8–0.9	2.5
0.7–0.8	5.1
0.6–0.7	10.2
0.4–0.6	15.2
<0.4	Use only stage 1 approach

Source: Boutwell, 1992

Example 10.1

A double-stage permeability test (Boutell, 1992) yielded the following data. Using this data estimate k_v and k_h.

Data for stage 1

h_1 (measured from the base of a 4-ft (122-cm) liner $= 194.86$ cm
h_2 $\qquad\qquad\qquad\qquad\qquad\qquad\qquad\quad = 182.9$ cm
$t_2 - t_1 = 54{,}000$ s
Change in temperature: 26 to 19°C
TEG dropped -2.8 cm, $c = -2.8$ cm
Standpipe $d = 1.27$ cm
Inside diameter (ID) of casing $= 4$ in. $= 10.16$ cm
$\qquad\qquad\qquad\qquad\qquad D = OD = 4.5$ in. $= 11.43$ cm
$b_1 = 24$ in. $= 60.96$ cm

Solution:

Average temperature $= 22.5$°C
From Eq. 6.4 and ASTM D 5084, correction factor $= 0.94$
From Eq. 10.2, $h'_2 = 182.9 + 2.8 = 185.7$ cm
From Eq. 10.5, $F = 0.94 \ln(194.8/185.7)/54{,}000 = 8.3 \times 10^{-7}$
Consider a permeable base at b_1, $a = -1$
From Eq. 10.13, $G1_1 = 3.14(1.27)^2(1 - 11.43/4 \times 60.96)/11 \times 11.43$
$= 0.038$
From Eq. 10.12, $k_1 = 8.3 \times 0.038 \times 10^{-7} = 3.2 \times 10^{-8}$ cm/s

Data for stage 2

After completion of the first stage, an extension of 6 in. (15.2 cm) was bored
and the test was performed with the following data after a nearly
steady-state condition:

$t_2 - t_1 = 30{,}600$ s
$h_1 = 166.1$ cm
$h_2 = 154$ cm
TEG rose 1.1 cm, $c = 1.1$ cm
$L = 6$ in. $= 15.2$ cm
$b_2 = 21$ in. $= 53.3$ cm
Temperature rose from 16 to 19°C
Average temperature: $T = 17.5$°C
From Eq. 6.4 and ASTM D 5084, correction factor $= 1.07$
From Eq. 10.2, $h' = 154 - 1.1 = 152.9$
$L|D = 6/4 = 1.5$
$4b_2|D = 4 \times 21/4 = 21$

Solution:

(a) Calculate k_2 from Eq. 10.16 and Eq. 10.17.

$$A' = (1.5 + (1 + (1.5)^2)^{0.5}) = 3.302$$

$$B' = (21 + 1.5 + (1 + (22.5)^2)^{0.5})/(21 - 1.5 + (1 + (21 - 1.5)^2)^{0.5})$$

$$= 45.02/39.02 = 1.15$$

From Eq. 10.15,

$$G2_1 = ((1.27)^2/16 \times 15.2(1 - 0.5623(\exp - 1.566 \times 1.5))$$
$$\times (2 \ln A' - \ln B')$$
$$= 0.007(2.39 - 0.164) = 0.0156$$

From Eq. 10.5,

$$F = 1.07 \ln (166.1/152.9)/30600 = 2.9 \times 10^{-6}$$

From Eq. 10.14,

$$k_2 = 2.9 \times 0.0156 \times 10^{-6} = 4.52 \times 10^{-8} \text{ cm/s}$$

$$k_2/k_1 = 4.52/3.2 = 1.46$$

(b) To estimate k_v and k_h, assume $m = 2.5$.
Consider $p = 1$

$$mL/D = 2.5 \times 1.5 = 3.75$$

$$4mb_2/D = 4 \times 2.5 \times 21/4 = 52.5$$

From Eq. 10.4,

$$G1_m = ((3.14 \times (1.27)^2/11 \times 2.5 \times 11.43))$$
$$\times (1 - 11.43/4 \times 2.5 \times 60.96)$$
$$= 0.0157$$

Calculate $G2_m$ from Eq. 10.7.

$$A = 3.75 + (1 + (3.75)^2)^{0.5} = 7.63$$

$$B = 52.5 + 3.75 + (1 + (52.5 + 3.75)^2)^{0.5}/$$
$$52.5 - 3.75 + (1 + (52.5 - 3.75)^2)^{0.5}$$
$$= 112.5/97.5 = 1.15$$

$$C = 3.75 + (1 + (3.75)^2)^{0.5}/7.63 = 1$$

$$G2_m = (1.27)^2(2 \ln 7.63 - \ln 1.15 + \ln 1)/16 \times 15.2(2.5)^2$$
$$(1 - 0.5623 \exp - 1.566 \times 1.5)$$
$$= 0.0044$$

From Eq. 10.18, $(G1_m/G1_1)(G2_1/G_m) = (0.0157/0.038)(0.0157/0.0044) = 1.47$, which is close to k_2/k_1 of 1.46; no more iterations are necessary.

From Eq. 10.3,

$$k_v = F \cdot G1_m = 0.0157 \times 8.3 \times 10^{-7} = 1.3 \times 10^{-8}$$

$$k_h = k_v m^2 = 6.25 \times 1.3 \times 10^{-8} = 8.1 \times 10^{-8} \text{ cm/s}$$

10.4   GEOMEMBRANES

10.4.1 Types

The use of geomembranes (also called flexible membrane liners) is mandatory. The commonly used geomembranes (EPA, 1993) are the high density polyethylene (HDPE), very low density polyethylene (VLDPE), polyvinyl chloride (PVC), and chlorosulfonated polyethylene with fabric reinforcement (CSPE-R). Details on the manufacture and properties of these polymers is given by Daniel and Koerner (EPA, 1993).

Table 10.6 (EPA, 1993) shows these commonly used geomembranes and their approximate weight relationships. Carbon black is added to HDPE resin for ultraviolet light stabilization. Other additives are added to the HDPE for the purpose of oxidation prevention, long-term durability, and ease of manufacturing. Applicable standards are ASTM D 1248 on resin quality, ASTM D 1238 on resin density (which typically varies from 0.934 to 0.94 g/cc), and ASTM D 1238 (melt flow index). HDPE sheets are typically 15 ft (4.5 m) wide but 30-ft (9-m) wide sheets are available. Sheet thicknesses range from 30 to 120 mils (0.75 to 3 mm).

Very low density polyethylene geomembranes are made up of the polyethylene resin, carbon black, and additives. Densities of VLDPE range from 0.89 to 0.912 g/cc. Applicable standards are ASTM D 792 or ASTM D 1505 for density. VLDPE sheets are typically 15 ft wide (4.5 m) in thicknesses of 30 to 120 mils (0.75 to 3 mm).

Both the HDPE and VLDPE also are available as "textured," which enhances the interface strength with soils or other geosynthetic products. Another variation of the HDPE geomembrane is a light colored surface layer coextruded onto a black base layer. The purpose of this variation is to reduce

Table 10.6
Types of commonly used geomembranes and their approximate weight percentage formulations

Type	Resin	Plasticizer	Filler	Carbon black or pigment	Additives
HDPE	95–98	0	0	2–3	0.25–1.0
VLDPE	94–96	0	0	2–3	1–4
PVC	50–70	25–35	0–10	2–5	2–5
CSPE	40–60	0	40–50	5–40	5–15

Source: After EPA, 1993

the surface temperature if the geomembrane is exposed for a long period of time; this in turn reduces expansion and formation of wrinkles.

PVC geomembranes belong to the thermoplastics class of polymers. The resin in a white powdered form is mixed with plasticizers in a liquid form to effect flexibility and modify other physical and mechanical properties. Applicable standards are usually specified by the manufacturer. PVC sheets are 10 to 60 mils in thickness and typically produced in 40 to 80 in. (100 to 200 cm) widths. They are transported in up to 1500 lb (6.7 kN) rolls to a panel fabrication facility (EPA, 1993); typically 5 to 10 rolls make up a panel.

CSPE (also known as hypalon) resin is a thermoplastic material with enough crystallinity to maintain flexibility without plasticizers. The "R" in CSPE-R stands for reinforcement scrim, which is a woven fabric made from polyester yarn. The designation of the scrim is based on the number of strands per inch of fabric. A 10 × 10 means 10 strands per inch in each direction. The spacing is far enough apart to allow the geomembrane to adhere to an opposing membrane in the field, as verified by the ply-adhesion test. ASTM D 792 may be used as a standard for density; a minimum density for CSPE is 1.2 g/cc. CSPE is available in thicknesses from 20 to 40 mils. CSPE-R sheet widths and roll weights are similar to PVC.

10.4.2 Characteristics of Geomembranes

10.4.2.1 Physical Properties   PVC is flexible, offers good workability, and withstands larger strains than do other polymers. A design strain value of 100% is not uncommon. HDPE is less flexible and strains are typically limited to 5%. CSPE-R is in the intermediate range of 10% strain. For the same thickness, the break tensile strength for HDPE is the highest. For a 60-mil (0.06-in.) HDPE, a typical yield strength is about 140 lb/in. and a typical yield strain is 13%. Strains up to 700% (ASTM D 882) are reported for some reinforced PVC products. PVC geomembranes are rarely used in modern liners because of the documented resistance of HDPE to a wide range of chemicals. Applicable standards are ASTM D 5199 for thickness, ASTM D 638 for tensile strength and elongation (for HDPE and VLDPE), ASTM D 882 for tensile strength and elongation (for PVC), and ASTM D 751 for tensile strength and elongation (for CSPE-R). Tear resistance (ASTM D 1004, Die C) may be specified for HDPE, VLDPE, and PVC. Puncture resistance (Federal Test Method (FTM) Std 101 C) may be specified for HDPE and VLDPE. Ply adhesion (ASTM D 413, machine method type A) may be specified for CSPE-R. A guidance on specification conformance is given in ASTM D 4759. Data on various products are periodically published by the Geotechnical Fabric Report (Industrial Fabric Association, St. Paul, MN).

10.4.2.2 Sensitivity to Organic Liquids and Vapors   Polymers are organic and thus are sensitive to organic fluids or vapors. When exposed to organic fluids or vapors, they can swell, leach, or shrink. Polyethylene has been shown to be resistant to many chemicals but CSPE has poor resistance to many chemicals. Butyl rubber has poor resistance to hydrocarbons. This problem is usually investigated with compatibility testing using EPA test method 9090.

10.4.2.3 Sensitivity to Temperature   Some polymers are sensitive to temperature. At low temperatures they become brittle and at high temperatures thermoplastic polymers become soft and plastic. Butyl rubber has good high and low temperature performance. PVC exhibits poor high and low temperature performance. Both low-temperature testing and tensile testing at elevated temperatures are covered by ASTM standards (Haxo and Waller, 1987).

Thermal coefficients for polymeric material are 10 times greater than metal or concrete (Haxo and Waller, 1987). ASTM D 2102 may be used to determine contraction or expansion coefficients. HDPE, for example, has a thermal expansion coefficient in the range of 8×10^{-5} to 12×10^{-5} in./0°F (Koerner, 1992). For a 100-ft length and a 100°F change in temperature, the expansion (or required slack) is 14.4 in. For stiff (high-modulus) membranes such as HDPE, cooling could cause high stress in the liner when the membrane is restrained. In softer membranes, such as PVC, the effect is not as important.

Sheet temperatures of 0°C and 50°C (32°F to 122°F) are the lower and upper limit where the geomembrane should not be unrolled or unfolded (EPA, 1993). Sufficient slack should be maintained in cold weather to avoid the development of tensile stresses in the geomembrane. The slack should be calculated (based on expected change in temperature) to avoid excessive slack that may result in folded membrane.

White geomembranes keep the membrane temperature low on hot days, which leads to a lower temperature difference, less expansion, and fewer wrinkles. Fewer wrinkles means a better contact with the underlying soil layer and, therefore, less leakage through potential flaws in the liner.

10.4.2.4 Creep   Geomembranes, drainage nets, and polymeric materials in general are subject to creep (changed dimension) under prolonged loading. Creep also occurs if the geomembrane is initially stressed because stress relaxation occurs over time. The most practical significance of creep is in the selection and design of drainage nets; creep could substantially reduce the capacity of the net to transmit flow. Creep should also be considered when geogrids are used for reinforcement. A stress relaxation in the geogrid could result in failure of the system reinforced.

Liner components (geomembrane, pipes, drainage nets, etc.) are under continuous stress as a result of landfill loading. In high-crystalline polymers (such as polyethylene), slippage occurs within the crystalline structure, which produces creep (Haxo and Waller, 1987). For low-crystalline polymers (such as polyester), much lesser slippage occurs and creep therefore is low. The creep behavior for a given polymer depends on several factors including temperature and the level of stress applied as a fraction of the yield stress.

10.4.2.5 Stress Cracking   This phenomenon occurs in some grades of polyethylene and polyester polymers (Haxo and Waller, 1987). It refers to a crack or raze developing after prolonged exposure to stress and strain. The result is a crack in the geomembrane caused by tensile stresses less than the short-term tensile strength (ASTM D 883).

Procedures for testing for stress cracking are described by Hsuan et al. (1993). Two types of cracks are recognized (Hsuan et al., 1993): slow crack growth (SCG) and rapid crack propagation (RCP). SCG refers to crack growth rates of less than 0.1 m/s at applied stresses less than yield stress. The rate depends on the applied stress, crack length, and ambient temperature. RCP refers to much higher (about 300 m/s) crack propagation rates and larger crack lengths (on the order of tens of meters) at higher stress levels.

Hsuan et al. (1993) proposed what they called the NCTL (notched constant tensile load) test as an alternative to the ASTM bent-strip test (ASTM D 1693). In the NCTL test, a notched specimen is subjected to a constant tensile stress and a surface wetting agent at 50°C is used to accelerate cracking. Field-exhumed samples failed at less than 100 hours. Failure was characterized as brittle within a slow crack growth mechanism.

10.4.2.6 Biaxial Behavior   Geomembranes are usually tested uniaxially. Performing the test biaxially (in a machine and transverse direction) simultaneously yields considerably lower elongation values (Haxo and Waller, 1987).

10.4.2.7 Other Environmental Factors   Polymers are usually resistant to biodegradation. However, ingredients such as plasticizers (i.e., in PVC) may be subject to microbial attack. ASTM 3083 may be used to assess the effects of soil burial. Some geomembranes such as PVC and LDPE (low density polyethylene) exhibit poor resistance to long-term exposure to ozone (ASTM D 518) and ultraviolet light (ASTM D 3334). CSPE exhibits good resistance. Burrowing animals are a threat to all soil-covered membranes.

10.4.3 *Geomembrane Placement and Seaming*

The subgrade in a composite liner system is usually a clay liner component. Obviously the surface of the subgrade must be graded smoothly to result in intimate contact with the liner. The upper 6 in. of the subgrade should have no particles greater than 0.25 in. and should be smooth rounded. Ruts in the subgrade receiving the geomembranes should not be larger than 1 in. Geomembrane panels (up to 5000 lb) are rolled into place using a front-end loader or other appropriate equipment. Subsequent to this, construction equipment is not permitted to ride over the geomembrane. Liner sheets are usually placed in accord with a sheet layout detail prepared by the installer based on the design drawings. Placement of the liner is usually on a slope and the anchor trench is prepared first. The geomembrane is not typically used as a reinforcement, and a typical anchor trench such as in Figure 10.8 is sufficient.

Figure 10.8
Anchor trench

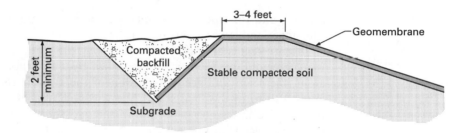

Seaming of the geomembrane sheets is a critical item that to a great extent controls the performance of the liner. Various seaming methods are shown in Figure 10.9 (EPA, 1993). Due to their high chemical resistance, polyethylenes cannot be solvent bonded and their surface hardness precludes taping. Thus, the methods used with polyethylene are extrusion welding and fusion seams.

Extrusion welding, which is similar to metallurgical welding, is used exclusively for polyethylene but is not applicable for PVC or CSPE-R. A molten ribbon of the polymer is extruded over the edge of or between the surfaces to be joined. The hot extruded polymer melts the surfaces of the sheets to be joined and the entire mass fuses together. The upper and lower sheets must be

Figure 10.9
Various methods available to fabricate geomembrane seams (EPA, 1993)

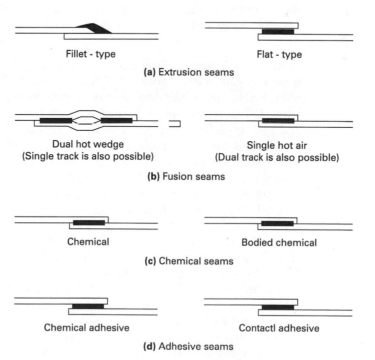

ground, and for thicknesses greater than 50 mil, the upper sheet is beveled. The upper and lower sheets may be heated.

Fusion seams are developed by melting a portion of the opposing faces of the geomembranes. This is accomplished by a hot shoe or wedge that travels along the seam and melts a portion of the two sheets. A roller pressure is applied to form the final seam. The "hot air" technology forces air between the two sheets to melt a portion of the two sheets. The thermal fusion technique (hot wedge or hot air) is applicable for all polymers.

Chemical fusion involves application of liquid chemical (bodied liquid contains 1% to 20% of the lining resin) between the two sheets and, after few seconds to soften the surface, the two sheets are pressed to form the seam. Chemical adhesive makes use of a dissolved bonding agent in a liquid chemical (or bodied liquid chemical) that is left in place after the seam is cured. Contact adhesives are applied to both surfaces of the sheets to a predetermined thickness after which one sheet is placed on top of the other, followed by application of a roller pressure. The chemical fusion and adhesive techniques are applicable to all polymers except polyethylene. Details of field seaming methods are further described in an EPA document (EPA, 1993).

10.4.4 *Seam Testing*

Seam testing involves testing of test strips, testing samples from the liner in place (destructive testing), and nondestructive testing. Seams are tested by applying tension across the seam (shear testing) and by tensioning the upper sheet and flap of the lower sheet in a 180° peeling action (peel testing). Details are given by Daniel and Koerner (EPA, 1993).

10.4.4.1 Trial Seams   Trial seams are tested during construction at specified times (typically every 4 hours) that may include the beginning of the day, afternoon break, and at the time of crew change. Test pieces 3 to 10 feet long are seamed. Two to six 1-in.-wide specimens are cut and field tested in both shear and peel. In the peel test, the specimen is tensioned until the seam pulls apart or the sheet material on either side of the seam fails (film tear bond or FTB). The seam efficiency is the seam peal strength divided by the sheet tensile strength. An efficiency of 62% is typically required for HDPE. The peel strength is typically specified at 1000 psi for VLDPE and 10 psi for PVC or CSPE-R. If any of the tests fail, more tests are made and corrective action is taken until the tests are successful. Required efficiency in shear (shear resistance of seam divided by the resistance of the unseamed sheet) is 95% for HDPE and 80% for PVC and CSPE-R; typically a seam shear strength of 1200 psi is required for VLDPE.

Considering a seam area of 1.0 in.^{2} and a shear strength of 1000 psi, the tensile strength required to fail the seam is 1000 lb/in., which is very high. Thus the strength of the geomembrane allows verification of the seam for only a small fraction of its shear strength. The same limitation applies to the peel (or tear resistance) test where the membrane strength is insufficient for adequate assessment of the bond strength between the geomembranes. Detailed discussions on this are provided by Peggs (1997).

10.4.4.2 Destructive Testing

Destructive testing is made on samples cut from the completed installation in the field. The frequency varies but typically a test is done every 300 to 500 linear ft at locations that are either predetermined or chosen by the engineer. Initially two 1-in.-wide, closely spaced specimens are cut from two locations 14 to 42 in. apart. Sample widths perpendicular to the seam are usually 12 in. The specimens are field tested for peel. If the tests are successful, then the complete seam (14 to 42 in.) is removed. A 14-in. seam may be cut into five shear and five peel specimens. A 28-in. seam may be cut in half. One half is divided into five peel and five shear specimens that are tested in the field or a laboratory, and the other half is divided into five peel and five shear specimens that are tested in a different laboratory. A 42-in. sample allows a 14-in. portion to be archived. The hole created to retrieve the test sample must be patched and tested as well (see EPA, 1993).

Applicable standards for seam peel tests are ASTM D 4437 (HDPE, VLDPE) and ASTM D 413 (PVC, CSPE-R). For seam shear testing, applicable standards are ASTM D 4437 (HDPE, VLDPE), ASTM D 3083 (PVC), and ASTM D 751 (CSPE-R). Sheet tests are covered under ASTM D 638 (HDPE, VLDPE), ASTM D 882 (PVC), and ASTM D 751 (CSPE-R).

10.4.4.3 Nondestructive Testing

To test the full length of the seam, nondestructive test (NDT) methods are employed. In the air lance method a jet of air at approximately 50 psi (350 kPa) is directed beneath the upper edge of the seam. If there is an opening in the seam, the air passes through and causes the membrane to locally inflate and flutter. The method works best on relatively thin (less than 45 mils (1.1 mm) thick) geomembranes and is therefore not recommended for the relatively stiff HDPE. The method is fast, relatively inexpensive, and applicable to all seams except the extrusion and fusion seams.

The vacuum chamber (box) method employs a box with a transparent top placed over the seam. A vacuum of 2.5 psi is applied. If the seam is defective, air enters the box from beneath the geomembrane and bubbles can be seen through the soapy solution originally placed over the seam. This method is applicable to all seams but progress is slow and cost is relatively high.

The pressurized dual system is used in conjunction with the dual hot wedge or the dual hot air seaming method (see Figure 10.9). The unbounded space is used to test the integrity of the seam. The space is pressurized with air and monitored for any drop in pressure that may be an indication of a leak. The test is fast and relatively inexpensive. Applicability is limited to the seams formed by the dual hot air or hot wedge technique.

A more recent technique is an electric leak detection system as illustrated in Figure 10.10 (Darilek et al., 1995). A voltage is impressed between one electrode in water (or soil) covering the liner and a grounded electrode (or an electrode in the leak detection system beneath the liner). Because geomembranes are electrical insulators, electrical flow will only occur if there is a

Figure 10.10
Principle of the
electrical leak location
method (Darilek et al.,
1995)

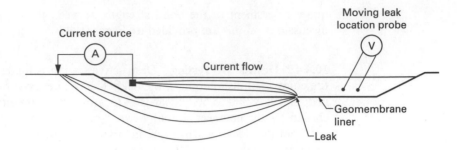

leak. Using this system, Darilek et al. (1995) reported 9.1 leaks/acre for a total of six uncovered liners. Leaks were mostly in extrusion welds. The four largest leaks were punctures and slits in liner panels. Double fusion welds had no leaks. Undamaged portions of the liners were those placed under good quality control and protected by layers of geonets and geotextiles before placing a pea gravel drainage layer. Areas experiencing leaks were those where gravel was placed directly over the liner.

10.4.5 Backfilling

The soil layer above the geomembrane is usually a drainage layer with relatively large particle size to meet drainage requirements. To avoid puncture of the geomembrane, a specified maximum size of 1/2 in. of rounded (no sharp edges) particles are placed in at least 6-in. lifts and spread (carefully and avoiding sharp turns) using low ground contact pressure equipment of 5 psi (35 kPa). It is preferable that the first lift be a relatively thick lift (12 in.) to help reduce damage to the geomembrane. As explained in Section 10.4.4, an intervening geotextile would help reduce damage to the geomembrane.

10.5   GEOSYNTHETIC CLAY LINERS (GCLs)

A geosynthetic clay liner (GCL) is a thin clay layer (usually sodium bentonite) supported by geotextiles or geomembranes. Figure 10.11 (EPA, 1993) shows the available GCLs.

The clay thickness varies from 0.16 to 0.32 in. (4 to 6 mm) and the weight per unit area varies from 0.66 to 1.2 psf (0.032 to 0.06 kN/m^2). The widths of factory-made GCLs vary from 7 to 17 ft (2.2 to 5.2 m). The requirements for subgrade preparations (minimal rutting, restrictions on oversized particles) are similar to those cited for geomembrane placements. GCLs are placed by hand or by lightweight equipment. Overlap distance varies from 6 to 12 in. (150 to 305 mm).

Advantages of GCLs are ease in placement and quality control, less volume compared to a 3-ft. clay liner, and more space for waste disposal. However, they are best suited for landfill plateaus or liners where slopes are

Figure 10.11
Cross section sketches of
currently available
geosynthetic clay liners
(GCLs) (EPA, 1993)

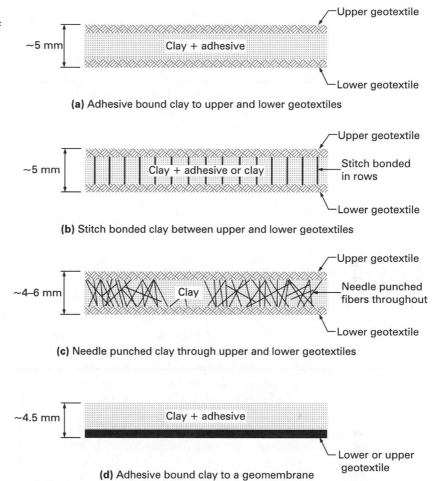

(a) Adhesive bound clay to upper and lower geotextiles

(b) Stitch bonded clay between upper and lower geotextiles

(c) Needle punched clay through upper and lower geotextiles

(d) Adhesive bound clay to a geomembrane

small (less than 5%). Mild slopes (generally less than 10% are preferred) because of the low shear strength of the hydrated bentonite clay. For geotextile-encased bentonite, the strength of the hydrated bentonite may govern the design; the effective friction angle is usually less than 10°. Also, the thin nature of GCLs requires substantially more control in the field (compared to a thick clay liner) to avoid damage by construction equipment.

Reinforced GCL offers higher shear resistance compared to GCLs without reinforcement (Gilbert et al., 1996). Reinforcement is provided by connecting the upper and lower geotextiles by needle-punched geotextile fibers. The shear strength depends on displacements, strength of the bentonite, and the pullout strength of the reinforcing fibers. For design application, the residual strength usually governs specially for cap design. Gilbert et al. (1997) conducted experiments that sheared GCL alone, and the interface with textured geomembrane and with geonets. Sheer displacements up to 48 mm were reached but the residual strength was not attained (i.e., strength continued to drop with increased displacement).

In general, the hydraulic conductivity of intact GCL is low (10^{-8} to 10^{-9} cm/s). Considering the small thickness of the GCL, the hydraulic gradient across the GCL will be larger than that of conventional clay liner—2 to 3 ft thick for a given head. For example, for a head of 12 in. over a 0.25-in. GCL, the gradient is $(12 + 0.25)/0.25$ whereas for a 3-ft clay liner, the gradient is $(1 + 3)/3$. Other factors to consider are the bearing capacity (bentonite should not be allowed to squeeze) and response to differential settlements. GCLs derive their tensile resistance by virtue of the geosynthetic that encases the bentonite and possible reinforcement. Other considerations are discussed in EPA (1993).

10.6   VERTICAL BARRIER WALLS

10.6.1 *Function and Types*

Vertical barrier walls are usually used to improve existing waste disposal facilities that have no modern liner and/or leachate collection system. Figure 10.12 shows an application of this technology. Table 10.7 shows the principal types of sealing mixtures used in cutoff wall construction (Manassero et al., 1995).

Figure 10.12 shows a section of an old landfill with a leachate mound resting on a somewhat pervious stratum which in turn overlies a low permeability material. A vertical wall of low permeability is installed around the facility and is keyed into the low permeability soil. So that the leachate will not back up and spill over the wall, a leachate collection system is placed at the inward (landfill) side of the wall. As illustrated in Figure 10.12, the invert of the leachate collection pipe is placed at a level below the lowest mean groundwater level on the outboard side of the wall. In this case, a hydraulic barrier is created and the flow direction would be inward (i.e., ingradient). Even if a defect exists in the wall, no leachate is released (but the flow to the collection drain would be higher).

The most common type of wall used in the United States is the soil-bentonite (SB) wall. The mixed in place (MIP) wall, cement-bentonite (CB) wall, plastic concrete (PC) wall, and others are also used. All but the MIP wall require a trench dug under a slurry of water and bentonite to stabilize the trench or, in the case of CB walls, to add cement and bentonite to form the wall after hydration. Conventional excavating equipment (a standard backhoe) may be used to excavate the trench for depths down to about 30 ft (10 m). For greater depths (down to 50 ft), an extended boom is fitted to the backhoe for deeper reach. A clamshell or a dragline is needed for depths greater than about 60 ft (see Figure 10.13). The width of the trench is typically 3 ft. Modern construction utilizes pug mills (see Figure 10.14) where the mix can be controlled to a large degree.

Figure 10.12
Ingradient design
using a vertical
barrier wall

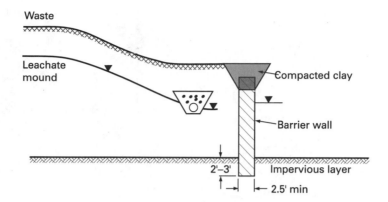

Table 10.7
New trends in backfill
sealing mixtures

Additives and main compounds	Improvements and advantages
Sodium metacrilates, alkaline silicato-alluminates and other dispersive chemical additives for CB and plastic concrete mixtures	Decreased hydraulic conductivity; increased resistance against acid compounds
Calcium bentonite, attapulgite in soil–bentonite mixtures	Constant or decreased with time hydraulic conductivity when permeated with organics
Filling materials (fly ashes, furnace slags, minerals, other by-products) for CB mixtures	Increased unit weight, decreased void ratio, decreased hydraulic conductivity, decreased diffusion coefficient, increased chemical resistance, decreased unit cost
Microfine cement in CB mixtures	Decreased hydraulic conductivity and diffusion coefficient
Treatment of clays with ammonium cations for CB and soil–bentonite mixtures. Use of other sorbent materials.	Increased sorption capacity

Source: After Manassero et al., 1995

Figure 10.13
Slurry wall excavation
operation

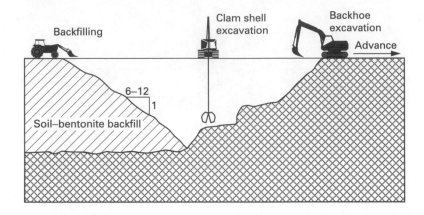

10.6.2 *Flow Across Slurry Walls*

Assuming constant but not equal heads on the inboard and outboard sides of
the wall (Figure 10.15), the volume of flow across the wall is estimated based
on the familiar Darcy's specific discharge equation as follows:

$$q = ((H_i - H_o)/s)k_w \qquad\qquad (10.19)$$

where q is the volume of flow per unit surface area of the wall, H_i is the
hydraulic head on the inboard side, H_o is the hydraulic head on the outboard
side, s is the thickness of the wall, and k_w is the hydraulic conductivity of the
wall.

Figure 10.14
Modern pug mill for
mixing wall ingredients
(courtesy Conti
Construction)

Figure 10.15
Flow across a vertical
barrier wall

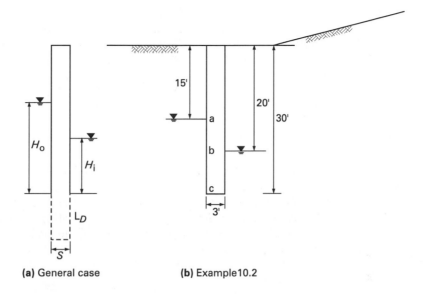

(a) General case   **(b)** Example10.2

Example 10.2

Consider a wall 30 ft (9.15 m) deep. The water level outboard is 15 ft (4.6 m) below the top of the wall and 20 ft (6.1 m) inboard. Estimate the flow across the wall if the hydraulic conductivity of the wall is 10^{-7} cm/s and the wall is 2000 m long 3 ft wide.

Solution: From Figure 10.15, at level a $(H_i - H_o) = 0$, $q = 0$ and at level b, $(H_i - H_o) = -5$ ft, $q = -(5/3)k_w$. The average flow per unit length through length a to b is $-(5/6)k_w$ (6.1 m − 4.6 m).
    For the flow from level b to level c $(H_i - H_o) = 5$, $q = -(5/3)k_w$.
    Thus for the total flow we have

$$\text{From a to b: } Q = -(5/6)(10^{-9}\text{ m/s})(1.5)(2000) = -2.5 \times 10^{-6}\text{ m}^3/\text{s}$$

$$\text{From b to c: } Q = -(5/3)(10^{-9}\text{ m/s})(3.05)(2000) = -10 \times 10^{-6}\text{ m}^3/\text{s}$$

$$\text{Total flow: } -12.5 \times 10^{-6}\text{ m}^3/\text{s}$$

If the average head difference is $(H_i - H_o)$ and the porosity of the slurry wall is assumed to be 0.5, then the seepage velocity across the wall is simply $2q$ and the breakthrough time is:

$$t = s^2/2 \cdot k_w(H_i - H_o) \tag{10.20}$$

The penetration of the wall into the low permeability soil is typically specified by state regulations (1 m to 1.5 m). Assuming that the average conductivity along the interface with the low permeability soil is k_b and a breakthrough time is specified to be as much as calculated across the wall, then the minimum depth of embedment is estimated as:

$$\frac{(2L_D + s)^2}{k_b} = \frac{2s^2}{k_w}$$

$$L_D = (s/2)((2k_b/k_w)^{0.5} - 1) \tag{10.21}$$

where L_D is the depth of embedment into the low permeability stratum.

As discussed in Chapter 6, transport by chemical diffusion could be dominant in low permeability soils. This may be the case for contaminant migration across slurry walls. For example, if the concentration of a chemical species in a leachate is c_0 inside the wall and c_1 outside the wall, then the steady-state flux of contaminants across the wall is given by (Shackelford, 1993)

$$J_D = -((c_1 - c_0)/s)D^*n \tag{10.22}$$

where J_D is the contaminant flux across the wall ($M/L^2 \cdot T$), c_0 and c_1 are the concentrations of the chemical species (M/L^3) inside and outside the wall, respectively, D^* is the effective diffusion coefficient (L^2/T), and n is the porosity.

Example 10.3

Consider the same wall described in Example 10.2, but with inward and outward fluid levels at ground surface. Assume that $c_0 = 100$ mg/l, $c_1 = 10$ mg/l, $n = 0.5$, and $D^* = 10^{-9}$ m^2/s. Compute the flux of contaminants (J_D) across the wall, the surface area of the wall, and the volume of contaminant passing through the wall.

Solution:

$$c_0 = 100 \times 1000 \text{ mg/m}^3, \; c_1 = 10 \times 1000 \text{ mg/m}^3$$

$$J_D = -((10 - 100)/0.91) \times 1000 \times 10^{-9} \text{ m}^2/\text{s} \times 0.5$$

$$= 49.45 \times 10^{-6} \text{ (mg/m}^2/\text{s)}$$

$$= 49.45 \times 10^{-6} \times (31.536 \times 10^6 \text{ s/yr}) = 1559.45 \text{ mg/m}^2/\text{y}$$

The surface area of the wall is 18,300 m^2 and the volume of contaminant passing through the wall is 1559.45 × 18,000 or 28.1 kg/y.

As illustrated in the above two examples, contaminant exits the wall by diffusion in a direction opposite to groundwater flow (ingradient). In the diffusion example, a retardation coefficient of 1.0 was assumed, which is applicable for chlorides.

10.6.3 Stability

The stability of the trench is evaluated by the methods of Chapter 13. The equilibrium of the mass tending to slide into the trench is considered together with the resisting force afforded by the slurry in the trench. Considering the general condition in Figure 10.16, the equation to be solved is

$$((\gamma H^2/2)\cot \alpha - P_w \cos \alpha)M + P_w \sin \alpha = \gamma_s H_s^2/2 \qquad \textbf{(10.23)}$$

where
$\quad M = (\tan \alpha - \tan \phi/F)/(1 + \tan \alpha \tan \phi/F)$
$\quad \phi =$ friction angle of the soil
$\quad \gamma =$ unit weight of the soil
$\quad \gamma_s =$ unit weight of the slurry
$\quad F =$ factor of safety
For the special case of dry soil $((P_w = 0), H_s = H)$,

$$(\gamma/\gamma_s) \cot \alpha = 1/M$$

Example 10.4

Consider a trench in dry sand with a friction angle of 30°, sand unit weight of 115 pcf, and slurry unit weight of 67 pcf. The wall is 30 ft deep. Assume that the wedge angle, α, is compatible with an active wedge $(45 + \phi/2 = 60°)$ and that the slurry is at the ground surface level. Estimate the factor of safety.

$$\gamma/\gamma_s \cot \alpha = (125/67)(0.58) = 1.08$$

Figure 10.16
Trench stability

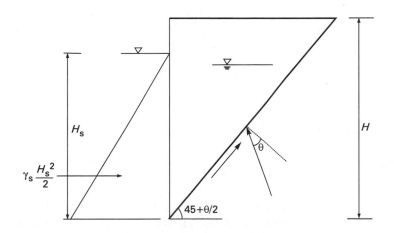

If we try $F = 1.8$, we get

$$M = (1.73 - 0.58/1.8)/(1 + 1.73 \times 0.58/1.8) = 0.9$$

$$1/M = 1.11 > 1.08$$

which is too large. If we try $F = 1.9$, we get

$$M = (1.73 - 0.58/1.9)/(1 + 1.73 \times 0.58/1.9) = 0.93$$

$$1/M = 1.075$$

which is a better estimate.   ———————

In the above calculations for dry sand, the calculated factor of safety is independent of the wall depth. If for the above problem the ground water is at a level 15 ft above the bottom of the wall (and $H_s = H$), then

$$((\gamma/\gamma_s) \cot \alpha - M' \cos \alpha) = (1 - M' \sin \alpha)/M$$

where

$$M' = P_w/(\gamma_s H_s^2/2)$$

$$= 62.4 \times 15(15/\sin 60/2)/(67 \times 900/2) = 0.27$$

The left-hand side (LHS) of the equation is equivalent to $1.08 - 0.27(0.5) = 0.945$. The factor of safety is computed by trial and error such that the right-hand side (RHS) is equivalent to 0.945. If we try $F = 1.5$, we get

$$M = (1.73 - 0.58/1.5)/(1 + 1.73 \times 0.58/1.5) = 0.806$$

$$1 - M' \sin 60/M = 1 - 0.27 \times 0.86/0.806 = 0.95$$

The LHS and RHS are numerically close and no further refinement is necessary.

10.6.4 *Slurry* Properties

As illustrated in the above calculations, the safety of the trench is dependent on the unit weight of the slurry used, which can vary from 65 to 90 pcf (10.2 to 12.5 kN/m³). The slurry is usually prepared in a mixing pond and pumped to the trench. A bag of bentonite typically weighs 100 lb. For a given mix (lb of bentonite/gal of water), the resulting unit weight can be estimated from the following expression:

$$\gamma_s = 7.48m_b + \gamma_w/(1 + (7.48m_b/G_s\gamma_w)) \tag{10.24}$$

where m_b is the bentonite content (lb of bentonite/gal of water), G_s is the specific gravity of bentonite (2.77), γ_w is the unit weight of water (lb/ft^3), and γ_s is the unit weight of slurry (lb/ft^3).

If three 100-lb bags of bentonite are mixed with 500 gal of water, then $m = 0.6$ and for $\gamma_w = 62.4$, the resulting unit weight is 65.2 lb/ft^3. The percent of bentonite by weight typically ranges from 4 to 8%. The unit weight of the slurry is measured using the procedure in ASTM D 4380.

Viscosity of the slurry is usually specified at 40 to 50 s as measured with a Marsh funnel and cup (API 13 B). The fluid is poured into a funnel to a prescribed level and the time required to discharge the slurry of 1 qt volume (946 ml) is measured. Water has a Marsh viscosity of 26 s at 23°C. The sand content by volume (ASTM D 4381) may be specified because sand may slough from the sides of the trench. In some cases, the slurry is "sanded" when trenching through porous refuse to avoid losing the slurry. In this case, quality control measures should ensure that the bottom of the trench is free of sand before introducing the SB mix. The sand content should not be more than 14% by volume or 5% by weight during excavation.

Filtrate loss is typically specified to assess the sealing capability. The test procedure involves dispersing 22.5 g of bentonite in 350 ml of deionized water, mixing it for about 20 min then allowing 16 to 24 h for hydration. Slurry is then introduced into a chamber with porous stone and filter paper at the bottom. A 100 psi (690 kPa) gas pressure is applied and filtrate collected in a cup. The test (per the API RP 13 B procedure) at 100 psi is continued until a cake of bentonite (the filter cake) develops. The volume of water that flows out of the cake in a 30-minute test is called the *filtrate*. Specifications on API filtrate vary but typically less than 20 ml is required. A limit of 25 ml is typically specified for cutoff wall construction. The pH during excavation is specified in the range of 7 to 10. The reasons for the various specification items are listed in Table 10.8.

Slurry trenches are often excavated adjacent or in close proximity to structures or utilities. Under good control, lateral displacement could vary from 0.3 to 0.5% of the depth of the wall. It is prudent to maintain a slurry depth several feet higher than the water table. If the water is under confined condition with piezometric heads near or above ground surface, it may be necessary to build a platform to compensate for increased water pressure; otherwise, the slurry wall option may not be feasible.

10.6.5 Soil Bentonite (SB) Walls

The slurry-stabilized trench is filled with a soil–bentonite mixture. To facilitate construction, the mix usually has a slump of about 6 in. (15 cm). The mix is designed to achieve a specified hydraulic conductivity requirement (usually 10^{-7} cm/s). The soil gradation controls the percentage of the added bentonite

Table 10.8
Reasons for
specification
parameters

Parameter	Reason
Density	Trench stability
Sand content	To avoid premature settling of sand particles
pH	Bentonite in acid water is not properly hydrated. High pH also diminishes hydration.
Viscosity	Characterizes how thick the slurry is and its ability to carry particles in suspension
Filtrate loss	Characterizes formation of a usually low permeability filter cake on trench wall

and hence the performance of the wall. In some specifications, the maximum particle size is limited to 3/4 in. (19 mm) or 1/2 in. (12.7 mm). The percentage of bentonite added depends on the fine content (percentage by weight passing a No. 200 sieve). The amount of bentonite by dry weight may vary from 1 to 5% and the moisture content for a 6-in. slump may range from 25 to 35%. The unit weight of the mix may range from 105 to 120 pcf (16.5 to 18.9 kN/m^3). The fine content is typically higher than 10% but lower than 50%. Figure 10.17 shows the hydraulic conductivity versus fine content without bentonite, and Figure 10.18 illustrates the influence of bentonite content.

A typical plan for SB installation consists of a slurry preparation area with a water source (in tanks or a lined pond) and a pond or a tank for storage and hydration of the slurry used in the backfill mixture. The mixing of the soil, bentonite, and water is best done using a pug mill (Figure 10.14). Performing the mix by tracking the ingredients along the sides of the trench requires a great deal of quality control. Free dropping the mix into the slurry is not recommended as it may cause segregation of the mix.

Excavating the trench requires an experienced backhoe operator. The length of the trench open at any time should be within the reach of the backhoe to allow continuous cleaning of the trench before introducing the SB mix. The SB mix from the pug mill may be transported in a regular concrete mixer truck. The toe of the SB backfill slope (Figure 10.19a) at the end of each shift should be at or near the excavation point. This slope is typically 1 vertical on 6 horizontal. When resuming excavation, the top of the SB backfill is cleaned with the excavation equipment. To start the operation, the SB mix is placed by tremie. Subsequently, the mix from the truck is placed 15 to 30 ft (6 to 9 m) from the slurry–backfill interface (Figure 10.19b).

Prior to introducing the SB mix, the depth of the trench is recorded after cleaning at a spacing of no more than 50 ft (15.25 m). The required depth of the trench is determined from interpretation of boring data and field observations during excavation by an experienced engineer and from classification of the materials excavated.

Figure 10.17
Permeability of backfill versus fine content (reproduced from Ryan, 1987, by permission)

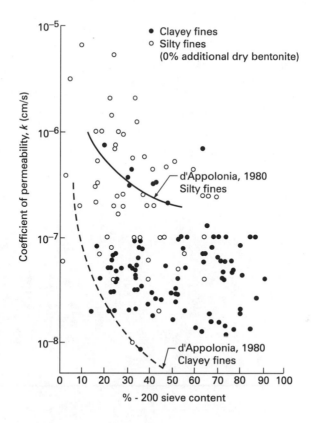

Figure 10.18
Permeability of SB backfill versus bentonite content (reproduced from d'Appolonia, 1980, by permission)

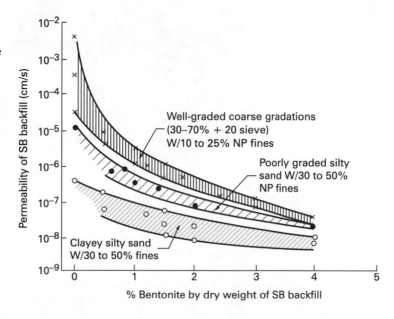

Figure 10.19
(a) Construction progress
(b) Construction of an
SB wall

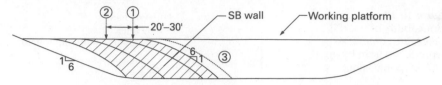

① Position of edge of backfill

② New backfill placed 20'–30' away from top edge

③ New front of wall as ② is placed

(a) Construction progress

(b) Delivery

10.6.6 *Other Wall Types*

In situations where some structural strength is required, consideration is given to cement–bentonite (CB) walls, plastic concrete walls, or other wall types. In the cement–bentonite wall, the slurry (in this case water, cement, and bentonite) is left to harden in place to form the wall and, unlike SB walls, excavated material is not used. The strength of the wall is related to the cement–water ratio (typically 10 to 20%). The higher this ratio is, the stiffer and more brittle the wall. Higher bentonite content (typically 4 to 6%) produces more ductile (i.e., higher strains at failure) but weaker walls (see Figure 10.20). The strength of a CB wall increases with time, as illustrated in Figure 10.20.

The hydraulic conductivity of CB walls is higher than that of SB walls (typically 10^{-5} to 10^{-6} cm/s). It also decreases with increased curing time, which places a burden on the designer to specify a hydraulic conductivity that can be verified during construction. The diffusion coefficient is comparable to compacted clay although the porosity is 2 to 3 times higher (Manassero et al.,

Figure 10.20
Strength deformability relationships for cement–bentonite slurries (reproduced from Millet et al., 1992, by permission)

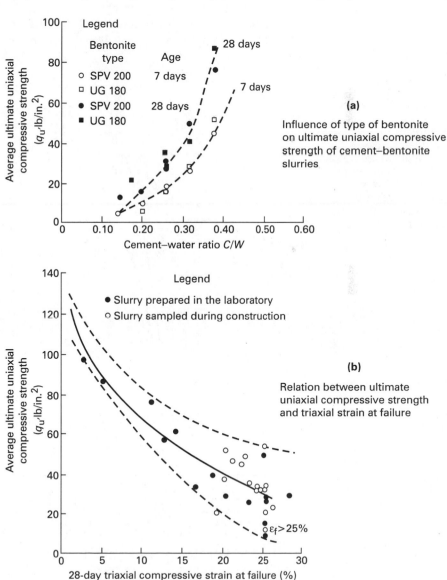

1995). Manassero et al. (1995) reported a diffusion coefficient for a CB mixture of 6.3×10^{-10} m²/s compared to 3.8×10^{-10} m²/s for clay. The retardation factor of 1.22 relative to ethanol for CB is small (essentially no sorption) compared to 2.1 for compacted clay. This is due to the much higher porosity of CB mixtures.

Plastic concrete walls are formed by a mixture of cement, bentonite, water, aggregates, and additives. This mixture, which varies in size from clay to fine-to-medium gravel, is introduced (tremied or pumped) to a slurry-filled trench and replaces the slurry. The aggregates impart more strength and ductility; hence the adjective *plastic*. As illustrated in Figure 10.21 (Davidson et al., 1992), ductility increases as strength decreases. The strength of the wall need not be higher than the surrounding soil. The dry portion of the mix is typically 75 to 85% by weight with fines (silt and clay) making up 30 to 40% of the dry

Figure 10.21
Representative unconfined compression test results on slurry wall backfill mixes (reproduced from Davidson et al., 1992, by permission)

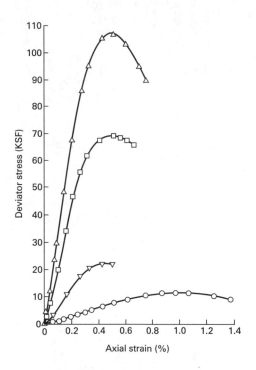

Legend

Mix	B/W	C/W	C/A	S/A	B/A	w	γd	Aggregate
O Cement bentonite	6	0.38	—	—	—	180	28.2	—
□ Mix A plastic conc.	4.7	0.6	16.7	23.1	1.3	17.5	112.7	45% CWR 55% NS
△ Mix B plastic conc.	3.5	0.68	16.3	9.8	0.8	15.6	116.5	45% CWR 55% NS
▽ Mix E plastic conc	6	0.6	16.7	30.1	1.7	25.3	97.6	Other proj. pit

run
B Bentonite weight   A Aggregate weight   γd Dry unit weight (lb/ft³)
W Water weight   S Bentonite-water weight   CWR Coarse waste rock weight
C Cement weight   w Water content   NS Natural sand weight

mixture. The largest particle size is usually less than 1/2 in. (12.5 mm) to maintain workability. The result is also low hydraulic conductivity (10^{-7} to 10^{-8} cm/s) after two to three weeks.

In situ soil mixing (Day and Ryan, 1994) is a relatively new technique for constructing a vertical barrier wall. Deep soil mixing (DSM) involves advancing multiple augers (2 to 8), with diameters of 0.8 to 1.0 m. Slurry or grout is pumped through the hollow-stem auger and mixing is accomplished by a counter — rotating while moving the tooling up and down through the soil.

10.6.7 Compatibility with Waste Fluid

A key design issue in developing a vertical barrier wall is the compatibility of the wall materials (cement, bentonite, and fine-grained soil) with the waste fluid. (See the procedures in Chapter 6.) Figure 10.22 shows the results of hydraulic conductivity tests on soil–bentonite samples recovered after 8 years from an existing SB wall. Tap water, leachate, and salty water were used as permeants. Apparently no adverse effect can be observed due to leachate or salt water.

Figure 10.22
Effect of various permeants on the hydraulic conductivity of an aged SB wall mix

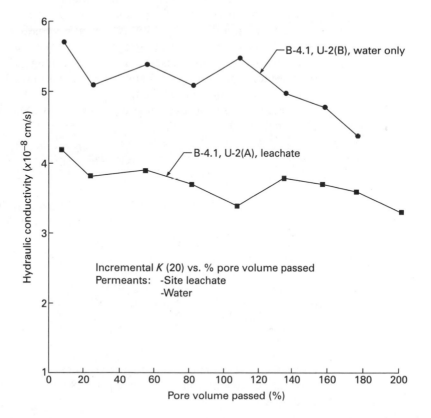

10.6.8 Wall Verification

The usual procedure for wall verification is to recover backfill samples from various depths and locations and conduct permeability tests. Recovering good quality samples requires drilling 4-in. holes with the possible need of drilling mud to stabilize the hole. This, in theory at least, could result in hydraulic fracturing although there are no reliable data to support this theory. Utilization of the piezocone (Chapter 7) could present a good alternative. The piezocone minimizes disturbance to the wall, eliminates the potential problem of hydraulic fracturing, and is less costly. An example of a piezocone profile in an SB wall is shown in Figure 10.23 (Oweis et al., 1994).

In Figure 10.23, the upper 5 to 10 ft is fill material with high point and sleeve resistance. Below that is the SB material with low point and sleeve resistance and excess pore water pressure increasing with depth. The occasional increase in tip and sleeve resistance with a corresponding drop in pore water pressure is indicative of thin sand layers. At or near the design depth, the point and sleeve resistance suddenly increases and the pore water pressure suddenly drops. This is not consistent with the clay-impervious layer that the wall was keyed into. In this case, the trench may have collapsed, or the bottom of the wall was not intersected. This could happen if a deep but relatively narrow wall is out of plumb, causing the cone to penetrate a portion of the wall but exit in a layer that was not the layer the wall is keyed into.

With the piezocone, dissipation tests may be conducted in the wall to assess its hydraulic conductivity. A typical dissipation test is shown in Figure 10.24. To achieve full dissipation of excess pore water pressure in low permeability soil, the test must be done over a long time period. However, the time needed can be shortened after establishing the initial portion of the curve (e.g., $t = 2000$ s in Figure 10.24). The dissipation ends (i.e., the curve will be horizontal) at ambient hydrostatic pressure. The coefficient of consolidation is estimated based on

$$c_h = T_{50} r^2 / t_{50} \tag{10.25}$$

where

c_h = coefficient of consolidation in the horizontal direction
T_{50} = dimensionless factor corresponding to 50% consolidation
r = radius of the piezocone probe
t_{50} = time elapsed from start of dissipation test to 50% dissipation of excess pore water pressure
T_{50} = 3.5 for a cone-sensing device or 5.5 for a side-sensing piezocone with a filter located immediately above the cone tip (Baligh and Levadoux, 1980)

Example 10.5

Given the dissipation curve of Figure 10.24 and a cone-sensing pore water pressure device with a cone diameter of 35.8 mm, determine the coefficient of consolidation.

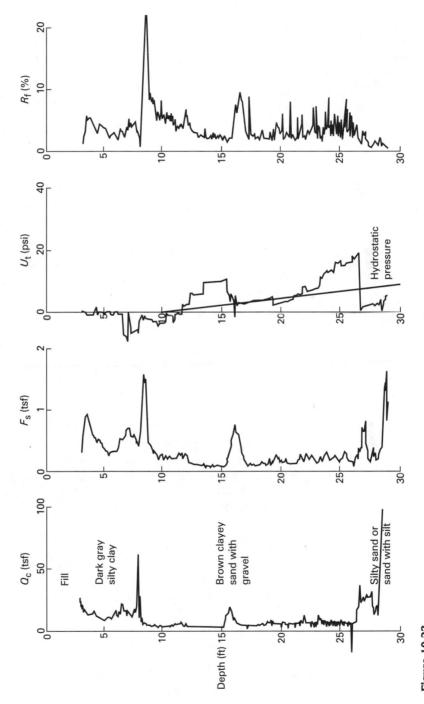

Figure 10.23
Example of a piezocone sounding in an SB wall (reproduced from Oweis et al., July 1994, pp. 740, 741, 742, by permission)

Figure 10.24
Typical dissipation
curve in an SB slurry
wall using the
piezocone (reproduced
from Oweis et al., July
1994, by permission)

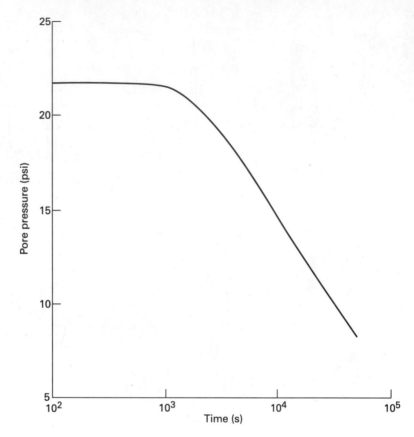

Solution: From Figure 10.24,
Initial pore water pressure = 21.8 psi
Ambient (final) pore water pressure = 7 psi
Pore water pressure at 50% dissipation:
7 + (21.8 − 7)/2 = 14.4 psi
t_{50} = 10,000 s = 167 min

Thus,

$$c_h = 3.5(3.58/2)^2/167 = 0.067 \text{ cm}^2/\text{min}$$

The hydraulic conductivity is defined in terms of the coefficient of consolidation as

$$k_h = c_h m_v \gamma_w \tag{10.26}$$

where k_h is the hydraulic conductivity in the horizontal direction, $m_v = 1/D$, the volume compressibility, D is the constrained modulus, and γ_w is the unit weight of the pore fluid.

The constrained modulus may be empirically estimated based on the point resistance q_c (Meigh, 1987). For low-plastic clays with q_c less than 7 tsf, the

Table 10.9
Field tests for slurry wall verification

Test	Advantage(s)	Disadvantage(s)
Field tracers	Full scale tests	Only major defects can be observed
Pumping tests inside a small confined area	Large scale tests	Expensive; only part of the confinement system is tested in a preliminarily defined location
Geophysical field measurements (electrical, thermal, etc.)	Large scale tests	Difficult setup: Tests are easily disturbed in polluted and/or highly industrial areas. Difficult correlations between different flow coefficients (e.g., hydraulic and electrical conductivity).
New interpretation approaches or equipments for infiltration tests: bored or preinstalled piezometers, B.A.T. systems, self-boring permeameters	Tests on the *in situ* slurry wall	Small volumes are involved. Expensive and time-consuming approach if a significant number of tests are requested. Attention must be paid to hydraulic fracturing.
Composite geosynthetic (geomembrane + geonet) inserted into the slurry wall	Large or full scale tests	Only feasible for composite slurry walls
Piezocone tests (CPTU)	Fast and cost effective. Continuous permeability log vs. depth can be obtained and calibrated by dissipation tests interpretation.	Penetration is not possible in hard backfilling materials. Further validation of interpretation procedure is still necessary.

Source: After Manassero et al., 1995

range of D/q_c is from 7 to 10. For high-plastic silt and clays, the range is from 2.5 to 7.5 (Meigh, 1987). For SB walls such correlations are not available. However, a reasonable estimate of D can be obtained from the consolidation test based on the compression index as follows:

$$D = \sigma'/0.435(C_c/1 + e_0) \tag{10.27}$$

where D is the constrained modulus, σ' is the average effective stress, C_c is the compression index, and e_0 is the initial void ratio.

Example 10.6

Given that $q_c = 6$ tsf, $C_c/(1 + e_0) = 0.15$, the sounding is at a depth of 26 ft, and the unit weight of the SB wall material is 110 pcf, estimate the hydraulic conductivity:

Solution:

At 50% dissipation, pore pressure $u = 14.4$ psi (from Example 10.5)
Effective stress $= (110 \times 26/144) - (14.4) = 5.46$ psi
Initial effective stress $= (110 \times 26/144) - (7) = 13$ psi
$\sigma' = (13 + 5.46)/2 = 9.23$ psi
$D = 9.23/(0.435 \times 0.15) = 141.5$ psi $= 20,370$ psf
$k_h/c_h = (1/20,370)(62.5 \text{ lb/ft}^3) = 3.1 \times 10^{-3} \text{ ft}^{-1} = 10^{-4} \text{ cm}^{-1}$
$c_h = 0.067 \text{ cm}^2/\text{min} = 1.12 \times 10^{-3} \text{ cm}^2/\text{s}$
$k = 1.12 \times 10^{-7} \text{ cm/s}$

Using an empirical $D = 7q_c$, then $D = 42$ tsf (840,000 psf) and the estimated hydraulic conductivity is:

$$k_h = 1.12 \times 10^{-7}(20,370/84,000) = 2.7 \times 10^{-8} \text{ cm/s} \quad \rule{2cm}{0.4pt}$$

Another way to assess the integrity of the wall is to conduct pumping tests in suspected areas. In this case, a well is installed outboard and piezometers are installed inward to monitor drawdowns during pumping. Large drawdowns inside the wall as a result of the pumping leads to the conclusion that the wall is ineffective. The suspect segment of the wall could be identified using piezocone soundings.

Additional procedures for wall verification are listed in Table 10.9 (Manassero et al., 1995).

10.6.9 Cracking

For walls subjected to drying, cracking and increase in hydraulic conductivity could result. Fluctuation of the water table tends to produce drying and shrinkage of the wall. In walls constructed to mitigate gas transmission, drying may occur as a result of gas pumping on the inboard or outboard side of the wall. The problem of drying is usually handled in design by constructing a cap several feet thick over the wall to keep the wall permanently wet. There is no set criteria for this procedure but maintaining a low fluctuation in temperature is necessary.

10.7   TRANSIT TIME

In assessing the thickness of clay liners, it is usually necessary to restrict the transit times to no less than 30 or 50 years. Contaminants in leachate can penetrate and exit the liner by advection or seepage of fluid through the liner under a hydraulic gradient, or by chemical diffusion, or both. In chemical diffusion, chemicals are transported under a concentration gradient (see Chapter 6).

For the liner case, both transport mechanisms occur. Due to conditions at the top and base of a liner, there exists a hydraulic gradient causing fluid flow. There also exists a concentration gradient causing a particular species in the landfill to transport from higher concentration (top of liner) to lower concentration (base of liner).

For transport by seepage through the liner (advection), the simplest way to predict the transit time is to assume that the fluid permeates the liner while the suction head at the wetting front is zero. The hydraulic gradient in this case is

$$i = (y_{max} + d)/d \qquad \text{(10.28a)}$$

and the specific discharge (Darcy velocity) is

$$q_1 = k_1 i \qquad \text{(10.28b)}$$

The seepage velocity from Eq. 6.7 is

$$v_s = q_1/n_e \qquad \text{(10.28c)}$$

where d is the liner thickness, n_e is the effective porosity available for flow, and y_{max} is the maximum leachate depth above liner.

Thus the transit time is

$$t = d/v = n_e d^2/(k_1(d + y_{max})) \qquad \text{(10.29)}$$

In Eq. 10.29, the effective porosity is usually less than the total porosity determined in the laboratory because soil channels with dead ends do not contribute to n_e.

If a suction is considered at the base of the liner, then

$$t = n_e d^2/k_1(y_{max} + d - h_s) \qquad \text{(10.30)}$$

where h_s is the suction head and is equal to or less than zero such that $-h_s$ is zero or larger.

The Green-Ampt model (Green and Ampt, 1911) describes moisture movements in unsaturated soil during ponded infiltration by assuming a wetting front moving down the soil column and a constant suction of capillary tension at the wetting front. Considering that z is the depth of penetration of the wetting front and h_s is the suction head (negative number) at the base of the wetting front, then for time dt the volume of water infiltrated is equated to the change in the volumetric water content. The flow equation is

$$(\theta_s - \theta_i)dz = k_u((y_{max} + z - h_s)/z)dt \qquad \text{(10.31)}$$

where θ_i is the initial volumetric water content, θ_s is the saturation volumetric moisture content ($=$ porosity), and k_u is the unsaturated hydraulic conductivity.

Eq. 10.31 can be readily integrated by parts over the thickness of the liner d for the time required for liner saturation:

$$t = ((\theta_s - \theta_i)/k_u)\{d - [y_{max} - h_s]\ln(y_{max} - h_s + d)/(y_{max} - h_s)\} \quad \textbf{(10.32)}$$

This formula requires estimating the suction and the unsaturated permeability of the liner, which may be determined experimentally (Olson and Daniel, 1981). ASTM D 2325 (capillary–moisture relationship) applies to coarse- and medium-grained soils whereas ASTM D 3152 applies to fine-grained soils. Empirical methods for assessing unsaturated permeability are discussed in Chapter 11.

Example 10.7

Consider a liner 90 cm thick with a saturated hydraulic conductivity of 10^{-7} cm/s and effective porosity of 0.3. The specified maximum fluid head above the liner is 30 cm.

a. Determine the transit time assuming a 5-cm suction.

b. Determine the transit time when the unsaturated conductivity is 10^{-8} cm/s, the initial volumetric moisture content is 30%, porosity is 0.45, and the suction is 38 cm.

Solution:

a. $t = 0.3 \times 90 \times 90/(10^{-7}(30 + 90 + 5)) = 1.944 \times 10^8$ s $= 6.2$ yr

b. $h_s = -38$ cm

$$t = (0.45 - 0.3) \times 10^8(90 - (30 + 38)\ln(30 + 38 + 90)/(30 + 38))$$

$$= 2.78 \times 10^8 \text{ s} = 8.8 \text{ yr}$$

In general, the mathematics of unsaturated flow problems is nonlinear. Details on the mathematical treatment of these problems and the quasi-analytical solutions are summarized elsewhere (Philip, 1958; EPA, 1988; GCA Corp., 1984). The advantage of the analytical or numerical treatment is that the assumption of the nature of the suction at the base of the liner can be avoided and the change of moisture content with depth is gradual.

As described above, the actual transport mechanism through the liner is either advection and diffusion or the dispersion mechanism described in Chapter 6.

An intact composite liner is characterized by a very low overall hydraulic conductivity because of the geomembrane component. However, this would not prevent diffusion through the geomembrane. Gray (1995) cites cases of low molecular weight solvents in dilute aqueous solution passing through intact HDPE liners with thicknesses of 0.76, 1.52, and 2.54 mm in about 1, 4, and 13 days, respectively. In tests reported by Haxo (1991), a complete breakthrough of trichloroethylene (TCE) and toluene through 33-mil (0.84-mm) LLDPE membrane occurred in about 10 days.

Because the transit time varies with the square of the liner thickness, it is clear that thicker liners afford increased protection. Most natural clays have poor sorption affinities for low molecular weight, nonionic, nonpolar organic solutes (Gray, 1995). Gray (1995) suggested adding small amounts (1 to 2%) of organic carbon (such as found in high-carbon fly ash) to increase the sorptive capacity.

10.8 LEAKAGE FROM LINERS AND COVERS

10.8.1 Regulatory Guidance for Liners

In its final double liner and leak detection systems rule for hazardous waste facilities, the EPA requires that the owner or an operator submit for approval an action leakage rate, or ALR. The ALR is defined as "the maximum design flow rate that a leak detection system (LDS) can remove without the fluid head on the bottom liner exceeding 1 ft." The action leakage rate must include an adequate safety margin to allow for uncertainties in the design (e.g., slope, hydraulic conductivity, thickness of drainage material); construction, operation, and location of the LDS; waste and leachate characteristics; likelihood and amounts of other sources of liquids in the LDS; and proposed response action (e.g., the action leakage rate must consider decrease in the flow capacity of the system over time resulting from siltation and clogging, creep of synthetic components of the system, overburden pressures, etc.) (EPA, 1992). The objective of the rule is to minimize the buildup of high heads on the bottom (second) liner and thereby "decrease the potential for migration of hazardous constituents out of the unit should a leak develop in the upper liner." The corresponding leakage rate (with a safety factor of 2 suggested) is termed as resulting from a "rapid and extremely large leak."

In 1992 an EPA supplemental background document offered empirical formulation for assessing the ALR. This guidance assumes that the hole in the linear is large enough to produce the calculated flow. Assuming a small leak from a liner defect, the width of the wetted zone may be derived based on the assumption that the gradient of the flow at the hole is equal to the slope of the LDS and that the depth of flow at the hole is equal to the thickness of the drainage layer. Hence

$$B = D/\sin \alpha = D/\tan \alpha \quad \text{(for small alpha values)} \quad \text{(10.33)}$$

where D is the leak detection system thickness, α is the slope of the leak detection system, and B is the average width of the flow in the LDS perpendicular to the flow. The flow through the LDS can be expressed as

$$Q = kh \tan \alpha \cdot B \quad \text{(10.34)}$$

where Q is the flow rate in the LDS, k is the hydraulic conductivity of the drainage layer, and h is the head on the liner beneath the LDS.

If the detection system is made of sand and gravel, the head h is less than the layer thickness D and the thickness of the flow is equal to h. Eq. 10.34 in this case reduces to

$$Q = kh^2 \qquad \text{for } h < D \tag{10.35}$$

In the case of a geonet, the head h is larger than the thickness of the geonet (D in this case). Hence the flow thickness is D and flow over a thickness ($h - D$) is deducted from Q in Eq. 10.36:

$$Q = kD(2h - D) \qquad \text{for } h > D \tag{10.36}$$

Example 10.8

A leak detection system consists of a sand layer having a hydraulic conductivity of 0.1 cm/s. A defect (hole) is expected on the average of one per acre. Propose an action leakage rate using a safety factor of 2.0.

Solution:

$$h = 30 \text{ cm} = 0.3 \text{ m}$$

$$k = 0.001 \text{ m/s}$$

From Eq. 10.35,

$$Q = 0.001(0.3)^2 = 0.00009 \text{ m}^3/\text{acre/s}$$

$$= 0.00009(3.28 \text{ ft/m})^3 \times (7.48 \text{ gal/ft}^3) \times (86{,}400 \text{ s/day})$$

$$= 2052 \text{ gal/acre/day}$$

$$\text{ALR} = Q/2 = 1026 \text{ gal/acre/day} \qquad \rule{2cm}{0.4pt}$$

Example 10.9

A geonet has a transmissivity (kD) of 3×10^{-5} m²/s. Assuming an allowable head buildup of 30 cm, determine the ALR if the design thickness is 6 cm and one leak per acre is expected.

Solution:

$$Q = 3 \times 10^{-5}(0.6 - 0.06) = 1.62 \times 10^{-5} \text{ m}^3/\text{s}$$

$$= 369 \text{ gal/acre/day}$$

$$\text{ALR} = 369/2 = 185 \text{ gal/acre/day} \qquad \rule{2cm}{0.4pt}$$

There may be sources of flow other than waste in an LDS, as illustrated in Figure 10.25 (EPA, 1992). If unusual volumes of fluids are detected in the

Figure 10.25
Sources of flow in a
leak detection,
collection, and removal
system (LDCRS) (EPA,
1992)

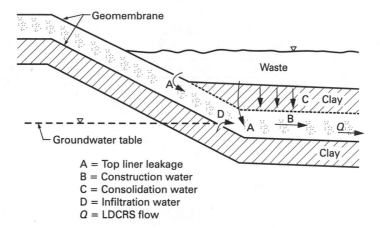

A = Top liner leakage
B = Construction water
C = Consolidation water
D = Infiltration water
Q = LDCRS flow

LDS, an investigation must be done to determine the source. This may include evaluating the quality of the fluid and comparing it with the expected leachate quality as well as reviewing various aspects of construction including the construction records and quality control.

10.8.2 *Flow through Imperfections in a Geomembrane*

10.8.2.1 Analytical Relationships The leakage mechanism presumes a gap between the membrane and soil (Figure 10.26), which cannot be prevented even in controlled laboratory experiments (Brown et al., 1987). Brown et al. (1987) developed analytical procedures for predicting the flow through a circular hole in a liner that is in contact with an underlying soil layer. The head acting on the liner was assumed to be small in comparison with the thickness of the underlying soil layer such that the hydraulic gradient through the soil layer is close to 1.0. Thus:

$$i = (h + d)/(d) = 1 \qquad \text{for } h \ll d \qquad (10.37)$$

where i is the hydraulic gradient, h is the head on top of the soil layer underlying the membrane, and d is the thickness of the soil layer.

In this procedure the vertical thickness of the gap is estimated. The radius of the gap is found by equating the sum of energy losses to the head acting on the top of the liner. The head losses are calculated as

$$H_1 = 0.6835(k^2/g)(R/r_0)^4 \qquad (10.38)$$

$$H_2 = (8\eta' k T_1/\rho \cdot g \ r_0^2)(R/r_0)^2 \qquad (10.39)$$

$$H_3 = [3\eta' k/\rho \cdot g \cdot t_g{}^3]r_0{}^2\{[(R/r_0)^2(2 \ln R/r_0 - 1)] + 1\} \qquad (10.40)$$

Figure 10.26
Forces acting on an
element of fluid

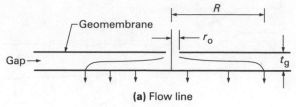

(a) Flow line

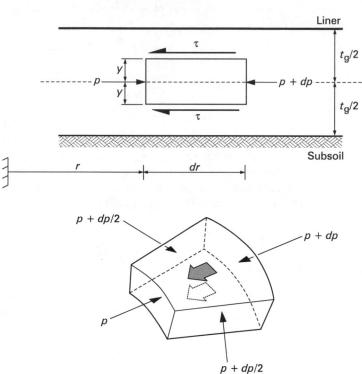

(b) Analytical model

where

H_1 = head loss due to convergence at the opening (cm)
H_2 = head loss due to viscous resistance to flow through the thickness of the liner
H_3 = head loss through the gap
r_0 = radius of the opening in the liner (cm)
R = radius of the gap beneath the liner (cm)
k = hydraulic conductivity of the soil beneath the liner (cm/s)
g = acceleration of gravity ($= 980$ cm/s^2)
ρ = density for water ($= 1.0$ g/cm^3)
η' = 8.95×10^{-3} poises for water
T_1 = membrane liner thickness (cm)
t_g = vertical thickness of the gap (cm)

Under the extreme conditions where the gap is too wide and/or the supporting soil is highly permeable, the head loss H_3 becomes very small. Considering that head loss H_2 is small in comparison with H_1, the problem reduces to a flow through an orifice and the head loss H_1 may be expressed as (Brown et al., 1987)

$$H_1 = 1.367v^2/2g = H \tag{10.41}$$

The flow rate through the opening is

$$Q = \pi \cdot R^2 v \tag{10.42}$$

where H is the total head on the liner (cm), v is the flow velocity through the opening (cm/s), and Q is the flow rate through the opening (cm^3/s).

Analyses similar to that by Brown et al. (1987) may be applied to the leakage with planar flow where the width of the imperfection is assumed to be $2b_0$ (see Figure 10.27). Considering the equilibrium of a fluid element within the gap beneath the liner, the following expression can be established:

$$\tau = -y\, dp/dx = -\eta'\, du/dy \tag{10.43}$$

where
    p = fluid pressure at a distance x from the center line
    y = distance from the middle plane of the fluid element
    τ = viscous shear on any plane at a distance y from the middle plane
    u = lateral flow velocity at a distance y from the middle plane
    η' = absolute viscosity of the fluid

Figure 10.27
Planar flow through a gap beneath a geomembrane

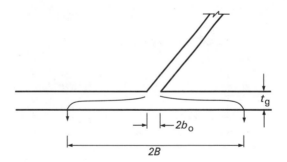

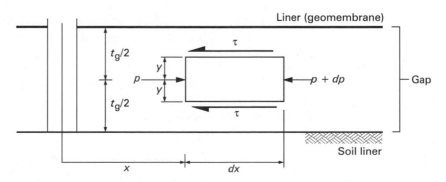

Integrating Eq. 10.43 gives

$$u = -(1/2 \cdot \eta')(dp/dx)(t_g^2/4) - y^2) \tag{10.44}$$

where t_g is the thickness of the gap beneath the liner.

If the total flow rate is Q_1^*/unit length, then the flow to one side of the imperfection is $Q_1^*/2$:

$$Q_1^* = 2 \int_{-t/2}^{t/2} u \cdot dy = -\frac{1}{\eta'} \frac{t^3}{6} \frac{dp}{dx} \tag{10.45}$$

The flow at a distance x is the flow $Q_1^*/2$ at $x = b_0$ minus the leakage:

$$Q^*/2 = Q_1^*/2 - kx \tag{10.46}$$

Noting that

$$Q_1^* = 2(B - b_0) \cdot k$$

integration results in

$$p_3 = (6\eta' k/t_g^3)(2B^2 + 2b_0^2 - 4Bb_0) \tag{10.47}$$

where p_3 is the fluid pressure drop associated with flow through the gap.

The quantities involving b_0 in Eq. 10.47 are usually small compared with B^2 and may be omitted. The head drop H_3 may be expressed as

$$H_3 = (12\eta' k/\rho \cdot g \cdot t_g^3)(B^2) \tag{10.48}$$

The head loss H_1 can be expressed by noting that the flow velocity is Bk/b_0 and then using Eq. 10.41 (Brown et al., 1987):

$$H_1 = 1.367v^2/2g \qquad \text{for } v = Bk/b_0 \tag{10.49}$$

The head loss through the membrane is usually small and can be ignored. However, it may be estimated using the Moody diagram (Brown et al., 1987).

10.8.2.2 Empirical Relations   Based on the work by Brown et al. (1987), Bonaparte et al. (1989) and Schroeder et al. (1994) cited the empirical relationships in Table 10.10 for leakage through a circular hole in a geomembrane under various contact conditions. The relationships are sensitive to the thickness of the gap along the imperfect contact. Brown et al. (1987) found that it is difficult to maintain a perfect contact even in the laboratory. They also concluded that larger gaps are associated with more rigid liners and more permeable soils. In developing guidance tables, they used the following empirical relationship relating gap thickness to permeability.

$$t_g = \log^{-1}(-0.811 + 0.292 \log k_s) \tag{10.50}$$

where t_g is expressed in cm and k_s in cm/s.

Table 10.10 Empirical relations for estimating flow from geomembrane flaws

Condition	Relationship
Geomembrane with flaws not in perfect contact with controlling soil	$R = C_1(a_0)^{b_1}(h)^{b_2}(k_s)^{b_3}$ $q_f = 0.877 k_s i_{av} \pi R^2$

	C_1	b_1	b_2	b_3
Excellent contact				
$10^{-4} \le k_s < 0.1$ cm/s	0.97	0.38	0.38	-0.25
$k_s < 10^{-4}$ cm/s	0.5	0.05	0.5	-0.06
Good contact	0.26	0.05	0.45	-0.13
Poor contact	0.61	0.05	0.45	-0.13

Condition	Relationship
Geomembrane with defect under free-flow condition (high permeability controlling soil)	$q_f = C_b n_f a_0 \sqrt{2gh}$
Geomembrane underlain by a geotextile over controlling soil	$R = \left(\dfrac{4 h t_f k_f}{k_s \left[2 \ln \dfrac{R}{r_0} + \left(\dfrac{r_0}{R} \right)^2 - 1 \right]} \right)^{1/2}$
Geomembrane in perfect contact	$q_p = \pi k_s h(2 r_0)/(1 - r_0/h_s)$ $i_{av} = 1 + (h/(2 h_s \ln R/r_0)$

where

- h = hydraulic head on the geomembrane (m)
- h_s = thickness of the controlling soil layer
- R = radius of interfacial flow around a geomembrane flaw (m)
- r_0 = radius of geomembrane (flaw (m)
- a_0 = flaw area (m²)
- q_f = interfacial flow leakage (m³/s) per flaw, flaw not in perfect contact (m³/s)
- t_f = thickness of geotextile
- q_p = leakage rate through a flaw in perfect contact with controlling soil (m³/s)
- k_s = hydraulic conductivity of controlling soil (m/s)
- k_f = hydraulic conductivity of geotextile (m/s)
- i_{av} = average hydraulic gradient on wetted area of controlling soil layer
- C_b = head loss coefficient (=0.6 for sharp-edged orifice)
- n_f = number of flaws
- g = acceleration of gravity (m/s²)

Condition	Relationship
Perfect contact	No gap or interface between geomembrane and controlling soil
Excellent contact	Exceptional geomembrane placement over well-prepared low permeability soil Cohesionless soil conforming to geomembrane GCL attached to geomembrane with good foundation
Good contact	Geomembrane installed with as few wrinkles as possible on adequately compacted low permeability soil with smooth surface
Poor contact	Geomembrane installed with a certain number of wrinkles on a poorly compacted low permeability soil that does not appear to be smooth

Example 10.10

A circular hole 0.08 cm in diameter is assumed to exist in a liner. Using a flexible liner thickness of 0.051 cm, determine the flow rate for $k = 3.4 \times 10^{-4}$ cm/s and 3.4×10^{-7} cm/s if the head above the liner is 10 cm. The thickness of the controlling soil layer is 60 cm. Use both the analytical and empirical methods and assume good contact. Assume also $\eta' = 8.95 \times 10^{-3}$ poise (1 poise = 1 dyne s/cm^2).

Solution:
Analytical method:
For $k_s = 3.4 \times 10^{-4}$ cm/s, from Eq. 10.50 we have

$$t_g = \log^{-1}(-0.811 - 3.47 \times 0.292) = 0.015 \text{ cm}$$

If we try $R = 16.04$ cm, we get

$$H_1 = 0.6835(3.4)^2 \times 10^{-8} \times (16.04/0.04)^4/980 = 2.08 \text{ cm}$$

$$H_2 = 8 \times 8.95 \times 10^{-3} \times 3.4 \times 10^{-4} \times 0.051 \times (16.04/0.04)^2/980(0.04)^2$$
$$= 0.127 \text{ cm}$$

$$H_3 = (3 \times 8.95 \times 10^{-3} \times 3.4 \times 10^{-4}(0.04)^2/980 \times 0.015^3)$$
$$\times ((16.04/0.04)^2(2 \ln (16.04/0.04) - 1) + 1) = 7.79 \text{ cm}$$

With $H = 10$ cm we can use $R = 16.04$ cm.

$$Q = 3.14 \times (16.04)^2 \times 3.4 \times 10^{-4} = 0.275 \text{ cm}^3/\text{s}$$

For $k_s = 3.4 \times 10^{-7}$ cm/s, from Eq. 10.50 we have

$$t_g = \log^{-1}(-0.811 + 0.292 \log 3.4 \times 10^{-7}) = 0.002 \text{ cm}$$

If we try $R = 26.7$ cm we get

$$R/r_0 = 26.7/0.04 = 667.5$$

$$H_1 = 0.6835(3.4)^2 \times 10^{-14}(667.5)^4/980 = 1.6 \times 10^{-5} \text{ cm}$$

$$H_2 = (8 \times 8.95 \times 10^{-3} \times 3.4 \times 10^{-7} \times 0.051/980 \times 0.04^2)(667.5)^2$$
$$= 3.5 \times 10^{-4}$$

$$H_3 = (3 \times 8.95 \times 10^{-3} \times 3.4 \times 10^{-7}(0.04)^2/980(0.002)^3)$$
$$(667.5^2(2 \ln 667.5 - 1) + 1) = 10$$

With $H \approx 10$ cm, we can use $R = 26.7$ cm.

$$Q = 3.14 \times 26.7^2 \times 3.4 \times 10^{-7} = 7.6 \times 10^{-4} \text{ cm}^3/\text{s}$$

Empirical method:

$$k_s = 3.4 \times 10^{-4} \text{ cm/s}$$

$$a_0 = 3.14(0.04 \times 10^{-2})^2 = 5.024 \times 10^{-7} \text{ m}^2$$

$$R = 0.26 \times (5.024 \times 10^{-7})^{0.05} \times (0.1)^{0.45} \times (3.4 \times 10^{-6})^{-0.13}$$

$$= 0.26 \times 0.48 \times 0.35 \times 5.14 = 0.22 \text{ m} = 22 \text{ cm}$$

$$i_{av} = 1 + 0.1/2 \times 0.6 \ln 22/0.04 = 1.013$$

$$q_f = 0.877 \times 3.4 \times 10^{-6} \times 1.013 \times 3.14(0.22)^2 = 0.46 \times 10^{-6} \text{ m}^3/\text{s}$$

$$= 0.46 \text{ cm}^3/\text{s}$$

$$k = 3.4 \times 10^{-7} \text{ cm/s}$$

$$R = 0.26 \times 0.48 \times 0.35 \times (3.4 \times 10^{-9})^{-0.13} = 0.55 \text{ m}$$

$$q_f = 0.877 \times 3.4 \times 10^{-9} \times 1.013 \times 3.14(0.55)^2 = 2.9 \times 10^{-9} \text{ m}^3/\text{s}$$

$$= 0.0029 \text{ cm}^3/\text{s}$$

Example 10.11

A membrane on a soil base with a hydraulic conductivity of 10^{-5} cm/s developed what was believed to be a long crack 0.1 cm wide. The head above the liner is 0.3 m. Estimate the leakage rate.

Solution: From Eq. 10.50, we estimate the gap thickness.

$$t_g = \log^{-1}(-0.811 + 0.292 \log 10^{-5}) = 0.00536 \text{ cm}$$

Assume H_1 is too small to be verified later. From Eq. 10.48, we have

$$B^2 = 30 \times 980 \times (0.00536)^3/(12 \times 8.95 \times 10^{-3} \times 10^{-5}) = 42.15 \times 10^2$$

$$B = 64.9 \text{ cm}$$

$$Q^* = 2 \times 64.9 \times 10^{-5} = 0.0013 \text{ cm}^3/\text{s/cm length}$$

$$v = 0.0013/0.1 = 0.013 \text{ cm/s}$$

$$H_1 = 1.367(0.013)^2/2 \times 980 = 1.18 \times 10^{-7} \text{(negligible)}$$

Slurry walls with defects will have higher effective permeability than will intact walls. A composite wall may be specified to effect very low hydraulic conductivity for the same reason a geomembrane is used in a composite liner. As is the case for the liner, the effectiveness of the geomembrane depends on the quality of the joints between the sheets of membrane. Figure 10.28 illustrates the adverse effect of a thin layer of high permeability material within a slurry wall. The flow through the wall could be several orders of magnitude

Figure 10.28
Wall without a
geomembrane
(reproduced from
Manassero, 1994,
by permission)

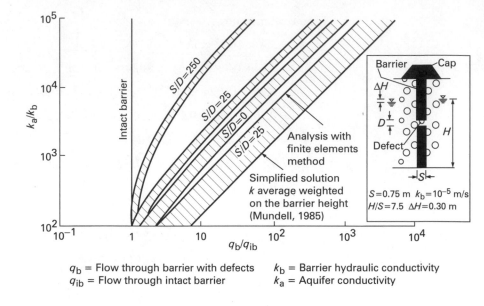

q_b = Flow through barrier with defects k_b = Barrier hydraulic conductivity
q_{ib} = Flow through intact barrier k_a = Aquifer conductivity

higher than that for an intact wall. Such pervious inclusions could develop as a result of cave-ins during construction or poor keying of the wall into a low permeability layer.

10.8.3 *Liner Efficiency*

McEnroe (1989) presented approximate formulas for estimating the liner efficiency in terms of the ratio of leakage through the liner to the infiltration volume (Figure 10.29). Considering that the liner thickness is much greater than the head above the liner, the infiltration rate through the liner becomes equivalent to the hydraulic conductivity of the liner. After developing an empirical equation from the analytical results of the thick liner approximation, a correction was made for the effect of the liner thickness using Wong's (1977)

Figure 10.29
Definition sketch for
liner efficiency

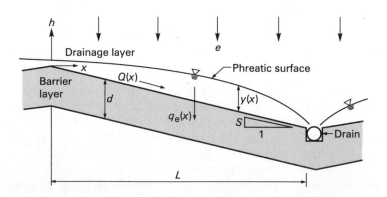

model. The relationship by McEnroe is

$$\frac{q_1}{q_0} = \left(1 + \frac{y_{av}}{d}\right)\left(\frac{k_1 t_0}{q_0} + 0.64 \frac{n_e \, y_{av}}{q_0} M_0{}^{0.91}\right)$$ **(10.51)**

where $M_0 = k_1 L/k_d \cdot S \cdot y_{av}$. Here,

$$y_{av} = y_s\left[1 - \exp\frac{(k_1 - e)t_0}{n_e y_s}\right]$$

where $y_s = (e - k_1)L/(2S \, k_d)$ for m less or equal to 1. Now,

$$y_s = \frac{\pi L}{4}\sqrt{\frac{e - k_1}{k_d}}\,(0.403)^{m'}$$

where
$m\ \ = (e - k_1)/0.4S^2 k_d$
$m'\ = (m)^{-0.55}$
$q_0\ =$ inflow volume per unit horizontal area (L) $= et_0$
$e\ \ =$ rate of inflow to the drainage layer (discharge per unit horizontal area) (L/T)
$t_0\ \ =$ duration of inflow period (T)
$q_1\ \ =$ leakage volume per unit horizontal area (L)
$y_{av} =$ average saturated depth during time t_0
$y_s\ \ =$ steady-state average saturated depth
$d\ \ =$ thickness of liner (L)
$k_1\ \ =$ vertical hydraulic conductivity of the liner (L/T)
$k_d\ \ =$ lateral hydraulic conductivity of the drainage layer (L/T)
$n_e\ \ =$ effective porosity of the drainage layer
$L\ \ =$ maximum horizontal drainage distance (L)
$S\ \ =$ slope of liner
The liner efficiency is defined as

$$E = 100(1 - q_1/q_0)$$ **(10.52)**

In a composite liner, the geomembrane is assumed to be impervious and flow can only occur through imperfections. The area over which flow occurs through the clay liner is therefore less than the area of the liner. If we consider a circular imperfection or a long imperfection, the radius or the width of the wetted area and the plan flow area is calculated using the methods of Section 10.8.2. The results are incorporated into the above analysis by defining an adjusted conductivity of the liner to be used in Eq. 10.51 based on

$$k_1' = k_1 \cdot A_w/A_1$$ **(10.53)**

where A_w is the wetted area beneath the geomembrane and A_l is the plan area of the liner.

For a soil liner without a geomembrane on top, the ratio A_w/A_l is 1.0.

Example 10.12

Determine the leakage ratio q_1/q_0 using the following parameters,

e = 0.5 cm/day over 10-day period
S = 0.02
$k_1 = 10^{-7}$ cm/s
$k_d = 0.01$ cm/s
d = 90 cm
n_e = 0.4
L = 100 m

Solution:

e = 0.5 cm/day = 57.87×10^{-7} cm/s
m = $56.87 \times 10^{-7}/0.4 \times 0.0004 \times 10^{-2} = 3.55$
m' = $(3.55)^{-0.55} = 0.5$
y_s = $(3.14 \times 100 \times 100/4)((56.87 \times 10^{-7})/10^{-2})^{0.5}(0.403)^{0.5} = 119$ cm
t_0 = 10 days = 0.0864×10^7 s
y_{av} = $119(1 - \exp - 56.87 \times 0.0864/0.4 \times 119) = 11.7$ cm
M_0 = $10^{-7} \times 100 \times 100/(10^{-2} \times 0.02 \times 11.7) = 0.43$
q_0 = $0.5 \times 10 = 5$ cm
k_1t_0 = 0.0864
q_1/q_0 = $(1 + 11.7/90)(0.0864/5 + (0.64 \times 0.4 \times 11.7/5)(0.43)^{0.91})$
= $1.13(0.0173 + 0.28) = 0.34$
E = $100(1 - 0.34) = 66\%$

If the inflow is applied instantaneously, then $t_0 = 0$.

y_{av} = $5/0.4 = 12.5$ cm
M_0 = $0.43(11.7/12.5) = 0.402$
q_0/q_1 = $(1 + 12.5/90)(0.64 \times 0.4 \times 12.5/5) \times (0.402)^{0.91} = 0.32$
E = $100(1 - 0.32) = 68\%$

SUMMARY

Liners are broadly classified as soil liners, geomembranes, or geosynthetic clay liners (GCL) where soil (bentonite) is sandwiched between two sheets of fabric. The soil liners are either compacted clay liners (CCL) or soil (silty sand) mixed with bentonite. Vertical liners (barrier wall) are designed as soil bentonite, cement bentonite, plastic concrete, or other low-permeability mixtures.

Low hydraulic conductivity is a requirement for all liners. A conductivity of 10^{-7} or lower can be achieved for soil liners. Geomembranes are virtually watertight and their hydraulic conductivity is controlled by defects in seaming and holes in the membrane. Polyethylene products are used extensively as geomembrane because of their resistance to many chemicals. Such products may experience stress cracking. Liquids, vapors, and gases can permeate geomembrane by way of molecular diffusion. GCL offers low conductivity and small thicknesses, but the shear strength of hydrated bentonite is low. This limits application of GCL to gentle slopes.

The performance of soil liners is controlled by moisture content, plasticity, and density. The plasticity index should be larger than 10 but less than 25, and the soil should be compacted wet of the optimum. The hydraulic conductivity is verified by testing cores recovered from the field, two-stage field permeability tests, and Sealed Double Ring Infiltrometer tests. Geomembrane performance is tested by seam testing in the field and laboratory and leak detection methods. Vertical walls are verified by testing undisturbed samples, pumping tests, piezocone soundings, and other techniques. The performance of a vertical barrier wall is controlled by the hydraulic conductivity of the mix, good tie into an impervious layer, and prevention of pervious windows or defects during construction.

Regulations place a limit on leakage through liners in terms of an action leaking rate (ALR). Relationships are provided in this chapter for estimating leakage rates through liners and liner efficiency.

NOTATIONS

b_0 half-width of imperfection in liner (L)

B half-width of gap (L)

C_b head loss coefficient

C_w undrained shear strength at moisture content w

c change in temperature effect gauge from time t_1 to t_2

c_0 concentration of chemical species inside

c_1 concentration of chemical species outside

c_h coefficient of consolidation in horizontal direction (L^2/T)

C_c compression index

D inside or outside diameter of casing (L)

D constrained modulus (F/L^2)

e_0 initial void ratio

D^* effective diffusion coefficient

D leak detection system thickness in Eq. 10.33

d inside diameter of standpipe (L)

d liner thickness

H_1 head loss due to convergence at opening (L)

H_2 head loss due to viscous resistance through hole (L)

H_3 head loss through gap (L)

H height of slurry wall trench (L)

F factor of safety

H_s height of slurry inside slurry wall trench (L)

h head (L)

h_1 head at time t_1

h_2 head at time t_2

h_s suction head at base of liner (L)

h_s thickness of controlling layer in Table 10.10

H_i hydraulic head — inboard side of a slurry wall (L)

H_0 hydraulic head — outboard side of a slurry wall (L)

J_D contaminant flux (M/L^2 · T)

k_h horizontal hydraulic conductivity (L/T)

k_v vertical hydraulic conductivity (L/T)

k_1 k_{v1} for $m = 1$

k_2 k_{v2} for $m = 1$

k_{v1} stage 1 hydraulic conductivity (L/T)

k_{v2} stage 2 hydraulic conductivity

k_s	hydraulic conductivity of smear zone (L/T)	D	inside or outside diameter of casing (L)	t_0	duration of inflow (T)
k_s	hydraulic conductivity of controlling soil in Table 10.10	r	radius (L)	v	velocity (L/T)
		q_f	leakage rate through a flaw not in perfect contact (L^3/T)	u	lateral flow velocity (L/T)
k_f	hydraulic conductivity of geotextile	q_p	leakage rate through a flaw in perfect contact (L^3/T)	y_{max}	maximum leachate buildup above liner (L)
k_w	hydraulic conductivity of slurry wall	q_0	flow volume per unit area (L)	y_{av}	average saturated depth above liner during time t_0 (L)
k_l	vertical hydraulic conductivity of liner	q_1	leakage volume through liner per unit area (L)	y_s	steady-state average saturated depth (L)
k_d	lateral hydraulic conductivity of drainage layer above liner	e	rate of inflow to drainage layer per unit area (L/T)	w_{opt}	optimum moisture content
L	length of stage 2 extension (L)	R_c	viscosity factor	γ	unit weight of soil (F/L^3)
m	$(k_h/k_v)^{0.5}$	R	radius of gap (L)	γ_s	unit weight of slurry (F/L^3)
n_f	number of flaws	p	k_h/k_s	σ'	average effective stress (F/L^2)
n	porosity	PI	plasticity index	θ_i	initial volumetric water content
n_e	effective porosity	s	slurry wall thickness (L)	θ_s	saturation volumetric water content
q	volume of flow per unit area (L/T)	T	thickness of smear zone (L)	η'	absolute viscosity (poise)
		T_1	membrane thickness (L)	τ	shear stress (F/L^2)
		t_s	thickness of gap (L)		
		t_f	thickness of geotextile (L)		

PROBLEMS

10.1 Develop a checklist for the parameters (and their values) required for accepting materials from a borrow area to use as soil liners.

10.2 Develop a checklist for the minimum requirements of a soil liner material as delivered and spread before compaction.

10.3 Explain why the use of an average reference density for compaction control could result in either undercompaction or overcompaction. Recommend procedures for mitigating this problem.

10.4 Explain why density alone is not a satisfactory criterion for field control in soil liner construction.

10.5 Explain why a reference density is not useful for assessing relative compaction if compaction is performed at a very high field moisture content.

10.6 If the unconfined compressive strength at optimum moisture content is 2000 psf, the optimum moisture content is 20%, and the field moisture content is 30%, determine the maximum pressure that a compactor can exert without sinking into the wet soil.

10.7 Using a two-stage permeability test and the following data, estimate the vertical and horizontal permeabilities.

Stage 1
$d = 1.27$ cm, $D = 10.2$ cm
$h_1 = 193$ cm, $h_2 = 180$ cm
$b_1 = 40$ cm
Average temp.: 21°C, $R = 0.97$
$c = 1$ cm
$t_2 - t_1 = 1.5$ hr

Stage 2
$h_1 = 199$ cm
$h_2 = 190$ cm
$b_2 = 50$ cm
$t_2 - t_1 = 0.75$ hr
Average temp.: 22°C, $R = 0.94$
$c = +0.3$ cm

10.8 A slurry wall is planned through 40 ft of silty sand and will be keyed into a thick clay layer. What is the minimum weight of the slurry required to maintain a stable cut at a safety factor of 1.4? The water table is 10 ft below the surface. Assume a friction angle of 35° for the sand.

10.9 For the slurry wall in Problem 10.8, determine the required bentonite content and the depth of the key assuming an interface conductivity of 10^{-6} cm/s and conductivity across the wall of 10^{-7} cm/s. Consider a 3-ft-wide wall.

10.10 A geonet with a thickness of 4.5 mm and a hydraulic conductivity of 2 cm/s is used for a leak detection under a geomembrane. Assuming that the head is not to exceed 0.3 m, what is the action leakage rate (use a safety factor of 3.0).

10.11 For Problem 10.10, use both the analytical and empirical methods to determine the number of circular holes in the liner that could generate the action leakage rate. Assume a thick base with a permeability of 10^{-5} cm/s, poor contact condition, a liner thickness of 0.15 cm, and a hole area of 1 cm^2.

10.12 A liner thickness of 0.15 mm overlies a leak detection system at a slope of 3%. A geonet or a sand layer 0.3 m thick are being considered. Determine the required hydraulic conductivity of the sand or the transmissivity of the geonet needed for the head in the LDS to not exceed 10 cm. The acting head on top of the liner is 30 cm and a hole in the liner with an area of 1 cm^2/acre is anticipated.

10.13 In a dual composite liner, a membrane 0.05 cm thick overlies a clayey base that is 0.6 m thick with a permeability of 10^{-5} cm/s.
   a. Considering holes with diameters of 0.08 cm, 0.16 cm, and 0.64 cm, plot the leakage rate versus hole area using the text guidance for the vertical thickness of the gap.
   b. Using gap thicknesses two and three times the value in part (a), plot the leakage rate versus hole area.
   c. Using the empirical method, determine the leakage rate for good and poor contact for the hole diameters in part (a).
   d. At which gap thicknesses are the empirical and analytical estimates of leakage in accord? In all cases assume an acting head of 30 cm and a thick base.

10.14 A liner 0.15 mm thick with a liquid head of 5 m is supported on a base with a hydraulic conductivity of 10^{-6} cm/s. Three possibilities exist for possible defects (a) a circular hole 1 cm^2 in area or (b) a longitudinal defect 0.3 m long and 2 mm wide, and (c) a longitudinal defect 3 m long and 3 mm wide. The thickness of the supporting soil is 1 m. Determine the leakage rate for each of the three cases, using both the analytical and empirical methods.

10.15 Consider a 3-ft soil liner with a hydraulic conductivity of 10^{-6} cm/s, a drainage layer with a hydraulic conductivity of 10^{-2} cm/s, a ratio of $e/k_d = 10^{-3}$, a slope of 0.02, a slope length $L = 400$ ft, and an effective porosity of 0.3. Considering infiltration over a 15-day period, determine the liner efficiency.

10.16 Repeat Problem 10.15 for liner permeabilities of 10^{-4} cm/s and 10^{-5} cm/s.

10.17 Determine the liner efficiency for the following conditions:

    Liner thickness = 1 m
    Liner slope = 0.05
    Slope length = 50 m
    Liner conductivity $k_1 = 10^{-7}$
    Inflow = 20 cm over a 20-day period
    Conductivity of drainage layer = 0.01 cm/s
    Effective porosity of drainage layer = 0.35

10.18 Repeat Problem 10.17 with the liner slope decreased to 1.5% and inflow applied instantaneously.

10.19 For Problem 10.15, if the liner thickness is 90 cm and its saturated hydraulic conductivity is 10^{-7} cm/s, determine the approximate transit time. Assume an effective porosity of 0.4 and no suction.

10.20 Assume that the liner in Problem 10.19 was below saturation at a volumetric water content of 35%. Determine the range of transit times, assuming a constant suction of 20 cm and a porosity of 0.45.

11 Leachate Generation and Collection

11.1 WASTE DECOMPOSITION

The waste in a sanitary landfill is a mixture of organic material (e.g., garbage, paper, cardboard, textiles, plastic, wood, rubber) and inorganic waste (which may include metals such as cans and wires). Refuse in a landfill undergoes oxidation and decomposition in the presence of oxygen, moisture, and appropriate temperature. Water, which is essential for decomposition, is derived partially from the waste itself and is about 10 to 20% by volume of refuse (Fenn et al., 1975). Another source of water is that percolating through the refuse as a result of precipitation, irrigation, or recirculation of leachate. Given a favorable environment, approximately 25 to 40% of municipal refuse is available for decomposition (Wall and Zeiss, 1995). The rate of decomposition is influenced by the content of refuse, ambient temperature, oxygen supply, and water content. Barlaz et al., 1990 estimated that refuse typically contains 40 to 50% cellulose, 12% hemicellulose, 10 to 15% lignin, and 4% protein on a dry weight basis. The cellulose and hemicellulose fractions of the waste account for 91% of the methane potential (Barlaz et al., 1990). Materials containing cellulose include paper, rags, fruit skin, etc. Rubber and most plastics are resistant to biodegradation.

Low moisture content (generally less than 20% on a wet weight basis) inhibits decomposition (Ham, 1993). Decomposition can be promoted by leachate recirculation, which is the process of reintroducing the leachate collected from the landfill back into the landfill. Before reintroducing the leachate back into the refuse, the pH and other indicators may be modified to enhance decomposition. Recirculation increases the moisture content and nutrient movement in the landfill, which has the beneficial action of enhancing biodegradation.

Biodegradation of waste may occur in the presence of oxygen (by aerobic bacteria), in an environment devoid of oxygen (by anaerobic bacteria), or with very little oxygen (by facultative anaerobic bacteria). Aerobic bacteria require oxygen to degrade organic material (and appropriate nutrients and moisture must be present). Anaerobic bacteria remain mostly dormant in the presence of oxygen. Facultative anaerobic bacteria are adaptable to the presence of oxygen.

In all cases organic waste is broken down by enzymes produced by bacteria in a manner comparable to food digestion. Considerable heat is generated by these reactions, which produce methane, carbon dioxide, and other gases as by-products. The heat produced from aerobic decomposition elevates

Figure 11.1
Gas composition
and evolution in
a typical landfill
(reproduced from
Gas Generation
Institute, 1981,
by permission)

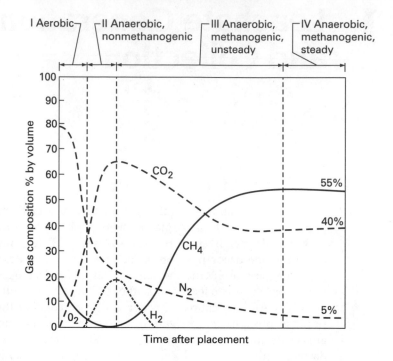

the initial ambient temperature. Peak temperatures of 160°F (71°C) can occur within a few days to a few weeks after application of the cover (Noble, 1976). The high increase in temperature (roughly 70°F) may cause combustion of dry waste and cause fires.

The consumption of nutrients and depletion of oxygen in the aerobic phase inhibits the aerobic process and initiates the anaerobic process (Figure 11.1). In the initial phase of anaerobic decomposition, the principal gas produced is carbon dioxide. With time, the amount of carbon dioxide decreases and methane increases, each reaching a plateau. The anaerobic process and gas generation could extend over many years. Methods of gas management are discussed in Chapter 14.

11.2   LEACHATE GENERATION

The percolation of water through the landfill generates leachate. Water that comes in contact with refuse as a result of surface runoff on exposed refuse is also called leachate. The source of water could be precipitation, irrigation, groundwater, or leachate recirculated through the landfill (see Figure 11.2). Leachate through a decomposing landfill contains a range of inorganic and organic chemicals, as illustrated in Tables 11.1 and 11.2 (Subtitle D study, EPA, 1986). These chemicals impact the groundwater quality beneath and beyond landfills. The major inorganic constituents of concern are typically lead and cadmium. The sources of cadmium in municipal refuse include nickel–cadmium batteries, plastics, nonfood packaging, and electronic

Figure 11.2
Leachate generation
parameters

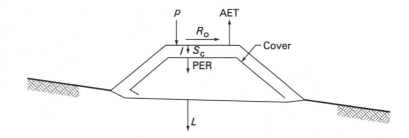

appliances. The sources of lead may include lead–acid batteries, plastics, cans, used oil, lightbulbs, and others (Franklin Associates, 1989).

The amount of leachate generated depends on the available water, landfill constituents, and the setting of the landfill with respect to groundwater and surface water. The available water is affected by the moisture content of the refuse, precipitation, surface runoff, recirculation practices, a rise in an otherwise low groundwater table, and moisture generated from the decomposition process.

Not all the water (from precipitation, irrigation, recirculation, or other sources) impinging on the surface of the landfill goes into leachate. Part of the water is lost to runoff if the landfill is covered with soil (see Figure 11.2) and

Table 11.1 Leachate constituents from municipal waste landfills

Constituent	Concentration (mg/l)	Constituent	Concentration (mg/l)
Chemical oxygen demand	50–90,000	Hardness (as $CaCO_3$)	0.1–36,000
Biochemical oxygen demand	5–75,000	Total phosphorus	0.1–150
Total organic carbon	50–45,000	Organic phosphorus	0.4–100
Total solids	1–75,000	Nitrate nitrogen	0.1–45
Total dissolved solids	725–55,000	Phosphate (inorganic)	0.4–150
Total suspended solids	10–45,000	Ammonia nitrogen	0.1–2000
Volatile suspended solids	20–750	Organic nitrogen	0.1–1000
Total volatile solids	90–50,000	Total Kjeldahl nitrogen	7–1970
Fixed solids	800–50,000	Acidity	2700–6000
Alkalinity (as $CaCO_3$)	0.1–20,350	Turbidity (Jackson units)	30–450
Total coliform bacteria (c.f.u./100 ml)	0–10	Chlorine	30–5000
Iron	200–5500	pH (dimensionless)	3.5–8.5
Zinc	0.6–220	Sodium	20–7600
Sulfate	25–500	Copper	0.1–9
Sodium	0.2–79	Lead	0.001–1.44
Total volatile acid	70–27,700	Magnesium	3–15,600
Manganese	0.6–41	Potassium	35–2300
Fecal coliform bacteria (c.f.u./1000 ml)	0–10	Cadmium	0–0.375
Specific conductance (mho/cm)	960–16,300	Mercury	0–0.16
Ammonium nitrogen	0–1106	Selenium	0–2.7
		Chromium	0.02–18

Source: Subtitle D study, 1986 (EPA, 1986)

Table 11.2
Organic constituents of
leachate from various
municipal waste
landfills

Constituent	Minimum (ppb)	Maximum (ppb)	Median (ppb)
Acetone	140	11,000	7500
Benzene	2	410	117
Bromomethane	10	170	55
1-Butanol	50	360	220
Carbon tetrachloride	2	398	10
Chlorobenzene	2	237	10
Chloroethane	5	170	7.5
bis(2-Chloroethoxy)methane	2	14	10
Chloroform	2	1300	10
Chloromethane	10	170	55
Delta BHC	0	5	0
Dibromomethane	5	25	10
1,4-Dichlorobenzene	2	20	7.7
Dichlorodifluoromethane	10	369	95
1,1-Dichloroethane	2	6300	65.5
1,2-Dichloroethane	0	11,000	7.5
cis-1,2-Dichloroethane	4	190	97
trans-1,2-Dichloroethane	4	1300	10
Dichloromethane	2	3300	230
1,2-Dichloropropane	2	100	10
Diethyl phthalate	2	45	31.5
Dimethyl phthalate	4	55	15
Di-*n*-butyl phthalate	4	12	10
Endrin	0	1	0.1
Ethyl acetate	5	50	42
Ethyl benzene	5	580	38
bis(2-Ethylhexyl)phthalate	6	110	22
Isophorene	10	85	10
Methyl ethyl ketone	110	28,000	8300
Methyl isobutane ketone	10	660	270
Naphthalene	4	19	8
Nitrobenzene	2	40	15
4-Nitrophenol	17	40	25
Pentachlorophenol	3	25	3
Phenol	10	28,800	257
2-Propanol	94	10,000	6900
1,1,2,2-Tetrachloroethane	7	210	20
Tetrachloroethane	2	100	40
Tetrahydrofuran	5	260	18
Toluene	2	1600	166
Toxaphene	0	5	1
1,1,1-Trichloroethane	0	2400	10
1,1,2-Trichloroethane	2	500	10
Trichloroethane	1	43	3.5
Trichlorofluoromethane	4	100	12.5
Vinyl chloride	0	100	10
m-Xylene	21	79	26
p-Xylene + *o*-xylene	12	50	18

Source: Subtitle D study, 1986 (EPA, 1986)

Table 11.3 Default soil, waste, and geosynthetic characteristics in HELP model

CLASSIFICATION		Total porosity vol/vol	Field capacity vol/vol	Wilting point vol/vol	Saturated hydraulic conductivity cm/s
USDA	USCS				
CoS	SP	0.417	0.045	0.018	1.0×10^{-2}
S	SW	0.437	0.062	0.024	5.8×10^{-3}
FS	SW	0.457	0.083	0.033	3.1×10^{-3}
LS	SM	0.437	0.105	0.047	1.7×10^{-3}
LFS	SM	0.457	0.131	0.058	1.0×10^{-3}
SL	SM	0.453	0.190	0.085	7.2×10^{-4}
FSL	SM	0.473	0.222	0.104	5.2×10^{-4}
L	ML	0.463	0.232	0.116	3.7×10^{-4}
SiL	ML	0.501	0.284	0.135	1.9×10^{-4}
SCL	SC	0.398	0.244	0.136	1.2×10^{-4}
CL	CL	0.464	0.310	0.187	6.4×10^{-5}
SiCL	CL	0.471	0.342	0.210	4.2×10^{-5}
SC	SC	0.430	0.321	0.221	3.3×10^{-5}
SiC	CH	0.479	0.371	0.251	2.5×10^{-5}
C	CH	0.475	0.378	0.265	1.7×10^{-5}
Barrier soil		0.427	0.418	0.367	1.0×10^{-7}
Bentonite mat (0.6 cm)		0.750	0.747	0.400	3.0×10^{-9}
Municipal waste (900 lb/yd³ or 312 kg/m³)		0.671	0.292	0.077	1.0×10^{-3}
Municipal waste (channeling and dead zones)		0.168	0.073	0.019	1.0×10^{-3}
Drainage net (0.5 cm)		0.850	0.010	0.005	$1.0 \times 10^{+1}$
Gravel		0.397	0.032	0.013	3.0×10^{-1}
L*	ML	0.419	0.307	0.180	1.9×10^{-5}
SiL*	ML	0.461	0.360	0.203	9.0×10^{-6}
SCL*	SC	0.365	0.305	0.202	2.7×10^{-6}
CL*	CL	0.437	0.373	0.266	3.6×10^{-6}
SiCL*	CL	0.445	0.393	0.277	1.9×10^{-6}
SC*	SC	0.400	0.366	0.288	7.8×10^{-7}
SiC*	CH	0.452	0.411	0.311	1.2×10^{-6}
C*	CH	0.451	0.419	0.332	6.8×10^{-7}
Coal-burning electric plant fly ash*		0.541	0.187	0.047	5.0×10^{-5}
Coal-burning electric plant bottom ash*		0.578	0.076	0.025	4.1×10^{-3}
Municipal Incinerator fly ash*		0.450	0.116	0.049	1.0×10^{-2}
Fine copper slag*		0.375	0.055	0.020	4.1×10^{-2}
Drainage net (0.6 cm)		0.850	0.010	0.005	$3.3 \times 10^{+1}$
High density polyethylene (HDPE)					2.0×10^{-13}
Low density polyethylene (LDPE)					4.0×10^{-13}
Polyvinyl chloride (PVC)					2.0×10^{-11}
Butyl rubber					1.0×10^{-12}
Chlorinated polyethylene (CPE)					4.0×10^{-12}
Hypalon or chlorosulfonated polyethylene (CSPE)					3.0×10^{-12}
Ethylene-propylene diene monomer (EPDM)					2.0×10^{-12}
Neoprene					3.0×10^{-12}

* Moderately compacted.
G: Gravel; S: Sand; Si: Silt (USDA); C: Clay; L: Loam (sand, silt, clay and humus mixture); Co: Coarse, F: Fine

can be collected as clean water. Another fraction is lost to evaporation and transpiration (water consumed by plants), which is commonly referred to as *evapotranspiration*. The remainder enters the landfill cover, and a portion of this is lost to storage in the cover. This storage is characterized by the moisture content at field capacity; *field capacity* is the maximum amount of water that the soil or refuse can retain in a gravitational field without percolation.

In theory, leachate can develop only after the field capacity moisture content is exceeded. A portion of the field capacity moisture is not available to plants. The upper limit of this moisture content is called the *wilting point*. Available moisture to plants is the field capacity minus the wilting point. These two parameters are related to the porosity, n, and soil grain size. Table 11.3 shows values for various parameters used for estimating leachate generation in the Hydrologic Evaluation of Landfill Performance (HELP) computer model (Schroeder et al., 1994).

The moisture content of municipal waste at placement varies from 10 to 20% (Fenn et al., 1975). If the field capacity is 30% by volume, then the refuse can hold an additional 10 to 20% by volume without leachate generation. Some leachate, however, will develop at moisture content less than field capacity because of the secondary conductivity (channeling) of refuse.

11.2.1 Water Balance Method

The water balance method was used by the U.S. Environmental Protection Agency (EPA) to estimate leachate flow from landfills (Fenn et al., 1975). The method assumes one-dimensional flow, conservation of mass, and transmission characteristics of soil and refuse. The basic mass balance equation (see Figure 11.2) is

$$P = I + R_0 \tag{11.1}$$

where P is the input water from precipitation, irrigation, recirculation, or surrounding surface runoff, I is the infiltration, and R_0 is the surface runoff.

Precipitation is the amount of water (in in. or mm) that accumulates on a sealed level surface. Data on this parameter are generally obtained from the U.S. Weather Bureau or from gauges installed near the landfill over a long period of time. Surface runoff is determined as a fraction of the precipitation as

$$R_0 = CP \tag{11.2}$$

where C is the empirical runoff coefficient, which depends on the type of surface, vegetation, and slope—as illustrated in Table 11.4 (Fenn et al., 1975).

A portion of the infiltrating liquid (I) will be lost to evapotranspiration (AET) and changes in the soil's ability to store moisture in the landfill cover (S_c); the remainder (PER_s) will percolate into the refuse. Thus,

$$PER_s = I - AET - S_c \tag{11.3}$$

Table 11.4
Runoff coefficients

Description of grass-covered soil	Slope of ground surface	Runoff coefficient (C)
Sandy soil	Flat (<2%)	0.05–0.1
	Mild (2–7%)	0.1–0.15
	Steep (>7%)	0.15–0.2
Clayey soil	Flat	0.13–0.17
	Mild	0.18–0.22
	Steep	0.25–0.35

Source: Fenn et al., 1975

where S_c is the change of moisture retention in the cover ($S_c = 0$ if the moisture in the cover is always at field capacity). Estimated values of the water-holding capacity at field capacity is given in Table 11.3. PER_s is the water percolating into the refuse.

Monthly potential evapotranspiration may be computed from the Thornthwaite equation. This equation uses the temperature efficiency index, TE, which is the sum of 12 monthly values of the heat index, I_t, which is given by

$$I_t = (t/5)^{1.514} \qquad (11.4)$$

where t is the mean monthly temperature in °C. The potential evapotranspiration (PET) is given by

$$PET \text{ (mm)} = 16(10t/TE)^a \text{ unadjusted} \qquad (11.5)$$

where

$$a = 6.75 \times 10^{-7}(TE)^3 - 7.7 \times 10^{-5}(TE)^2 + 1.792 \times 10^{-2}TE + 0.49239$$

$$(11.6)$$

where TE is the sum of 12 monthly values of heat index I_t. The moisture content held by the soil after evapotranspiration is characterized by the wilting point (Table 11.3).

The above value of unadjusted evapotranspiration PET (for a 12-hour standard duration of sunlight) is corrected for unequal day lengths and months. Table 11.5 (Chow, 1964) presents the correction factors for various latitudes. The factors for 50° are used for poleward from 50°.

If the annual $(I - PET)$ is positive, the moisture retention, S_T, available for evapotranspiration is characterized

S_T = (field capacity − wilting point) times the root penetration distance, which is limited by the thickness of the cover

$$(11.7)$$

If $(I - \text{Adj } PET)$ is negative, the soil moisture retention will be less.

Table 11.6 is an abbreviated version of a table (Thornthwaite and Mather, 1957) for estimating the soil moisture retention after the potential evapotranspiration has occurred. The deficiencies are summed up on a running basis. A

Table 11.5 Adjustment factors for potential evapotranspiration computed by the Thornthwaite equation

Latitude	Jan.	Feb.	Mar.	Apr.	May	June	July	Aug.	Sep.	Oct.	Nov.	Dec.
0	1.04	0.94	1.04	1.01	1.04	1.01	1.04	1.04	1.01	1.04	1.01	1.04
10	1.00	0.91	1.03	1.03	1.08	1.06	1.08	1.07	1.02	1.02	0.98	0.99
20	0.95	0.90	1.03	1.05	1.13	1.11	1.14	1.11	1.02	1.00	0.93	0.94
30	0.90	0.87	1.03	1.08	1.18	1.17	1.20	1.14	1.03	0.98	0.89	0.88
35	0.87	0.85	1.03	1.09	1.21	1.21	1.23	1.16	1.03	0.97	0.86	0.85
40	0.84	0.83	1.03	1.11	1.24	1.25	1.27	1.18	1.04	0.96	0.83	0.81
45	0.80	0.81	1.02	1.13	1.28	1.29	1.31	1.21	1.04	0.94	0.79	0.75
50	0.74	0.78	1.02	1.15	1.33	1.36	1.37	1.25	1.06	0.92	0.76	0.70

sum of zero is assigned to the last month having a positive value of $(I - \text{Adj } PET)$ because at the end of the wet season the soil moisture is at field capacity. For negative months, Table 11.6 is used to estimate moisture retention. After the dry period when $(I - \text{Adj } PET)$ becomes positive, the retention is that for the previous month plus $(I - \text{Adj } PET)$ for that month; however, the value

Table 11.6 Soil moisture retention after potential evapotranspiration has occurred

	S_T (mm)[b]								
$\Sigma NEG\ (I - PET)$[a]	25	50	75	100	125	150	200	250	300
0	25	50	75	100	125	150	200	250	300
10	16	41	65	90	115	140	190	240	290
20	10	33	57	81	106	131	181	231	280
30	7	27	50	74	98	122	172	222	271
40	4	21	43	66	90	114	163	213	262
50	3	17	38	60	83	107	155	204	254
60	2	14	33	54	76	100	148	196	245
70	1	11	28	49	70	93	140	188	237
80	1	9	25	44	65	87	133	181	229
90	1	7	22	40	60	82	127	174	222
100		6	19	36	55	76	120	167	214
150		2	10	22	37	54	94	136	181
200		1	5	13	24	39	73	111	153
250			2	8	16	28	56	91	130
300			1	5	11	20	44	74	109
350			1	3	7	14	34	61	92
400				2	5	10	26	50	78
450				1	3	7	20	41	66
500				1	2	5	16	33	56
600					1	3	10	22	40
700						1	6	15	28
800						1	4	10	20
1000							1	4	10

[a] $NEG(I - PET)$ is lack of infiltration water needed for vegetation. See Table 11.7 (row 12) for an example.
[b] S_T is the soil moisture storage at field capacity.
Source: After Thornthwaite and Mather, 1957

cannot exceed S_T. Where a positive $(I - \text{Adj } PET)$ occurs between two negative values, retention is calculated by direct addition of $(I - \text{Adj } PET)$ to the previous retention.

The actual evapotranspiration, AET, is equivalent to $(\text{Adj } PET)$ if the soil is at field capacity and adequate water supply is available for evapotranspiration. This occurs during wet months. For dry months (months with negative $(I - \text{Adj } PET)$), PET drops to below $(\text{Adj } PET)$ and cannot exceed the infiltration plus the absolute value of the change of storage, S_c.

As water percolates through the refuse, a portion will be lost to increasing storage in refuse unless refuse is at field capacity. In this case the change in storage (S_r) is zero. Additional water generated by refuse decomposition (W_d) and groundwater intrusion (W_g) are added for estimating the total volume of leachate (L_1) to be collected.

$$L_1 = PER_r - S_r + W_d + W_g \qquad (11.8)$$

The volume W_d is relatively small and is usually ignored. Modern landfills are constructed well above the seasonal high groundwater table and for such landfills W_g is zero. For old landfills, W_g may be significant.

Example 11.1

Use the water balance method shown in Table 11.7 to evaluate the landfill site in Lyndhurst, New Jersey.

Solution: The site has a grassed sandy soil cover with a field capacity (vol/vol) of 0.108, a wilting point of 0.03, and a cover thickness of 0.64 m. Row 1 is the average monthly temperature data obtained from the U.S. Weather Bureau. Row 2 is the computed monthly heat index from Eq. 11.4. Row 3 is the potential evapotranspiration from Eq. 11.5. Row 4 is the adjustment factor from Table 11.5 (latitude of 40°). Row 5 is the adjusted evapotranspiration (row 3 × row 4). Row 6 is the precipitation data obtained from the U.S. Weather Bureau. Row 7 is the runoff coefficient. A runoff coefficient of 0.25 is used. For January, a higher runoff coefficient of 0.5 is used to account for frozen ground (below freezing temperature). Row 8 is the runoff from Eq. 11.2 (row 6 × row 7). Row 9 is the infiltration from Eq. 11.1 (row 6 − row 8). Row 10 is the infiltration impinging on the cover minus the adjusted evapotranspiration (row 9 − row 5).

In row 11, the running sum of moisture deficiency indicates water loss. Summation is started by assigning zero to the last month with positive $(I - \text{Adj } PET)$ since at the end of the wet season soil moisture is at its field capacity.

For row 12, the monthly moisture stored is determined as follows.

a. Determine the initial moisture storage from the soil type and the plant root depth.

b. Assign this value to the last month with $(I - \text{Adj } PET) > 0$.

c. For months with $(I - \text{Adj } PET) < 0$, use Table 11.6.

d. After the dry period when $(I - \text{Adj } PET) > 0$, the storage available for evapotranspiration, S_c, in any month = storage in the previous month + $(I - \text{Adj } PET)$ for the month ≤ storage capacity.

Table 11.7 An example of the use of the water balance method

Row parameter		Jan	Feb	Mar	Apr.	May	June	July	Aug.	Sep.	Oct.	Nov.	Dec.	Annual total
1	Temperature (°C)	−0.39	0.22	4.90	10.80	16.70	21.70	24.70	23.80	19.80	13.80	7.80	1.72	
2	I_t	0.00	0.01	0.97	3.21	6.21	9.23	11.23	10.61	8.03	4.65	1.96	0.20	56.31
3	PET	0.00	0.18	13.21	39.21	71.44	102.45	122.44	116.34	90.31	54.95	25.05	3.13	638.71
4	Adj Fac	0.84	0.83	1.03	1.11	1.24	1.25	1.27	1.18	1.04	0.96	0.83	0.81	
5	Adj PET	0.00	0.15	13.61	43.52	88.59	128.06	155.50	137.28	93.92	52.75	20.80	2.53	736.71
6	P (mm)	84.60	73.90	99.60	88.90	92.50	84.10	97.30	105.70	96.00	76.00	87.40	83.60	1069.60
7	C (R/O)	0.50	0.25	0.25	0.25	0.25	0.25	0.25	0.25	0.25	0.25	0.25	0.25	
8	R_0 (mm)	42.30	18.48	24.90	72.23	23.13	21.03	24.33	26.43	24.00	19.00	21.85	20.90	288.55
9	I	42.30	55.43	74.70	66.68	69.38	63.08	72.98	79.28	72.00	57.00	65.55	62.70	781.05
10	I − Adj PET	42.30	55.27	61.09	23.15	−19.21	−64.99	−82.52	−58.01	−21.92	4.25	44.75	60.17	44.34
11	NEG (I − Adj PET)				0.00	−19.21	−84.20	−166.72	−224.73	−246.65				
12	S_T (Table 11.6)	50.00	50.00	50.00	50.00	35.00	9.00	2.00	1.00	1.00	5.25	50.00	50.00	
13	dS_c	0.00	0.00	0.00	0.00	−15.00	−26.00	−7.00	−1.00	0.00	4.25	44.75	0.00	0.00
14	AET (mm)	0.00	0.15	13.61	43.52	84.38	89.08	79.98	80.28	72.00	52.75	20.80	2.53	539.06
15	PERC (mm)	42.30	55.27	61.09	23.15	0.00	0.00	0.00	0.00	0.00	0.00	0.01	60.17	241.99

Row 13 is the change in storage from month to month, dS_c, which equals the storage from this month less the storage from the previous month. Row 14 is the actual evapotranspiration:

a. $(I - \text{Adj } PET) > 0$, $AET = \text{Adj } PET$ (wet months)
b. $(I - \text{Adj } PET) < 0$, $AET = (I + |\Delta S_c|)$ (dry months)

Row 15 is the percolation from cover impinging on refuse, calculated as follows.

a. $PER = 0. (I - \text{Adj } PET) < 0$ (dry months)
b. $PER = I - AET - \Delta S_c$ (wet months)

As mentioned earlier, there exists a very versatile quasi-two-dimensional hydrologic computer model for conducting water balance analysis of landfills, cover systems, and other solid waste containment facilities. It is known as the Hydrologic Evaluation of Landfill Performance, or HELP, model, authored by Schroeder et al. (1994). The configuration of HELP is shown in Figure 11.3. The HELP model accepts weather, soil, and design data, and uses solution techniques that account for the effects of surface storage, snowmelt, runoff, infiltration, evapotranspiration, vegetative growth, soil moisture storage, lateral subsurface drainage, leachate recirculation, unsaturated vertical drainage, and leakage through soil, geomembrane, or composite liners.

The curve number CN used by HELP for mild slopes is given in Figure 11.4. The CN number from Figure 11.4 is adjusted for various slopes (0.04 to 0.5 ft/ft) and slope lengths (50 to 500 ft) using

$$CN_S = 100 - (100 - CN)(L^2/S)^m \tag{11.9}$$

where

CN: curve number from Figure 11.4 (standard for slope of 4%, 500 ft long)
CN_S: curve number adjusted for slope angle and slope length
S = nondimensional slope = actual slope (ft/ft)/0.04. If the actual slope is $1v$ on $3h$, then $S = 0.33/0.04 = 8.25$.
l = nondimensional length (actual slope length (ft)/500 ft)
$m = (CN)^{-0.81}$

The default soil data used by HELP are given in Table 11.3. HELP assigns default soil parameters based on the following relationships.

$$FC = 0.1535 - 0.0018(\%S) + 0.0039(\%C) + 0.1943n \tag{11.10}$$

$$WP = 0.0370 - 0.0004(\%S) + 0.0044(\%C) + 0.0482n \tag{11.11}$$

$$k_s = (g/v)(n^3/(1 - n)^2)(d_g^2/1.8 \times 10^4) \tag{11.12}$$

where:

FC = field capacity (vol/vol)
WP = wilting point (vol/vol)
n = total porosity = $e/1 + e$
e = void ratio

Figure 11.3
Schematic of landfill
profile illustrating
typical landfill
features (Schroeder
et al., 1994)

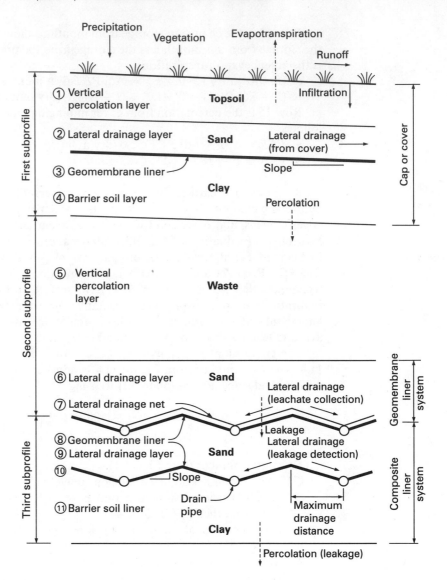

%S = sand (particle size 0.05 to 2.0 mm) expressed as a percentage of total weight

%Si = silt (particle size 0.002 to 0.05 mm) expressed as a percentage of total weight

%C = clay (particle size < 0.002 mm) expressed as a percentage of total weight

d_g = geometric mean of soil particles (mm) = $\exp(-1.151 - 0.07713(\%C) - 0.03454(\%Si))$

k_s = saturated hydraulic conductivity (cm/s)

Figure 11.4
Relation between SCS curve number and soil texture number for various levels of vegetation (Schroeder et al., 1994)

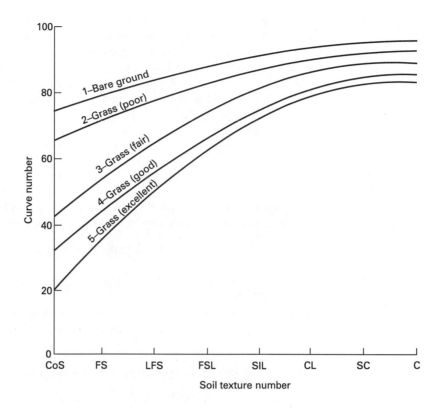

g = acceleration of gravity = 981 cm/s^2

v = kinematic viscosity of water = 1.14×10^{-2} cm^2/s at 15°C.

HELP computes the unsaturated hydraulic conductivity using the default values for the field capacity and wilting point coupled with Brooks and Corey's (1964) parameters. Using these parameters, the unsaturated hydraulic conductivity is calculated from Eqs. 11.13, 11.14, and 11.15 after intermediate steps.

$$\frac{\theta - \theta_r}{n - \theta_r} = (P_b/P_c)^\lambda \qquad (11.13)$$

where

θ = volumetric water content (at field capacity or wilting point) (vol/vol)

θ_r = residual saturation volumetric water content (vol/vol), which is the amount of water that remains in the soil under infinite capillary suction. It may be related to the wilting point using

$$\theta_r = 0.014 + 0.25 WP \qquad \text{for } WP \geq 0.04$$

$$\theta_i = 0.6 \, WP \qquad\qquad \text{for } WP < 0.04 \qquad (11.14)$$

P_c = capillary pressure (bars); 0.33 at field capacity, 15 at wilting point
P_b = bubbling pressure (bars), which is the pressure at which a continuous network of flow channels are formed
λ = pore-size distribution index

With $(\theta)_r$ known from Eq. 11.14, the values of *FC* and *WP* are taken from Eqs. 11.10 and 11.11 and substituted for θ at capillary pressures of 0.33 bars and 15 bars, respectively in Eq. 11.13. This results in two equations and two unknowns (λ and P_b) to solve for. With these pressures, the unsaturated hydraulic conductivity is calculated as

$$k_u = k_s((\theta - \theta_r)/(n - \theta_r))(3 + 2/\lambda) \tag{11.15}$$

In terms of the pore size distribution index, porosity, and bubbling pressure the saturated hydraulic conductivity is expressed as

$$k_s = 21 \text{ cm}^3/\text{s}((n - \theta_r)/P_b)^2(\lambda^2/(\lambda + 1)(\lambda + 2)) \tag{11.16}$$

The saturated hydraulic conductivity decreases with a decrease in porosity and pore size distribution index.

For a vegetated soil (which results in root channels), HELP allows increasing the saturated hydraulic conductivity in terms of the leaf area index (*LAI*) as follows:

$$(k_s)_v/(k_s)_{uv} = 1 + 0.5966LAI + 0.132659LAI^2 + 0.1123454LAI^3$$
$$- 0.0477762LAI^4 + 0.004325035LAI^5 \tag{11.17}$$

where $(k_s)_v$ is the hydraulic conductivity for vegetated soil and $(k_s)_{uv}$ is the hydraulic conductivity of unvegetated soil.

A liner structure may include geosynthetic components. Specifications for the geomembrane include pinhole (defect area = 7.84×10^{-7} m^2) density expressed as the number of pinholes per acre, installation defects (number of defects per acre using a defect area of 0.0001 m^2), and geomembrane liner placement quality. Other data are the geomembrane saturated hydraulic conductivity and geotextile transmissivity. The leakage rate depends on the contact conditions based on the empirical formulas presented in Chapter 10. Table 11.8 provides default data used for geotextile hydraulic properties.

Table 11.8
Needle-punched, nonwoven geotextile properties used in EPA HELP model

| | | IN-PLANE FLOW | | CROSS-PLANE FLOW |
Applied compressive stress (kPa)	Resulting geotextile thickness (cm)	Geotextile transmissivity (cm²/s)	Horizontal hydraulic conductivity (cm/s)	Vertical hydraulic conductivity (cm/s)
1 to 8	0.41	0.3	0.7	0.4
100	0.19	0.04	0.2	—
200	0.17	0.02	0.1	—

11.3   LEACHATE COLLECTION

11.3.1 *Steady-state Volume*

A leachate collection system is required above a liner in order to avoid buildup of the leachate head above the 1-ft (30.5-cm) level stipulated in the regulations. The analysis of flow considers the Dupuit assumption to be valid. This means that flow is horizontal toward the drain and the drainage layer has infinite hydraulic conductivity in the vertical direction. The liner is usually sloped toward the drain to enhance drainage (see Figure 11.5). In some designs (Figure 11.5), the liner is sloped uniformly from one end to the other. A schematic of a leachate collection system is shown in Figure 11.6.

Considering a vertical section at a distance x from the top of the slope (Figure 11.5), the flow quantity (McEnroe, 1989) is expressed as

$$Q = ex = -ky \, dh/dx = -k_d y \, d(y - Sx)/dx \qquad (11.18)$$

where S is the tan α, k_d is the hydraulic conductivity of the drainage layer, and e is the percolation impinging on the drainage layer. From Eq. 11.18

$$k_d y \, dy/dx - k_d yS + ex = 0 \qquad (11.19)$$

In the above expressions, it is assumed that the leakage through the layer beneath the drainage layer is zero.

For a flat ($S = 0$) condition, the solution to Eq. 11.19 for the boundary condition ($y = 0$ at $x = L$) is

$$y = (c(L^2 - x^2))^{0.5} \qquad (11.20)$$

and the maximum value of y, y_{max} at $x = 0$ is given by

$$y_{max} = L(c)^{0.5} \qquad (11.21)$$

Figure 11.5
(a) Collection with L-shaped liner geometry
(b) Collection on a uniformily sloped liner

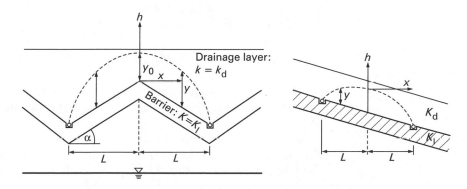

Figure 11.6
Schematic leachate
collection system
(Washington State
Department of
Ecology)

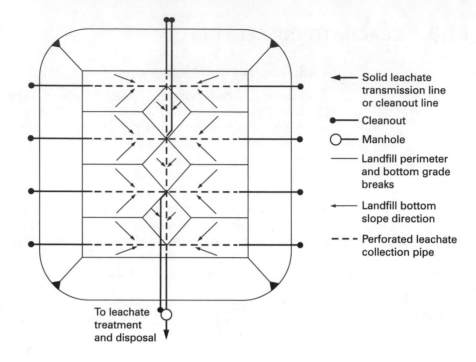

Solid leachate
transmission line
or cleanout line

● Cleanout

○— Manhole

Landfill perimeter
and bottom grade
breaks

Landfill bottom
slope direction

– – – Perforated leachate
collection pipe

To leachate
treatment
and disposal

where

$$c = e/k_d \tag{11.22}$$

For a liner with a slope, the following equation was given by Moore (1983) for the maximum value of y.

$$y_{max} = L((c + S^2)^{0.5} - S) \tag{11.23}$$

The formulation of Eq. 11.23 is not stated in Moore (1983). However, the formula can be derived by assuming that the second term of Eq. 11.19 is constant and equal to $k_d(y_{max})S$ and that $y = 0$ at $x = L$. With these assumptions, Eq. 11.19 is readily solved as

$$y^2 + 2y_{max} S(L - x) - c(L^2 - x^2) = 0 \tag{11.24}$$

If we further assume that $y = y_{max}$ at $x = 0$, then Eq. 11.24 reduces to the Moore (1983) equation by solving the quadratic equation for the positive root of y_{max}.

An analytical solution of Eq. 11.19 is obtained by noting (McEnroe, 1989) that the slope dy/dx for $S = 0$ is about -1 at the downstream boundary (bottom drain) and assuming that for small slope S, this approximation is valid. Then the head at the downstream y_L for a functional drain is obtained from Eq. 11.20 for $S = 0$:

$$y_L = Y = cL \qquad (11.25)$$

The solution of Eq. 11.19 depends on the ratio of the parameter c to the slope S (McEnroe, 1989).

If drain system is not functioning properly, saturated depth may exceed y_L in Eq. 11.25. The solution to Eq. 11.19 then is given in dimensionless form.

Case 1   For $c < S^2/4$,

$$x_* = (D)(E \cdot F/(E + 2A)(F - 2A))^{S/2A} \qquad (11.26a)$$

Case 2   For $c = S^2/4$,

$$x_* = ((S - 2Y_*)/(S - 2u_*)) \exp \{(2S(Y_* - u_*)/(S - 2Y_*)(s - 2u_*))\} \qquad (11.26b)$$

Case 3   For $c > S^2/4$,

$$x_* = D \exp\{(-S/2(\tan^{-1}((2u_* - S)/B) - \tan^{-1}(2Y_* - S)/B))\} \qquad (11.26c)$$

where
$$A = (S^2 - 4c)^{0.5}$$
$$B = (4c - S^2)^{0.5}$$
$$D = ((c - SY_* + Y_*^2)/(c - S u_* + u_*^2))^{0.5}$$
$$E = 2Y_* - S - A$$
$$F = 2u_* - S + A$$
$$Y_* = c$$
$$u_* = y_*/x_*$$
$$x_* = x/L$$
$$y_* = y/L$$

Figure 11.7 (McEnroe, 1989) shows the maximum normalized y (y_{max}/L) as a function of c ($=e/k$) for various slopes S.

Peyton and Schroeder (1990) presented a solution that permits an approximation for estimating the average saturated depth above the soil liner based on empirical data of numerical results. The solution, based on an empirical fit of numerical results, takes the form (all length dimensions in inches)

$$Q_d = 2C_1 k_d y_{av}(SL + y_0)/L^2 \qquad (11.27a)$$

where
Q_d = lateral drainage rate per unit area of the liner
$y_0 = (y_{av})^{1.16}/(SL)^{0.16}$ = saturated depth above liner at $x = 0$
(crest of the drainage layer) $\qquad (11.27b)$

Figure 11.7
Dimensionless
maximum saturated
depth (reproduced
from McEnroe, 1989,
by permission)

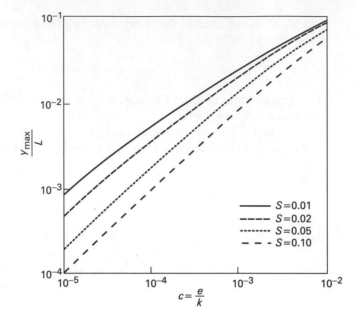

$$C_1 = 0.51 + 0.00205SL \tag{11.27c}$$

y_{av} = average saturated depth

The volume of leakage through the liner is

$$Q_1 = L_f k_1 (y_{av} + d)/d \tag{11.27d}$$

where

Q_1 = leakage through the liner

k_1 = hydraulic conductivity of the liner

L_f = fraction of the horizontal area of the soil through which percolation is occurring under a leaky synthetic liner. ($L_f = 1.0$ in the absence of a synthetic liner)

d = thickness of the liner

It is also true that

$$Q_d + Q_1 = e \tag{11.27e}$$

For the case of long and/or steep slopes, $y_0 = 0$ and, assuming a thick liner, $Q_1 = L_f k_1$ and

$$y_{av} = (e - L_f k_1)L/(2SC_1 k_d) \tag{11.27f}$$

If the saturation boundary is assumed to be a parabola, then the maximum saturated depth is $1.5y_{av}$. If k_1 is very small, then

$$(y_{max}/L) = 0.75c/SC_1 \qquad (11.27g)$$

For the geometry in Figure 11.5, the flow rate per unit length to the drain is

$$2L \cdot e \qquad (11.27h)$$

Example 11.2

Assume that the calculated percolation through the landfill cover into saturated refuse is 40 cm/yr. The designer specified a maximum head of 10 cm above the liner. The minimum thickness of the drainage layer for leachate collection is 30 cm and the specified hydraulic conductivity is 0.01 cm/s.

a. Determine the required spacing of the drains. The liner slopes uniformly at 5% slope.

b. If the drainage layer has a conductivity of 0.001 cm/s, what is the maximum thickness of saturation in the drainage layer.

Solution

a.

$k = 0.01$ cm/s $= 315,360$ cm/yr
$c = e/k_d = 40/315,360 = 1.27 \times 10^{-4}$

From Figure 11.7,

$y_{(max)*} = 2 \times 10^{-3} = y_{max}/L = 10$ cm/L
$L = 50$ meters

Drain spacing $= 2L = 100$ m

From Eq. 11.23,

$y_{max}/L = (0.000127 + 0.0025)^{0.5} - 0.05$
$= 1.254 \times 10^{-3}$

$L = 79.75$ m

Drain spacing $= 2L = 159.5$ m

From Eqs. 11.27c and 11.27g,

Assume $L = 36$ m $= 1417$ in.
$C_1 = 0.51 + 0.00205 \times 1417 \times 0.05 = 0.65$
$y_{max}/L = 0.75 \times 0.000127/0.05 \times 0.65 = 2.93 \times 10^{-3}$
$y_{max} = 10$ cm $= 3.94$ in.
$L = 1344.71$ in. $= 34$ m, close to 36, no more iterations

Drain spacing $= 68$ m

b. If $k_d = 0.001$ cm/s, and for the spacings calculated above, y_{max} by various methods is

From Figure 11.7,

$c = 1.27 \times 10^{-3}$
For $S = 0.05$, $y_{max}/L = 1.54 \times 10^{-2}$
$y_{max} = 50 \times 1.54 \times 10^{-2} = 0.77$ m $= 77$ cm

From Eq. 11.23,

$L = 79.75$ m

$$y_{max}/L = (0.0025 + 0.00127)^{0.5} - 0.05 = 0.0114$$
$$y_{max} = 0.91 \text{ m} = 91 \text{ cm}$$

From Eqs. 11.29c and 11.29g,

$$C_L = 0.51 + 0.00205 \times 0.05 \times 34 \times 3.28 \text{ ft/m} \times 12 \text{ in./ft} = 0.65$$
$$y_{max} = 34 \times 0.75 \times 0.00127/(0.05 \times 0.65) = 0.996 \text{ m} = 99.6 \text{ cm}$$

In all cases, the maximum thickness of the saturation is more than the thickness of the drainage layer. For a proper design, the drainage layer thickness should be larger than the saturation thickness. Thus the solution presented in Figure 11.7 is the most rigorous solution. The HELP model may also be used for sizing the drain spacing and design of the leachate collection system.

11.3.2 Transient Flow

In old landfills without modern liners, a leachate mound develops within the landfill. The height of the leachate mound depends on the climate (rainfall, temperature, etc.), the hydraulic conductivity of the final cover, the extent of the vegetation on the landfill, the geometry of the landfill, and the hydraulic conductivity of the soils beneath the landfill.

Figure 11.8 shows a typical leachate mound in the Hackensack Meadows district in New Jersey. The landfill is underlain first by a layer of sand and then by a thick layer of clay. An effective remediation scheme is to build a containment wall around the facility and install a leachate collection drain inside the wall with invert below the prevailing water level outside the landfill. In this case a constant head is maintained along the perimeter of the landfill. The hydraulic barrier thus created would preclude migration of leachate across the barrier if the barrier is keyed into a stratum with a low hydraulic conductivity (Oweis and Marturano, 1990; Oweis et al., 1994). If an effective cap is placed, the leachate mound will drain over time at a progressively reduced rate. If the cap is not effective, the lack of reduction in leachate collection may be an indication of a failed cap (Oweis and Biswas, 1993). Considering the geometry shown in Figure 11.9, the steady-state solution for the

Figure 11.8
Typical leachate mound in the Hackensack Meadows landfill

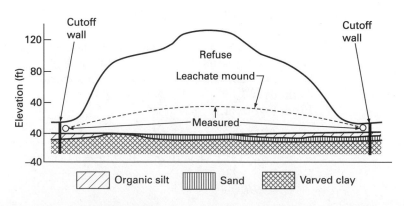

leachate mound height at a distance x from the toe may be estimated (Oweis and Biswas, 1993) based on

$$h_x = (d^2 + 2ex(L - x/2))^{0.5} \qquad \text{(11.28a)}$$

and the maximum leachate mound at $x = L$ is

$$h_{max} = (d^2 + eL^2)^{0.5} \qquad \text{(11.28b)}$$

Under such conditions the flow to the perimeter drain will simply be equal to eL.

The time to achieve a steady state is dependent on the refuse conductivity and other parameters. Oweis and Biswas (1993) presented formulas for the transient case of leachate mound buildup.

A reduction in leachate generation due to an impervious cap would lead to a reduction in the height of the leachate mound and a reduction in leachate flow to the drain. The following approximate relationships can be used to predict the decline of the height of the leachate mound from an initial maximum steady-state value followed by a decline in percolation. The analytical procedure (Oweis and Biswas, 1993) requires the following steps.

1. Select the period at the end of which $h'_{x,t}$ is desired.
2. Make an initial estimate of h and solve for h' using Eq. 11.29

$$h' = 0.5(h_t + h'_t)$$
$$\eta = kh'/S_y \qquad \text{(11.29)}$$

where h_t' is the value of h determined for the previous time $t' < t$, h_t is the estimated h value at time t, and S_y is specific yield for refuse (about 0.1; Oweis et al., 1990).

3. Solve for $\phi_{x,t}$ using Eq. 11.30, then solve $h_{x,t}$.

Figure 11.9
Idealization of leachate mound development

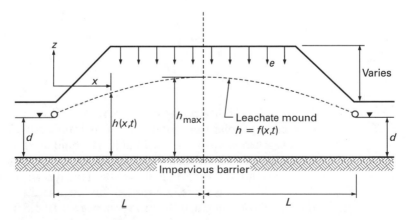

$$\phi_{x,t} = \phi_0 + ex(L - x/2) + \sum_{n=1}^{\infty} [\exp(-\beta_n^2 \eta t)] A_n \sin \beta_n x \qquad \textbf{(11.30)}$$

where

$$\phi_{x,t} = k(h_{x,t})^2/2$$

$$\beta_n = (2n - 1)\pi/2L \qquad \textbf{(11.30a)}$$

The coefficients A_n are obtained from initial conditions where

$$A_m = (2/L) \int_0^l g_x \sin(m\pi x/2L) \, dx, \qquad m = 1, 3, 5, \ldots \qquad \textbf{(11.30b)}$$

$$g_x = (\phi_{max} - \phi_0)\sin \pi x/2L - ex(L - x/2) \qquad \textbf{(11.30c)}$$

or

$$A_n = (\phi_{max} - \phi_0)\delta_n - 16eL^2/(2n - 1)^3\pi^3 \qquad n = 1, 2, 3 \qquad \textbf{(11.30d)}$$

where $\delta_n = 1$ for $n = 1$; $\delta_n = 0$ for $n \neq 1$; and $\phi_{0,t} = kd^2/2 = \phi_0$, $h_{0,t} = d$ **(11.30e)**

$$\phi_{max} = k(h_{max})^2/2 \qquad \textbf{(11.30f)}$$

$$h_{x,t} = \sqrt{(2 \cdot \phi/k)} \qquad \textbf{(11.30g)}$$

4. Compare step 3 to step 2 and repeat if necessary using the improved value of $h_{x,t}$ determined in step 2.

5. Stop the iteration when $(h \text{ (step 4)} - h \text{ (step 2)}/h \text{ (step 2)})$ is less than 3%. The 3% limit was selected to limit the number of iterations to about 3.

The leachate flow into the drain may be estimated using

$$q_{0,t} = -\left[eL + \sum_{n=1}^{\infty} (\exp(-\beta_n^2 \eta t)A_n \beta_n) \right] \qquad \textbf{(11.31)}$$

Example 11.3

Consider a landfill prior to closure with a total width of 2000 ft and a perimeter leachate collection system. A low hydraulic conductivity clay layer is 10 ft beneath the base of this landfill. The fluid level in the drain is at a constant head of 10 ft above the clay layer. The maximum leachate level within the landfill is 35.37 ft above the clay layer in the middle of the landfill. A regulated cap is placed on the landfill; assume that the cap is functional and the percolation through the cap is no more than 1.5 in./yr (0.125 ft/yr). The hydraulic conductivity of the refuse is assumed to be 1000 ft/yr (10^{-3} cm/s) and the specific yield is 0.1. Estimate what the maximum leachate head decline will be $\frac{1}{2}$ yr and 1 yr following cap installation. The following calculations are made for convenience.

Solution:

$k = 1000$ ft/yr, $S_y = 0.1$, $e = 0.125$ ft/yr

$h_{max} = h_{L,0} = 35.37$ ft

$\phi_{max} = (35.37)^2 \times 1000/2 = 0.6255 \times 10^6$ ft³/yr

From Eq. 11.30f, for $h_{max} = 35.37$ ft, $k = 1000$ ft/yr

$\phi_0 = \frac{1}{2} \times 1000 \times (10)^2 = 0.05 \times 10^6$ ft³/yr

From Eq. 11.30e, for $h = 10$ ft, $k = 1000$ ft/yr

$\phi_{max} - \phi_0 = 0.5755 \times 10^6$ ft³/yr

$n = 1, \beta_n = 1.57 \times 10^{-3}$ ft⁻¹

From Eq. 11.30e, for $n = 1$, $L = 1000$ ft

$n = 2, \beta_n = 4.71 \times 10^{-3}$ ft⁻¹

$n = 3, \beta_n = 7.85 \times 10^{-3}$ ft⁻¹

From Eq. 11.30d,

$$A_n = 0.5755 \times 10^6 \delta_n - \frac{16 \times 0.125 \times (1000)^2}{(2n-1)^3 \, \pi^3}$$

$$= 0.5755 \times 10^6 \, \delta_n - \frac{0.0645 \times 10^6}{(2n-1)^3}$$

For $n = 1$, $A_n = 0.511 \times 10^6$

For $n = 2$, $A_n = 0.002389 \times 10^6$

For $n = 3$, $A_n = -0.000516$

From Eq. 11.30,

$$\beta_n^2 = (\pi/2L)^2(2n-1)^2 = 2.467 \times 10^{-6}(2n-1)^2$$

$n = 1, \beta_n^2 = 2.467 \times 10^{-6}$

$n = 2, \beta_n^2 = 22.206 \times 10^{-6}$

From Eq. 11.29,

$\eta t = kh' \cdot t/S_y = (1000/0.1) \, (h't) = (0.01 \, h't) \times 10^6$

mound height h estimate for $t = \frac{1}{2}$ yr:

Trial 1

$t = \frac{1}{2}$ yr, assume $h_t = 32'$, $h' = (32 + 35.37)/2 = 33.69$

$\eta t = (1000/0.1)h't = (1000/0.1) \, (33.69) \, (\frac{1}{2}) = (0.01)(33.69)(\frac{1}{2})(10)^6$

$= 0.1684 \times 10^6$ ft²

Calculate

$$[\exp(-\beta_n^2 \eta t)]A_n \, \sin(2n-1)\pi/2$$

$n = 1, \beta_n^2 \eta t = (2.467 \times 10^{-6})(0.1684)(10^6) = 0.4154$

$n = 2, (22.206 \times 10^{-6})(0.1684)(10^6) = 3.74$

$$\phi_0 = 0.5 \times 10^5 \text{ ft}^3/\text{yr} \text{ [Eq. 11.30e, } h = 10 \text{ ft, } k = 1000 \text{ ft/yr]}$$

$n = 1, e^{-0.4154} \times 0.511 \times 1.0 \times 10^6 = 3.373 \times 10^5$

$n = 2, e^{-3.74} \times 0.0024 \times 10^6 \qquad = \underline{0.00057 \times 10^5}$

Sum $\qquad\qquad\qquad\qquad\qquad\quad = 3.373 \times 10^5$

From Eq. 11.30, calculate ϕ at $x = L = 1000$ ft

$$\phi = 0.5 \times 10^5 + 0.125(1000)(500) + 3.373 \times 10^5$$

$$= 1.125 \times 10^5 + 3.373 \times 10^5 = 4.498 \times 10^5 \text{ ft}^3/\text{yr}$$

$$h = \left(\frac{2\phi}{k}\right)^{0.5} = \left(\frac{2 \times 4.498}{1000} \times 10^5\right)^{0.5} = (200 \times 4.498)^{0.5} = 30 \text{ ft}$$

$$100\left(\frac{h_{\text{calculated}} - h_{\text{assumed}}}{h_{\text{assumed}}}\right) = \left(\frac{30 - 32}{32}\right)(100) = 6.25\% > 3\%$$

Try again! Note that the contribution of terms for $n > 1$ is very small and could be ignored for the mound dissipation problem in this case.

Trial 2

$$h = 30, h' = \frac{30 + 35.37}{2} = 32.685 \eta t = 0.1634 \times 10^6$$

$n = 1, [\exp(-\beta_n^2 \eta t)]A_n \sin(2n - 1)\pi/2$

$\quad = e^{-0.403} \times 0.511 \times 10^6$

$\quad = 3.415 \times 10^5$

$\phi = (3.415 + 1.125)10^5 = 4.54 \times 10^5 \text{ ft}^3/\text{yr}$

$h = (4.54 \times 200)^{0.5} = 30.13 \text{ ft}$

$$100\left(\frac{h_{\text{calculated}} - h_{\text{assumed}}}{h_{\text{assumed}}}\right) = \left(\frac{30.13 - 30}{30}\right)100 = 0.43\% < 3\%$$

Use $h = 30.1$ ft.

Thus the calculated leachate mound decline in $\frac{1}{2}$ yr is 5.27 ft (i.e., $35.37' - 30.1'$).

Calculate the flow volume from Eq. 11.31 using only the first term of the series as higher terms are very small.

$e = 0.125, L = 1000', h' = (30.1 + 35.372/2 = 32.735)$

From Eq. 11.29, $\eta = \dfrac{1000 \times 32.735}{0.1} = 0.327 \times 10^6$

$$n = 1, \beta_n^2 = 2.467 \times 10^{-6}, \beta_n^2 \, \eta t = 0.403$$
$$A_n = 0.511 \times 10^6, \beta_n = 1.57 \times 10^{-3}$$
$$\eta t = 0.327 \times 0.5 \times 10^6 = 0.1635 \times 10^6$$
$$q = -(0.125 \times 1000 + e^{-0.403} \times 0.511 \times 10^6 \times 1.57 \times 10^{-3}$$
$$= -661.2 \text{ ft}^3/\text{yr}/\text{ft}$$

Calculate leachate mound dissipation after 1 yr.

$$t = 1 \text{ year}$$
Try $h = 27.0$ ft, $h' = (27 + 30.1)/2 = 28.55$ ft,
$$\eta t = 1000/0.1 \times 28.55 \times 1 = 0.2855 \times 10^6$$

Note that the value of h at the end of $\frac{1}{2}$ yr is used for calculating h' (Eq. 11.29).

$$\beta_n^2 \eta t = (2.467 \times 10^{-6})(0.286)(10^6) = 0.704$$
$$n = 1, [\exp(-\beta_n^2 \eta t)]A_n \sin(2n-1)(\pi/2)$$
$$= e^{-0.704} \times 0.511 \times 10^6$$
$$= 2.527 \times 10^5$$
$$\phi = (2.527 + 1.125)(10^5) = 3.652 \times 10^5 \text{ ft}^3/\text{yr}$$
$$h = (3.652 \times 200)^{0.5} = 27.03 \text{ ft}$$

Use $h = 27.0$ ft.

Calculate the flow volume using the first terms of series in Eq. 11.31.

$$h' = (28.55)\eta t = (1000/0.1) \times 28.55 \times 1 = 0.2855 \times 10^6$$
$$\beta_n^2 \eta t = 2.467 \times 10^{-6} \times 0.2855 \times 10^6 = 0.704$$
$$\text{Discharge} = -(0.125 \times 1000 + e^{-0.704} \times 0.511 \times 10^6 \times 1.57 \times 10^{-3})$$
$$= -521.8 \text{ ft}^3 \text{ yr}/\text{ft}$$

11.3.3 *Leachate Collection Pipe*

A schematic leachate collection system layout was shown in Figure 11.6. The leachate collection pipe is sized to handle the anticipated flow. A collection pipe is usually 6 in. (15 cm) in diameter, although some designers prefer an 8-in. (20-cm) diameter as a minimum.

The hydraulics of the pipe is evaluated using the Manning formula:

$$Q = 1.486AR^{2/3}S^{0.5}/n \qquad \textbf{(11.32a)}$$

where Q is the flow rate, A is the cross-sectional area, R is the hydraulic radius (wetted area/wetted perimeter), S is the slope, and n is the Manning roughness coefficient.

Assuming a circular plastic pipe ($n = 0.01$) running full, the Eq. 11.32a may be expressed as

$$D = 0.237(Q/S^{0.5})^{0.375} \qquad \textbf{(11.32b)}$$

where D is the required pipe diameter.

Example 11.4

If the drains in Example 11.2 slope at a uniform 0.05% slope to a common header sloping at 0.1%, determine the design flow rate and the minimum size for the drain and the common header. The area covered by the drainage system is 300 m × 800 m. Assume the method of Figure 11.7 applies.

Solution: The flow rate and minimum size for the drain are calculated using
$e = 40$ cm/yr $= 1.27 \times 10^{-6}$ cm/s
$Q = 100$ m $\times$ 300 m $\times$ 1.27×10^{-8} m/s $= 3.81 \times 10^{-4}$ m^3/s
From Eq. 11.32b,

$$D = 0.237(3.81 \times 10^{-4}/0.0005^{0.5})^{0.375} = 0.051 \text{ m} = 5.1 \text{ cm}$$

Use a minimum of 15 cm.

The flow rate and minimum size for the header are calculated as
$Q = 800 \times 300 \times 1.27 \times 10^{-8}$ m/s $= 3.05 \times 10^{-3}$ m^3/s
$D = 0.237(0.00305/0.001^{0.5})^{0.375} = 0.098$
$D = 9.8$ cm
Use a minimum of 15 cm.

The structural design of the leachate collection pipe depends on the vertical load on the pipe and bedding conditions. Flexible pipes are usually used in a leachate collection system. Terminology such as trench width, haunching, bedding, springline, and others are illustrated in Figure 11.10. The load-carrying capacity of such pipes is derived mostly from interaction with the surrounding bedding. Plastic, steel, and ductile iron pipes are all flexible pipes. Steel and ductile iron are susceptible to corrosion by leachate and perforations cannot be easily made. Thus, plastic pipes are more common and may include polyvinyl chlorine (PVC), polyethylene (PE), acrylonitrile-butadiene-styrene (ABS), and others.

Rigid pipes such as concrete pipe (CP) derive most of their strength from their inherent structure but are subject to chemical attack and perforations are hard to make. Corrosion-resistant rigid pipes such as vitrified clay also lack adequate bending strength. Because of all these limitations, rigid pipes are not normally used in a leachate collection system.

The techniques developed for solid pipes can be used for perforated pipes by arbitrarily increasing the actual design load. If l_p is the cumulative length of perforations per unit length of pipe, then the design load used is

$$\text{Design load} = \text{actual design load} \times (1/(1 - l_p)) \qquad \text{(11.33)}$$

The load on a pipe depends on whether the pipe is in a trench (trench condition) or embedded in an embankment above the base of the embankment (positive projecting condition), as illustrated in Figure 11.11. For a pipe in a trench, the load on the top of the pipe is a fraction of the load generated by

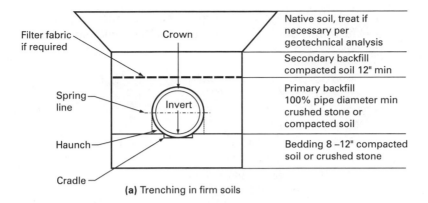

(a) Trenching in firm soils

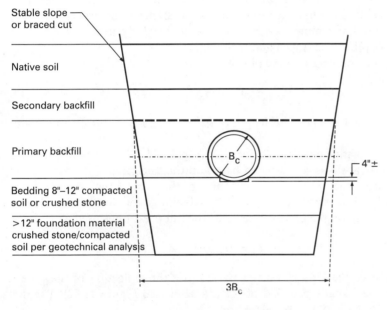

(b) Trenching in soft soils

Figure 11.10 Pipe trenching terminology

Figure 11.11
Pipe installation (EPA, 1983)

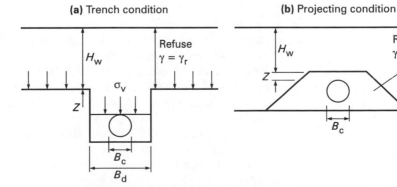

the material above the trench plus the load fraction generated by the trench backfill minus the frictional resistance of the trench wall. The vertical stress on the pipe is

$$\sigma_v = B_d \gamma_f C_d + (\gamma_r H_w) C_w \qquad (11.34)$$

where

$C_d = (1 - C_w)/2k_f$
$C_w = \exp(-2k_f(Z/B_d))$
γ_r = average unit weight of the refuse and soil cover
γ_f = unit weight of the backfill
H_w = height of the waste
B_d = width of the trench at a level corresponding to the top of the pipe
k_f = coefficient characterizing the side resistance of the trench
k_f = 0.165 maximum for sand and gravel, 0.19 for gravel
Z = depth of the trench above the pipe
The load per unit length of the pipe is

$$W = B_c \sigma_v \qquad (11.35)$$

where W is the force per unit length of the pipe and B_c is the outside diameter of the pipe.

For the positive projecting condition (Figure 11.11b), the parameter k_f is 0, and the vertical stress is computed based on the thickness of the backfill above the pipe and the height of refuse above that. In actual design practice, designers often conservatively ignore the advantage due to arching in computing the pressure on the pipe. In this case, the full pressure is used as

$$\sigma_v = H_w \gamma_r + \gamma_f Z \qquad (11.36)$$

The section of the pipe is derived by limiting the deflection of the pipe to a specific value, usually 5 to 7% of the diameter (see Figure 11.12). The deflection is computed (EPA, 1983) based on

$$\delta_y/B_c = \sigma_v \cdot D_e K_b/(EI/r^3 + 0.061E') \qquad (11.37)$$

where

δ_y = horizontal and vertical deflection of the pipe (in.)
D_e = empirical time factor to account for soil/pipe time-dependent behavior (conservatively may be assumed 2.0)
E = Young's modulus of the pipe material (lb/in.^{2})
E' = modulus of passive soil resistance (lb/in.^{2}) (typically 400 lb/in.^{2} for average compaction). Howard (1981) conducted a comprehensive study involving several soil types
K_b = bedding constant, which varies with the angle of bedding. The typically used value is 0.1 for plastic pipe. This parameter varies from 0.11 for a point bottom support to 0.083 for full support.
r = radius of the pipe (in.)
I = moment of inertia of the pipe section (in.^{4}/in.)

Figure 11.12
(a) Example of the effect of trench geometry and pipe sizing on ring deflection (EPA, 1989) (b) Vertical ring deflection versus vertical soil pressure for 18-in. corrugated polyethylene in high-pressure soil cell

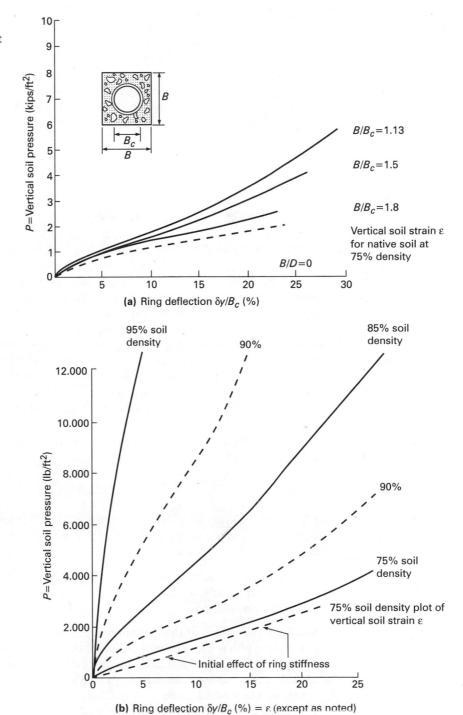

(a) Ring deflection $\delta y/B_c$ (%)

(b) Ring deflection $\delta y/B_c$ (%) = ε (except as noted)

The ratio EI/r^3 is called the *ring stiffness*. ASTM method ASTM D 2412 is used to determine this ratio based on measured percentage deflection. ASTM D 2321 is used as the recommended installation practice. Pipes are typically selected by schedule numbers. A schedule number (typically 40 or 80) indicates the approximate value of the expression $1000 \times P/S$, where P is the service pressure and S is the allowable stress (both are expressed in lb/in.^{2}). SDR (standard dimension ratio) is the ratio of the specified outside diameter of a plastic pipe to its wall thickness.

In addition to earth loads, the pipe could be subjected to concentrated or distributed loads. For concentrated loads the load per unit length of the pipe is given by

$$W_{sc} = PF/C_1 \cdot L \qquad (11.38)$$

Here, W_{sc} is the load on the pipe (lb/ft), P is the concentrated load (lb), and L is the effective length of the pipe (ft). Three ft is used for pipes over 3 ft long; the actual length is used for shorter pipes. The effective length is defined as the length over which the average load caused by the surface wheels produces nearly the same stress in the pipe wall as does the actual load, which varies from point to point. C_1 is the load coefficient that is a function of $B_c/2H$ and $L/2H$, where H is the height of fill from the top of the pipe to the ground surface. Figure 4.15 (Chapter 4) can be used to compute C_1 by entering $B_c/2H$ and $L/2H$ as m and n. C_1 is $4I$, where I is the influence value from Figure 4.15. Lastly, F is the impact factor, which reaches 1.3 for cover thicknesses less than 1 ft, 1.2 for thicknesses of 1.1 to 2.0 ft, 1.1 for thicknesses 2.1 to 2.9 ft, and 1.0 for thicknesses 3 ft or larger.

For distributed loads such as truck load, the load on the pipe, W_{sd}, may be determined from

$$W_{sd} = C_z pFB_c \qquad (11.39)$$

where p is the intensity of the distributed load (lb/ft^2) and C_z is the load coefficient that is a function of $D/2H$ and $M/2H$, where D and M are the width and length, respectively, of the loaded area.

Example 11.5

If the depth of waste in Example 11.1 is 50 ft and the average unit weight (accounting for daily, intermediate, and final cover) is 60 lb/ft^3, determine the required pipe strength.

Solution: Assume that the minimum depth of the trench above the pipe is 2.5 ft and the width is 2 ft. The unit weight of the washed stone backfill is 135 lb/ft^3. Use a creep factor of 2.0, a modulus of passive resistance of 400 psi for backfill, and a bedding constant $K_b = 0.083$ using Eq. 11.35.

$k_f = 0.19$ washed stone
$C_w = \exp(-2 \times 0.19 \times 2.5/2.0) = 0.62$
$C_d = (1 - 0.62)/2 \times 0.19 = 1.00$
$\sigma_v = 2 \times 135 \times 1.0 + 50 \times 60 \times 0.62 = 2265$ lb/ft^2 = 14.8 lb/in.^{2}

Assume that the perforations are 15% of the pipe surface area.

Corrected $\sigma_v = 14.8/0.85 = 17.4$ psi

Design for $\delta y/B = 0.05$

From Eq. 11.37,

$0.05 = 17.4 \times 2 \times 0.083/((EI/r^3) + 0.061 \times 400)$

$EI/r^3 = 33.34$

If arching is conservatively ignored and the pipe is assumed projecting, then

$$\sigma_v = 50 \times 60 + 2.5 \times 135 = 3337.5 \text{ psf} = 23 \text{ psi}$$

(27.05 psi accounting for the perforations)

$$EI/r^3 = 65.4$$

From the above calculations, the pipe schedule or SDR is specified based on the manufacturer's literature or specific tests.

11.4 DRAINAGE AND FILTER REQUIREMENTS

11.4.1 *Soil Filters*

In addition to providing support, backfill should allow leachate to flow into the pipe. Thus, backfill must be a relatively free-draining material. It is also desirable to provide redundancy in the design of the backfill to meet the drainage requirements even if the pipe becomes clogged. The following characteristics are recommended for the drainage stone (Washington State Department of Ecology).

U.S. standard size/sieve no.	% passing by weight
$1\frac{1}{2}$ in.	100
No. 50	<5
Coefficient of uniformity $C_u (D_{60}/D_{10})$	4 or greater
Coefficient of curvature $C_c = (D_{30})^2/(D_{10})(D_{60})$	1.0 to 3.0

D_{10}, D_{30}, and D_{60} are the sieve sizes where 10, 30, and 60% (respectively) of the material by weight is finer.

In addition to the above requirements, the carbonate content should be limited to limit the potential for dissolution and clogging. Schmuker and Buffalini (1995) cited a 1994 Ohio standard that limits loss of weight to less than 5% using an acidic solution (pH = 4).

When water flows through soil into a drainage system, the soil through which water moves is called *base soil*. In between the base soil and the drainage layer (or a free face), a filter layer is often needed to intercept cracks or openings in the base soil and to prevent movement of soil particles by water passing through the openings. The filter is graded so that the particles cannot pass through the filter voids and are caught at the filter face; this prevents continued movement of soil particles and concentrated flow through cracks and openings (SCS, 1986). The U.S. Soil Conservation Service (SCS, 1986) established filter requirements based on four soil categories as follows.

Category	% by weight finer than No. 200 standard U.S. sieve (0.075 mm)
1. Fine silts and clays	>85
2. Sands, silts, clays, silty and clayey sands	40–85
3. Silty and clayey sands, gravels	15–39
4. Sands and gravels	<15

The category designation for soil containing particles larger than a No. 4 sieve (4.75 mm) should be determined from the gradation curve of the base soil that has been adjusted for 100% passing the No. 4 sieve.

The filter criteria is established in terms of d_{15}, d_{85}, D_{10}, D_{15}, D_{85}, and D_{90}; d_{15} and d_{85} are the particle sizes (mm) corresponding to 15 and 85% (respectively) for the base material. D_{10}, D_{15}, D_{85}, and D_{90} are the particle sizes (mm) corresponding to 10%, 15%, 85%, and 90% (respectively) finer by weight of the filter material.

The filter requirements for the four base categories are shown in the following chart.

Category	Filter criteria
1	$D_{15} < 9 \times d_{85}$ $D_{15} = 0.2$ mm minimum
2	$D_{15} = \leq 0.7$ mm
3	$D_{15} = \leq ((40 - A)/(40 - 15))(4d_{85} - 0.7 \text{ mm}) + 0.7$ mm $A = \%$ passing a No. 200 sieve (0.075 mm) after regrading
4	$D_{15} \leq 4d_{85}$. If D_{85} is <0.7 mm, use 0.7 mm.

For all categories except 4, the gradation curve for the base soil containing particles larger than a No. 4 sieve should be adjusted to 100% passing a No. 4 sieve (4.75 mm). For all categories, the fines (passing a No. 200 sieve) must be nonplastic. The maximum particle size is limited to 3 in. (75 mm), and the minimum, D_{15}, should be equal to $4d_{85}$ or 0.2 mm, whichever is greater.

Filters should have a relatively uniform grain size distribution curve, without absence of some sizes (as indicated by gaps in the gradation curve). The following limits are suggested (SCS, 1986) for preventing segregation of filters.

Minimum D_{10} (mm)	Maximum D_{90} (mm)
Less than 0.5	20
0.5–1.0	25
1.0–2.0	30
2.0–5.0	40
5.0–10	50
10–50	60

For filters surrounding perforated pipes the D_{85} for the filter should be no smaller than the perforation diameter.

11.4.2 *Geotextile Filter*

Geotextiles (also called *filter fabrics*) are made from synthetic polymers (resins) such as polypropylene, polyester polyethylene, and nylon. Approximately 75% of all geotextiles are made of polypropylene resin; 20% are made of polyester; and 5% are made of polyethylene, nylon, and other special-purpose resins (EPA, 1993). In addition to the resin type and additives (carbon black and others), the fabric type and geotextile type are used to characterize various fabrics. An informative but condensed description of fabric and geotextile types are described in EPA (1993) and are summarized here.

The resin and additives are introduced to an extruder that supplies heat, mixing action, and filtering. The molten material is forced through a die containing small orifices. The exiting fibers (yarns) are stretched and cooled. *Yarn* is a generic name for any continuous strand (fiber, filament, or tape) used to form a textile fabric. The yarns described here are joined together to make a fabric or geotextile. Woven geotextiles are in a "basket type" uniform weave: Each yarn goes over and under an intersecting yarn on an alternate basis. The result is a face with defined windows (openings).

Nonwoven, needle-punched geotextiles go through a needling process where barbed needles penetrate the fabric and entangle numerous fibers transverse to the plane of the fabric. Nonwoven, heat-bonded type geotextiles are formed by passing the unbonded fiber mat through a source of heat. Some of the fabrics are melted at various points. The mat is compressed and fibers are joined at their intersection by melt bonding.

The usual specification items are the mass per unit area (ASTM D 5261), grab tensile strength (ASTM D 4632), trapezoidal tear strength (ASTM D 4533), burst strength (ASTM D 3786), puncture strength (ASTM D 4833), thickness (ASTM D 5199), apparent opening size (ASTM D 4751), permittivity (ASTM D 4491), and ultraviolet light resistance (ASTM D 4355).

Seaming of geotextiles is required when used for filtration. Various types of seaming (flat, J, butterfly) are described in EPA (1993). ASTM D 4884 covers the seam strength.

The selection of the geotextile is based on permittivity and filter requirements. The permittivity is defined as

$$\psi = k_n/t \tag{11.41}$$

where ψ is the permittivity, k_n is the hydraulic conductivity normal to the plane of the fabric, and t is the thickness of the fabric. If the leachate head above the fabric is h and the inflow rate per unit area is Q_d, then the required permittivity is

$$(\psi)_{required} = k_n/t = Q_d/h \tag{11.42}$$

Example 11.6

a. The leachate generation is 20 in./yr and the maximum allowed head buildup is 12 in. Calculate the required permittivity. b. If the flow rate in the laboratory is 600 mm^3/s at a head of 50 mm and the area normal to the flow is 4185 mm^2, calculate the permittivity.

Solution:

a. $Q_d = 0.00456$ ft/day

$$(\psi)_{required} = 0.00456/1 = 0.00456 \text{ day}^{-1} = 5.3 \times 10^{-8} \text{ s}^{-1}$$

b. $(\psi)_{geotextile} = 600/(50 \times 4185) = 2.9 \times 10^{-3} \text{ s}^{-1}$

The factor of safety, or the design ratio (DR), is defined as the ratio of the geotextile permittivity to the required permittivity. For Example 11.6, the DR is 54,717.

Several criteria have been proposed for sizing the geotextile as filter when in contact with the base soil. The 95% opening size, O_{95}, of the fabric is related to the soil to be retained in the following type of relationship:

$$O_{95} < \text{base size} (d_{50}, C_u, D_r) \tag{11.43}$$

Here, O_{95} is the opening size expressed in terms of U.S. standard sieve, the size of particles of which the fabric will retain 95%. O_{95} is approximately equal to AOS (apparent opening size), which is defined as the size of a uniform glassed bead for which 5% or less pass through the fabric. d_{50} is the particle size (mm) of the base material corresponding to 50% by weight finer, C_u is the uniformity coefficient of the base material, and D_r is the relative density of the base material.

The retention criteria, given by Giroud (1982), is expressed as

$$O_{95} < \beta C_u d_{50} \qquad \text{for } 1 < C_u < 3 \tag{11.44}$$

where
  β depends on the relative density
  $D_r < 35\%, \beta = 1.0$
  $35\% < D_r < 65\%, \beta = 1.5$
  $D_r > 65\%, \beta = 2.0$

$$O_{95} < \beta d_{50}/C_u \qquad \text{for } C_u > 3 \tag{11.45}$$

where $\beta = 9$, 13.5, and 18 for $D_r < 35\%$, $35\% < D_r < 65\%$, and $> 65\%$, respectively.

Other criteria proposed (NYDEC, 1993) are

$$O_{95}/d_{85} < 2 \text{ or } 3, K_f > 10K_s$$

where K_f is the geotextile permeability (hydraulic conductivity) and K_s is the soil permeability (hydraulic conductivity). The AASHTO (1990) criteria is $O_{95} < 0.6$ mm for soil with less than 50% passing a No. 200 sieve, $O_{95} < 0.3$ mm for soil with more than 50% passing a No. 200 sieve, and $K_f > K_s$.

11.4.3 Geonets

Geosynthetic drainage nets (geonets) have been used by some designers as a substitute for soil drainage layers because space can be saved. Geonets require a geotextile filter above them. Long-term creep (Smith and Kraemer, 1987) and clogging are major design considerations.

Geonets are formed by layers of intersecting ribs such that liquid can flow within the open spaces. Available geonets in thickness vary from 4.0 and 6.9 mm (EPA, 1993) and are usually made of polyethylene in the natural density range of 0.934 to 0.94 g/cc. Geonets always function with a geomembrane or a geotextile on their two planar surfaces; the net can therefore be protected from clogging by adjacent soil material (see Section 11.4.4). The quality control testing can utilize applicable ASTM standards (ASTM D 1505 or ASTM D 792 for density, ASTM D 5261 for mass per unit area, and ASTM D 5199 for thickness). The transmissivity requirement may be determined by considering the flow system in Figure 11.13.

Considering that the maximum head allowed is h_{max} (usually 30 cm), the distance to the drainage layer is L, and the flow rate per unit area is Q_d, then from Darcy's law we obtain

$$\text{Gradient } i = (h_{max} + L \tan \alpha)/(L/\cos \alpha)$$
$$= (h_{max} \cos \alpha + L \sin \alpha)/L \tag{11.46}$$
$$eL = k_p l i = iT$$

Figure 11.13
Transmissivity
requirement of a geonet

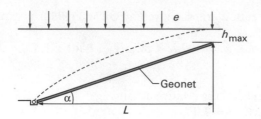

where k_p is the hydraulic conductivity in the plane of the geonet, t is the thickness of the geonet, and T is the transmissivity $= k_p t$. For small angle α, $\cos \alpha$ can be assumed to be 1.0 and

$$T = eL^2/(h_{max} + L \sin \alpha) \qquad (11.47)$$

Example 11.7

Given that $Q_d = 20$ in./yr, $L = 40$ ft, and the slope angle $\alpha = 4°$, determine the required transmissivity of the drainage layer for a maximum head no greater than 12 in.

Solution:
$$e = 20 \text{ in./yr} = 1.61 \times 10^{-6} \text{ cm/s}$$
$$L = 40 \text{ ft} = 1220 \text{ cm}$$
$$h_{max} = 30 \text{ cm}$$
$$T = (1220)^2 \times 1.61 \times 10^{-6}/(30 + 1220 \times 0.069)$$
$$= 0.022 \text{ cm}^2/\text{s} = 2.2 \times 10^{-6} \text{ m}^2/\text{s}$$

The transmissivity of geonets may be evaluated using ASTM D 4716. A large factor of safety is applied on the required T to allow for reduction in T due to creep, clogging, and other design factors (Koerner, 1990).

11.4.4 *Clogging*

Clogging refers to the process where a drainage medium (sand, gravel, geotextiles, drainage nets, or perforations in a drainage pipe) becomes clogged and thereby is reduced in its ability to transmit water. Clogging can occur from sedimentation, chemical, and biological causes. Clogging in soil filters depends on the amount of suspended solids that blind the filter. The soil filter criteria aim at controlling the migration of fine particles from the base through the filter.

Raghava and Atwater (1994) cited early research (Berend, 1967; Behnke, 1969) that demonstrated clogging of sands by turbid runoff and waste water. Settling of suspended solids initiates clogging at and below the infiltration surface and complete clogging of recharge basins of medium to fine sand can occur within a week with suspended solids as low as 50 ppm (by weight). A geotextile or permeable soil in contact with refuse could be clogged by suspended solids in the leachate stream.

Puig et al. (1986) cite a case of iron clogging a geotextile by chemical oxidation of ferrous iron and biological oxidation of ferrous iron in the presence of autotrophic bacteria. It was found that drain clogging was more severe in areas rich in ferrous iron and organic matter. The concentration of ferrous iron in the soil water was 0.2 mg/l.

While chemical oxidation beneath the landfill could be hampered by oxygen deficiency, high levels of biological activity (BOD) could give rise to biological clogging via formation of slime and sheath material, biomass formation, cohering, Fe S precipitation, and carbonate precipitation (Harborth and Hannert, 1987).

Geotextiles used as filters in contact with municipal waste are highly prone to clogging (Rollin and Denis, 1987). Experiments suggest significant clogging of geotextile under a leachate environment (Cancelli and Cazzuffi, 1987; EPA, 1992).

Stronger leachates (those with higher BOD, COD, and TS) have greater clogging impact. Drainage materials such as coarse sands and fine gravels are less likely to clog than are fine or silty sands. The U.S. Army Corps of Engineers (COE, 1977; Haliburton and Wood, 1982) gradient ratio method may be used to assess clogging. The gradient ratio is defined as the hydraulic gradient through the geotextile and 1 in. (25.4 mm) of soil immediately above it divided by the hydraulic gradient across the next 2 in. (51 mm) of soil above it. The results are interpreted as follows:

$GR < 1$ piping of adjacent soil through geotextile

$GR = 1$ geotextile does not inhibit flow

$GR > 3$ severe clogging (not acceptable)

Koerner et al. (1993) reported results from exhumed drainage media from three landfills. The permittivity of the geotextile exhumed from site 1 (a landfill receiving domestic and light industrial waste) was found to have decreased to 8.2×10^{-4} from 1.1 s^{-1} (a reduction factor of about 7364). The fabric was used as a sock around a perforated pipe and was clogged with particulate and biomass. The geotextile in site 2 (same landfill) experienced less clogging (clogged permittivity of 0.033 s^{-1} from 0.9). The fabric had a 7% open-area, nonwoven monofilament geotextile. The fabric was used to line a pipe trench to intercept leachate seeping from the side of the landfill.

At site 3, the clogged geotextile was wrapped around a 4-in. perforated, socked pipe. The permittivity of the geotextile was determined (after laboratory simulations) to have decreased by 5 orders of magnitude. The areal geotextile at the landfill base (needle-punched, nonwoven 16 oz/yd^2) decreased in permittivity by 2 orders of magnitude. The relatively better performance of the areal fabric versus the sock fabric was attributed to higher entrance velocities for the sock fabric.

At site 4, needle-punched, nonwoven geotextile was wrapped around a 4-in., schedule 40 PVC perforated well casing used for leachate recirculation and gas extraction at a vacuum of 3 to 6 in. of mercury. Geotextile specimens were exhumed from locations at 10, 25, and 50 ft below the landfill cover. The permittivity at the 50 ft depth decreased by about 4 orders of magnitude (3.4×10^{-4} s^{-1} from 1.1). The samples at the 25 and 50 ft levels were black with organic matter and caked with fine sediments. Heavy deposits were observed where the fabric spanned the pipe perforations.

Clogging of geosynthetic drainage systems is not restricted to geotextiles. Raghava and Atwater (1994) reported clogging of a composite geonet drainage system during in-plane leachate flow. Clogging due to surface sealing is well known in spray irrigation practices. Microbial clogging is common in pumping well applications and agricultural drains.

There is no criteria (other than the filter criterion discussed in this chapter) to quantify the clogging potential. Poor practices such as socking perforated pipes and well screens in refuse should be avoided. Even the need for a geotextile to separate refuse from drainage stone is questionable. In the landfills with successful performance, a drainage layer with a large open area and drainage pipes with large openings help reduce the potential of clogging. Maintaining low flow velocity also reduces the potential for clogging. Also, key elements of the drainage system should be accessible for cleaning. Cleanout for drainage pipes is now a common design feature.

If drainage geonets are to be used for economy and preserving landfill space, a large factor of safety should be applied to account for a decrease in transmissivity because of creep and clogging. It is preferable to use geonets in the secondary collection (detection) system and open gravel with high permeability for the primary drainage system.

Koerner et al. (1993) recommended that the permittivity of a geotextile surrounding a stone-filled trench (18 in. × 12 in. in size) be reduced (to account for clogging) by factors of 10, 20, 30, and 40 for drain spacings of 50, 100, 150, and 200 ft, respectively. Such a geotextile may not even be needed if the gravel is in direct contact with refuse or the drainage layer is an open gravel layer. If the drainage layer is composed of high permeability sand, then the gravel size should be designed as a filter. In this manner, the use of a geotextile and a potentially clogged medium can be avoided. Koerner et al. (1993) also suggested that corrugated pipes are less prone to clogging than are smooth pipes.

The leachate quality may be used as another parameter for assessing the potential for clogging. Table 11.9 shows a suggested criterion to avoid clogging hazard associated with drip irrigation systems using wastewater

Table 11.9
Wastewater quality criteria for evaluation of drip irrigation clogging hazard (Nakayama and Bucks, 1991)

	HAZARD RATING		
Clogging factors	Minor	Moderate	Severe
Physical (mg/l)			
Suspended solids	<50	50–100	>100
Chemical (mg/l)			
pH	<7.0	7.0–8.0	>8.0
Dissolved solids	<500	500–2000	>2000
Manganese	<0.1	0.1–1.5	>1.5
Total iron	<0.2	0.2–1.5	>1.5
Hydrogen sulfide	<0.2	0.2–2.0	>2.0
Biological (mo/ml)			
Bacterial number	<10,000	10,000–50,000	>50,000

(Nakayama and Bucks, 1991). In all these cases, the stone drain was enveloped by a geotextile in contact with refuse and seeping leachate.

11.5 LEACHATE RECIRCULATION

Leachate recirculation refers to the process of reintroducing collected leachate back into the landfill. This has been effective in enhancing biodegradation because of increased moisture and movement of moisture (which carries nutrients). RCRA Subtitle D regulations allow leachate management by recirculation for modern landfills with a regulated composite liner and a leachate collection system. A successful application of leachate recirculation requires attention to the following.

1. Refuse vertical hydraulic conductivity is typically less than the horizontal permeability. Leachate therefore may have to be injected through vertical walls. Figure 11.14 shows a leachate recirculation system that is built as the refuse is placed (Al-Yousfi, 1992); this system has operated successfully at a landfill in Delaware. For a closed landfill, vertical wells can be drilled through the refuse (Oweis et al., 1990).

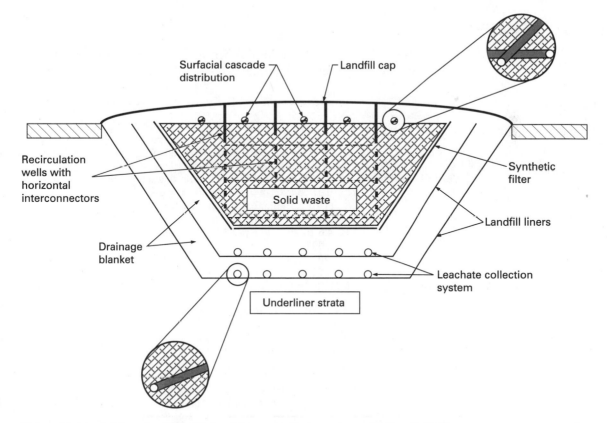

Figure 11.14   Sanitary landfill with leachate recirculation system (Al-Yousfi, 1992)

2. For well-compacted refuse with a substantial component of soils or waste of low hydraulic conductivity, recirculation may not be feasible.

3. For landfills with caps where leachate is recirculated through horizontal pipes beneath the cap, clogging of the pipe perforations may develop. Leachate should be allowed to cascade on a bed of open stone. Furthermore, the leachate should be spread uniformly over the landfill surface through a network of piping. The mounding between pipes may be estimated using Eq. 11.20; k in this case would be the hydraulic conductivity of the base beneath the cap and e would be the rate of leachate recirculation. It is possible that only a portion of the collected leachate can be circulated before leachate exits from the side of the landfill. This may occur if a daily or intermediate cover is of low conductivity.

4. Other than horizontal or vertical wells, leachate spraying (similar to spray irrigation) has been tried on some landfills. However, leachate odor and blowing of leachate to the operating zone limits the use of spraying. Another limitation is prolonged rain and freezing weather. Other techniques — such as prewetting of refuse and surface ponding — have been used with mixed success. Table 11.10 presents a summary of some case histories (Reinhart and Carson, 1993).

11.6   SUMPS

In essentially all designs, a pipe header carrying leachate will terminate in a sump from which the leachate is pumped. Figure 11.15 shows a variety of

Table 11.10
Full-scale leachate recirculation hydraulic application rates

Recirculation method	Application rates
Prewetting	0.2 m³/tn compacted waste (593 kg/tn) (48 gal/tn (1000 lb/yd³))
Vertical injection wells	A. 0.23 to 0.57 m³/hr per 6.4 cm diameter well 0.07 to 0.17 m³/m² landfill area/day (1 to 2.5 gpm/2.5 in diameter well, 1.7 to 4.1 gpd/ft² landfill area) B. 4.6 to 46 m³/hr per 1.2 m diameter well 0.005 to 0.09 m³/m² landfill area/day (20 to 200 gpm/4 ft diameter well, 0.12 to 2.3 gpd/ft² landfill area)
Horizontal trenches	0.31 to 0.62 m³/m of trench length/day at 14 to 23 m³/hr (25–50 gpd/ft of trench length at 60 to 100 gpm)
Surface ponds	0.0053 to 0.0077 m³/m²/day (0.13–0.19 gal/ft²/day)
Spray irrigation	A. 0.73 m³/m² of landfill area/day (18 gpd/ft² of landfill area) B. 0.001 to 0.0032 m³/m² of landfill area/day (0.025 to 0.078 gpd/ft² of landfill area)

Source: Reinhart and Carson, 1993

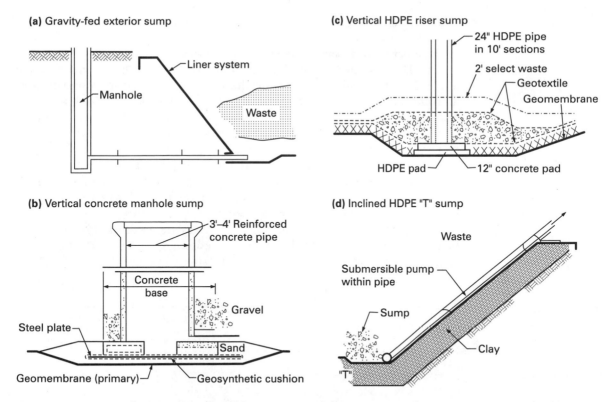

(a) Gravity-fed exterior sump

(b) Vertical concrete manhole sump

(c) Vertical HDPE riser sump

(d) Inclined HDPE "T" sump

Figure 11.15 Sump designs (reproduced from Koerner and Koerner, 1994, by permission)

sump designs that have evolved over many years (Koerner and Koerner, 1994). Figure 11.15a is a simple design but requires penetration of the liner. The advantage, however, is that leachate drains by gravity to the sump that is outside the toe of the landfill and not covered by waste. The system in Figure 11.15b avoids penetrating the liner but negative skin friction imposed by settling refuse requires a strong riser and imposes large stresses on the liner. The detail in Figure 11.15c is similar to that in Figure 11.15b except that a lighter plastic riser is used. The detail in Figure 11.15d requires no riser and no liner penetration; the problem with this design is potential clogging and lack of access to maintain the sump. The system in Figure 11.15d also is more sensitive to differential settlement.

11.7 QUALITY CONTROL

Parameters affecting the performance of soil drainage layers are grain size, hydraulic conductivity, and carbonate content. Carbonates are soluble in water. Drainage media rich in carbonate are more likely, therefore, to clog. Handling of the material in the field increases the fine content; approximately 0.5% additional fines are generated each time the material is handled (Daniel and Koerner, 1993). Therefore, overworking the material should be limited or avoided.

Koerner and Koerner (1994) provide useful guidance on construction quality. Drainage material on a slope should be placed from low grade to high grade. It is preferable to place the drainage soil on a geomembrane in the morning when the membrane is cool and the likelihood of wrinkles is small. The membrane should not be allowed to wave or wrinkle as the drainage soil is placed. Table 11.11a shows suggested criteria for testing on soil drainage material (Koerner and Koerner, 1994). Table 11.11b shows suggested quality control tests for geonets (Koerner and Koerner, 1994). During installation, geonets are overlapped a minimum of 6 in. at their roll ends and a minimum of 3 in. at their roll edges. When damaged geonets are patched, the patch should extend at least 12 in. beyond the tear.

The applicable standards for geotextile filters are ASTM D 5261 (mass per unit area), ASTM D 4833 (puncture resistance), ASTM D 4533 (tear resistance), ASTM D 4632 (grab tensile strength), ASTM D 3786 (burst

Table 11.11a Suggested tests and minimum testing frequency for specification conformance and quality assurance for natural soil drainage material

Test method	Frequency for spec. conformance	Frequency for quality assurance
At borrow source		
Grain size (ASTM D 422)	1 per 1000 cy	1 per 2500 cy
Hydraulic conductivity (ASTM D 2434)	1 per 1000 cy	1 per 2500 cy
Carbonate content (ASTM D 4373)	1 per 1000 cy	1 per 2500 cy
On site — after placement		
Grain size (ASTM D 422)		
Drainage layer	1 per acre	1 per 2.5 acres
Other uses	1 per 300 cy	1 per 650 cy
Hydraulic conductivity (ASTM D 2434)		
Drainage layer	1 per 3 acres	1 per 7.5 acres
Other uses	1 per 1000 cy	1 per 2000 cy
Carbonate content (ASTM D 4373)	1 per 1000 cy	1 per 2500 cy

Table 11.11b Suggested tests and minimum testing frequency for specification conformance and quality assurance for geonet drains

Test method	Frequency for spec. conformance	Frequency for quality assurance
Density (ASTM D 1505)	1 per acre	1 per 4 acres
Mass/unit area (ASTM D 5261)	1 per acre	1 per 4 acres
Thickness (ASTM D 5199)	1 per acre	1 per 4 acres
Compressive strength (ASTM D 1621)	1 per acre	1 per 4 acres
Ply adhesion* (ASTM D 413 mod.)	1 per acre	1 per 4 acres
Transmissivity (ASTM D 4716)	—	Project-specific
Direct shear (ASTM D 5321)	—	Project-specific

*Only for bonded geocomposites

resistance), ASTM D 5199 (apparent opening size), and ASTM D 4491 (permittivity).

Plastic pipes typically specified are either Schedule 80 polyvinyl chloride (PVC) or high density polyethylene (HDPE) with a standard dimension ratio (outside diameter divided by wall thickness) of 11. Pipes are usually delivered in 20 ft lengths. Koerner and Koerner (1994) recommend that one pipe sample be tested for every 1000-ft length of pipe. ASTM D 212 provides physical dimensions of PVC and HDPE pipes. ASTM D 792 is recommended for testing the density of PVC pipes and ASTM D 1505 for the density of HDPE pipes. ASTM D 2412 (plate bearing tests) and ASTM D 2444 (impact resistance) are recommended for both PVC and HDPE pipes (smooth or corrugated). AASHTO (American Association of State and Transportation Officials) M294-90 and M252-90 are recommended for HDPE corrugated pipes 12 to 36 in. in diameter and 3 to 10 in. in diameter, respectively.

SUMMARY

Solid wastes in landfills undergo oxidation and decomposition in the presence of moisture. The decomposition produces liquids, gas, and solid products, which move along percolating water or with refuse producing what is termed *leachate*. Because of its potential adverse impact on surface waters and groundwater, leachate needs to be properly collected and treated.

Leachate generation estimates are made using some type of water balance technique. The method explained in this chapter allows for recording site-specific rainfall and temperature data, and analyzing site-specific soil cover over the waste. A more detailed analysis can be made using the HELP model.

In old landfills without a leachate collection system beneath the liner, a leachate mound will develop. The height of the mound depends on the conductivity of underlying soil and the dimension of the landfill. A simple method is presented in this chapter for estimating leachate generation from the mound for transient and long-term conditions.

Leachate is collected and conveyed through a network of pipes and sumps. The formulas used for pipe design at a normal site are used at landfill sites as well. The potential for the clogging of pipes and drainage soil or fabric is an important design consideration as are safety factors when calculating drainage requirements.

NOTATIONS

A	cross-sectional area (L^2)	B_c	outside diameter of the pipe (L)	B_d	width of trench at the top of the pipe (L)
AET	evapotranspiration (L)				

C runoff coefficient

C_d, C_w, C_l load coefficients for calculating vertical stress or load on a pipe

C_u uniformity coefficient

C_c coefficient of curvature

d head at drain level at landfill perimeter (L)

D_e empirical time factor

E Young modulus (F/L^2)

E' modulus of passive soil resistance (F/L^2)

e percolation impinging on the drainage layer (L/T)

h head of the leachate mound inside the landfill (L)

h' estimated head of the leachate mound in iterative calculation (L)

H_w depth of waste above the trench (L)

I infiltration (L)

I moment of inertia of a pipe section per unit length (L^4/L)

I_t heat index

K_b pipe bedding constant

k_d hydraulic conductivity of the drainage layer

k_u unsaturated hydraulic conductivity

k_n hydraulic conductivity normal to the plane of the fabric (L/T)

k_s saturated hydraulic conductivity

$(k_s)_v$ hydraulic conductivity for vegetated soil

$(k_s)_{uv}$ hydraulic conductivity for unvegetated soil

L half-drain spacing (L)

L nondimensional length

L effective pipe length

L_1 leachate generation (L)

R hydraulic radius

R_0 surface runoff (L)

r pipe radius (L)

S_c change in soil storage (L)

S_r change of storage in refuse (L)

t fabric thickness (l)

t time (T)

n Manning roughness coefficient

P input water from precipitation, irrigation and surrounding runoff (L)

P concentrated live load (F)

p distributed live load (F/L^2)

P_c capillary pressure (bars)

P_b bubbling pressure (bars)

PER_s water percolating into refuse (L)

Q flow quantity in a pipe (L^3/T)

Q flow quantity per unit length (L^3/T/L)

Q_d lateral drainage rate (or inflow) per unit area of linear (L/T)

S slope

S nondimensional slope

S_y specific yield of refuse

y depth of leachate above the linear (L)

Z depth of the trench above the pipe (L)

W_d liquid from refuse decomposition (L)

W_g liquid from groundwater intrusion

W_{sc} load on a pipe per unit length from concentrated live load (F/L)

W_{sd} load on a pipe per unit length from a distributed load (F/L)

Θ volumetric water content

Θ_r residual saturation volumetric water content

λ pore size distribution index

γ_r average unit weight of the landfill (refuse and soil cover) (M/L^{2}T^2)

γ_f unit weight of the backfill in a trench (M/L^{2}T^2)

δ_y deflection of a pipe wall (L)

σ_v vertical stress on a pipe (F/L^2)

Ψ fabric permittivity (L^2/T)

PROBLEMS

11.1 For a certain landfill, the latitude is 45°, and the root zone is 10″. Assume that the available moisture = 50 mm. Determine percolation, giving your answers in mm.

Parameter	Jan	Feb	Mar	Apr	May	Jun	Jul	Aug	Sep	Oct	Nov	Dec
Temp (°F)	11.1	16.7	27.8	44	56.3	64.8	69.1	67.2	58.3	47.8	32.2	18.7
Precpt (in.)	0.74	0.77	1.71	2.73	3.58	4.11	4.09	3.92	3.65	2.14	1.52	1

11.2 Consider a ratio of $c = e/k_d$ of 10^{-5}, 10^{-4}, 10^{-3}, and 10^{-2}, a liner slope of 0.05, and $L = 15$ m, 30 m, 45 m, and 60 m. Plot y_{max}/L. Obtain the answer using all methods discussed in this chapter. Assume a liner with negligible hydraulic conductivity.

11.3 Consider a landfill where the hydraulic conductivity k_d of the drainage layer is 0.01 cm/s, $c = 10^{-2}$, the maximum allowable saturated thickness in the drainage layer above the liner is 15 cm, and the length of the drain is 200 m.
   (a) Calculate the design flow rate into the drains if the drains are placed within a uniform slope of 5%.
   (b) Calculate the design flow rate if drains are placed in a V-shaped arrangement at a slope of 5%.
   (c) If the length of the liner is 600 m and all drains slope to a single header, determine the number of drains and the design flow rate for the header.
   (d) If the sump where leachate is collected is to handle storage for 2 weeks while the pumping system is not operable, determine the required storage volume of the sump.

11.4 Repeat Problem 11.3 if the liner slope is 2%.

11.5 Repeat Problem 11.3 if k_d is 10^{-3} cm/s.

11.6 A landfill is 100 ft high, 1000 ft wide from toe to toe, and 3000 ft long. The refuse in the landfill is underlain by a sand layer 30 ft deep overlying plastic clay with negligible hydraulic conductivity. The conductivity of the sand and the refuse is 5×10^{-3} cm/s. A slurry wall and a leachate collection drain surrounds the landfill such that a constant head is maintained at 5 ft below the top of the sand layer. Consider an operating condition with a percolation rate of 40 cm/yr.
   (a) Determine the maximum height of the leachate mound inside the landfill and the total volume of leachate expected to be collected.
   (b) Suppose a cap was placed to limit percolation to 10 cm/yr. Estimate the height of leachate mound 6 months, 12 months, and 18 months after cap construction.
   (c) For the capped landfill, calculate the expected leachate generation after 18 months from cap placement.

11.7 If 6-in. plastic pipes are available for the lateral drains and 8-in. pipes for the header in Problem 11.3, what is the minimum drain slope required for both?

11.8 Suppose some lateral drains are designed as 6-in. pipes embedded in a 3-ft-wide trench that is 4 ft deep. The bedding is washed, crushed stone. Determine the structural requirements assuming three types of pipe: ABS, PVC, and polyethylene. Determine the requirements assuming:

(a) 100 ft of waste at an average unit weight of 55 lb/ft³.

(b) 50 ft of waste at an average unit weight of 55 pcf.

In all cases assume a unit weight of backfill of 135 pcf. For all pipe types assume a permissible deflection as 0.05 times the outside diameter. (The manufacturers will need to be contacted for data on pipes.)

11.9 For the pipes selected in Problem 11.8, evaluate whether they can survive the load from a Caterpillar D-8 with only 1.5 ft of cover over the pipe. The D-8 has a track load of 16,425 lb, spread over a ground contact of 18 in. × 7 ft 9 in.

11.10 For Problem 11.8, determine whether the pipe with only 1.5-ft cover can survive the passage of a loaded scraper with a wheel load of 45,470 lb.

11.11 Determine the filter requirements using a soil filter or a geotextile for a medium-dense silty sand with gravel having the following gradations:

Sieve size	% passing by weight
3 in.	100
1 in.	90
3/8 in.	82
No. 4	78
No. 10	72
No. 20	66
No. 40	54
No. 100	32
No. 200	20

11.12 Consider a leachate inflow rate of 0.0075 ft/day and a maximum head of 12 in. Determine the required permittivity of a geotextile to pass the flow.

11.13 Consider a liner overlain by a geonet for leachate collection. The maximum drainage distance is 50 ft at a slope of 3%. Using a leachate inflow rate of 0.005 ft/day and a maximum allowable head of leachate of 12 in. over the liner, determine the required transmissivity of the geonet.

11.14 A below-grade landfill has side slopes of $1V$ on $4H$. The depth of refuse is 20 ft and the leachate inflow is 50 in./yr. A geonet collects the flow from the side slopes. Determine the transmissivity of the geonet such that there is no buildup of fluid above the top of the geonet.

11.15 The results from a gradient ratio test showed the following:

Head at geotextile interface $\quad H_1 = 4$ in.

Head 1 in. above interface $\quad H_2 = 2.4$ in.

Head at 3 in. above interface $\quad H_3 = 3.2$ in.

Determine the potential for clogging the geotextile filter in contact with the soil tested.

11.16 Consider a landfill in a midwestern city. The cover is clay layer, the depth of the root zone is 0.6 m, and the cover slope is 3%. The precipitation (in mm) starting in January and ending in December is 60, 90, 85, 80, 100, 115, 107, 92, 70, 60, 85, and 84. The potential evaportranspiration values (in mm) for the respective months are 0, 1, 18, 55, 100, 129, 160, 140, 95, 50, 18, and 4. Estimate the leachate generation. Assume that the refuse is at field capacity.

12 Caps

12.1 REGULATORY GUIDANCE

The RCRA (Resource Conservation and Recovery Act) guidance on hazardous waste landfill final cover design attempts to achieve the following five goals.

1. Minimize infiltration from precipitation into the landfill, and hence minimize leachate generation.

2. Develop a cap that is not more permeable than the liner system, and hence allow no more precipitation to infiltrate the landfill than can escape through the bottom liner.

3. Promote drainage from the surface with minimal erosion.

4. Accommodate settlements and subsidence.

5. Operate with minimum maintenance.

A recommended design for an RCRA final cover is shown in Figure 12.1 (Landreth, 1992). The vegetation cover is needed to minimize erosion but plant species should not have a deep root system that reaches the barrier layer. The recommended thickness will accommodate most nonwoody cover plants. Top slopes between 3 and 5% (after settlement) are recommended to prevent pooling but minimize erosion. For slopes exceeding 5%, the maximum erosion rate should not exceed 2 tn/acre/yr as estimated by the universal soil loss equation (USLE). A surface drainage system must accommodate runoff to avoid rills and gullies. The drainage layer functions as the leachate collection and removal system (LCRS) in the base liner and must, therefore, pass the flow in a lateral direction; a minimum thickness of 12 in. (30.5 cm) and 2% grades are recommended. To avoid clogging, a separation filter is recommended between the drainage layer and the vegetative support layer. The 20-mil (minimum thickness) geomembrane must be protected by a bedding of sand (SP) free of stone or sharp objects 6 in. (15 cm) above and below the barrier unless the clay below and the drainage layer above serves as bedding. Both the geomembrane and the underlying low permeability soil must be below the average depth of frost penetration.

There is no specific RCRA guidance on methods for settlement- and subsidence-resistant final covers. Because storage of liquid waste in drums has been banned, corrosion and collapse of drums should no longer produce significant settlements in hazardous waste landfills. For municipal solid waste landfills (MSWLFs), however, settlement will continue for years and the cap integrity is more of a problem. In old MSWLFs with mostly soil components, projected settlements can be tolerable and an RCRA cap can be expected to function well with proper maintenance (Zamiskie et al., 1994).

Figure 12.1
USEPA-recommended
landfill cover design
(Landreth, 1992)

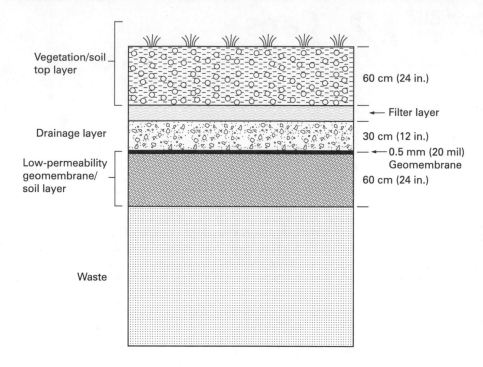

Vegetation/soil
top layer

60 cm (24 in.)

← Filter layer

Drainage layer

30 cm (12 in.)

← 0.5 mm (20 mil)
Geomembrane

Low-permeability
geomembrane/
soil layer

60 cm (24 in.)

Waste

12.2   FUNCTION

The key function of a cap is to limit percolation into the refuse and the development of leachate (see Chapter 11). Other functions are gas control, vector control, future site use, and aesthetics. The relative suitability of soils for various functions is illustrated in Table 12.1 (Lutton et al., 1979). The geotechnical elements of the design are cover stability against sliding, resistance to erosion, and structural resistance to differential settlements. A cover with a multicomponent structure was shown in Figure 12.1. The vegetative cover promotes evapotranspiration (see Chapter 10) and reduces percolation of water through the cap. In addition, vegetation is an important ingredient for erosion control. The drainage layer beneath the top soils diverts infiltrating water to a collection and removal system, reduces the hydraulic head on the barrier layer beneath, and limits percolation. The drainage layer, which could be either a drain stone layer or a geonet (see Chapter 11), promotes future utility of the site by enhancing mobility.

The hydraulic barrier has a function similar to a liner (see Chapter 10). The barrier layer could be either a low permeability soil, a geomembrane, a geosynthetic clay liner (GCL), a combination of these three, or some other low permeability barrier.

The foundation separation layer provides the platform for constructing the barrier layer and a transition to the underlying gas control layer. The gas control layer is composed of drain stone or a geosynthetic. Waste materials such as broken glass have been used in some landfills.

Table 12.1 Ranking of soil types according to performance of some cover function

Soil type USCS symbol	Impedance of water percolation	Hydraulic conductivity (approx.) (cm/s)	Support vegetation	Impedance of gas migration	Resistance to water erosion	Frost resistance	Crack resistance
GW	X	10^{-2}	X	X	I	I	I
GP	XII	10^{-1}	X	IX	I	I	I
GM	VII	5×10^{-4}	VI	VII	IV	IV	III
GC	V	10^{-4}	V	IV	III	VII	V
SW	IX	10^{-3}	IX	VIII	II	II	I
SP	XI	5×10^{-2}	IX	VII	II	II	I
SM	VIII	10^{-3}	II	VI	IV	V	II
SC	VI	2×10^{-4}	I	V	VI	VI	IV
ML	IV	10^{-5}	III	III	VII	X	VI
CL	II	3×10^{-8}	VII	II	VIII	VIII	VIII
OL			IV		VII	VIII	VII
MH	III	10^{-7}	IV		IX	IX	IX
CH	I	10^{-9}	VIII	I	X	III	X
OH			VIII				IX
PT			III				

Ranking: I (best) to XIII (poorest).
Source: Lutton et al., 1979

12.3 MATERIALS

The use and criteria for liners are discussed in Chapter 10. Liners are usually built on a stable foundation where settlement is only on the order of inches. Closed landfills could settle by as much as 50% of their original thickness (Jaros, 1990). Designing a cap to accommodate the associated differential settlement is possible but the cost is prohibitive.

12.3.1 Clay

Differential settlements could cause cracking of a clay cap (Oweis, 1989). Other causes of cracking are desiccation, freeze–thaw cycles, and buildup of gas pressure beneath the cap. Differential settlements cause tensile-bending stresses. Because of the low tensile resistance of soils a crack develops, which reduces the effective thickness of the cap and decreases the effectiveness of the cap in limiting percolation. Tension tests on compacted clay suggest that the elastic Young's modulus, E, in tension is comparable or higher than that in compression (Leonards and Narain, 1963; Ajaz and Parry, 1975; Cheng et al., 1993). Table 12.2 shows strains at failure for compacted clay and other barriers (LaGatta et al., 1997).

The ratio of the tensile strength of rock (as determined by the point load test) typically ranges from 1/10 to 1/15 of the unconfined compressive strength. Cheng et al. (1993) reported a ratio of about 1/13 for nonplastic (ML) soil and a ratio of 1/5 for (ML-CL) soil with a plasticity index of 9. The tensile strain at failure can therefore be estimated using the following steps.

Table 12.2
Compilation of
published
tensile strains at
failure for cover
materials

Type or source of soil	Water content	Plasticity index (%)	Maximum tensile strain (%)
Natural clayey soil	19.9	7	0.80
Bentonite	101	487	3.4
Illite	31.5	34	0.84
Kaolinite	37.6	38	0.16
Portland dam	16.3	8	0.17
Rector Creek dam	19.8	16	0.16
Woodcrest dam	10.2	Nonplastic	0.18
Shell Oil dam	11.2	Nonplastic	0.07
Willard test dam embankment	16.4	11	0.20
Gault clay	19–31	39	0.1–1.7
Balderhead city	10–18	14	0.1–1.6
Clay	—	32	1.3–2.8
Kaolin	21–30	16	2.8–4.8
Clay A	16–29	31	1.5–4.1
Clay B	19–33	49	1.6–3.6
Clay C	18–26	32	1.7–4.4
Needle-punched GCL			5–16
Geotextile encased stitch-bonded GCL			5
Bentonite geomembrane composite GCL			30
Geomembrane biaxial bending			20–100
Sand-bentonite mixture—no overburden, no significant cracking			4

Source: After La Gatta et al., 1997

1. Conduct an unconfined compression test and plot the stress–strain curve.

2. Extend the initial segment of the curve to a point corresponding to the tensile strength and read the corresponding strain value. The tensile strength may be determined based on the point load test or estimated as 0.1 or 0.15, the unconfined compression strength.

Cracking of a clay cap can develop when the cap spans over a void or by cap compliance with the settlement of the landfill. In the first case, the cap can be treated as a simply supported beam with unit width and a uniform load. The combined deflection due to moment and shear (Timoshenko and MacCullough, 1949) is expressed as

$$\delta = 5M \frac{(2l)^2}{48EI} \left(1 + 9.6 \frac{EQ}{G(2l)^2} \right) \tag{12.1}$$

where

δ = deflection at the center

l = half-span length

E = Young's modulus

G = shear modulus

I = moment of inertia about the neutral axis

Q = static moment of the cross-sectional area from the neutral axis to the extreme fiber

The moment, M, is expressed as

$$M = \varepsilon \cdot EI/C \tag{12.2}$$

where ε is the strain and C is the distance from the neutral axis to the extreme fiber.

Assuming that the field strains are at failure condition and that such strains are produced by biaxial bending such that $\varepsilon = (1/\sqrt{2})\,\varepsilon_f$ (where ε_f is the strain at the extreme fiber), then, substituting Eq. 12.2 into Eq. 12.1,

$$\delta/l = (14.14 \cdot \varepsilon_f \cdot l/48C)(1 + 14.4IE/tG(2l)^2) \tag{12.3}$$

For a homogeneous material with $E/G = 2.8$, the limiting deflection ratio δ/l for crack development is

$$(\delta/l)_f = (0.60\varepsilon_f \cdot l/t)(1 + 3.36(t/2l)^2) \tag{12.4}$$

for the neutral axis through the middle and

$$(\delta/l)_f = (0.30\varepsilon_f \cdot l/t)(1 + 13.44(t/2l)^2) \tag{12.5}$$

for the neutral axis at top or bottom.

With the range of failure strains for clay in Table 12.2 (0.1 to 0.8%), a span of 10 ft ($=2l$), and a thickness of 2 ft, the limiting δ/l may range from 0.1 to 1.4%. If the cap is unsupported over the 10 ft length and is 2 ft thick, the self-weight stress is about 240 psf and the maximum moment (assuming a simple span) is 3000 ft-lb. For the maximum resultant biaxial strain to be less than 0.8%, the modulus E in tension will have to be about 800,000 psf or more. The computed tensile stress for a 3000 ft-lb bending moment is about 4500 psf, which is much larger than the compressive strength for a clay cap. Thus, cracking of clay caps on a settling landfill is almost certain even if the unsupported length is only a few feet (Oweis, 1989).

Figure 12.2 shows the results of a finite element analysis (Bredario et al., 1995), which shows the dependence of crack development on the stiffness and strength of the cap. The beam analysis described above only applies to the initiation of cracking; it would not be realistic for predicting the distortion required for a complete failure or a full development of a crack.

In practice, the settlements often are calculated and then the question is posed as to whether the cap will crack. If two points are at a distance $2l$ and

Figure 12.2
Cracking versus
distortion as a function
of material properties,
60 cm thick CCL
spanning a 3.5 m
diameter void
(reproduced from
Bredario et al., 1995,
by permission)

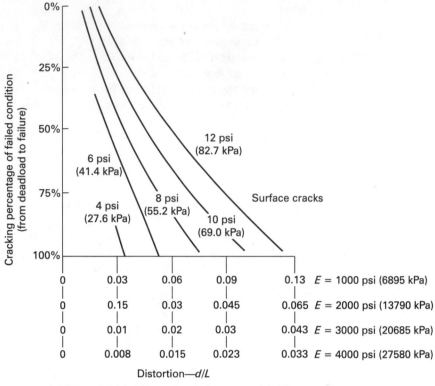

24" (61 cm) thick CCL, 10' (3 m) diameter void, $h/L = 0.4$

the midpoint settlement, δ, is measured from a straight line connecting the two points, the radius of curvature, R, is

$$R = (l^2/2\delta) + \delta/2 \qquad (12.6)$$

and the bending strain, ε, accounting for the biaxial deformation condition is

$$\varepsilon = \sqrt{2}(t/2R) \qquad (12.7)$$

If a plane is deformed into a spherical segment, the strain due to stretching from the biaxial deformation condition is

$$\varepsilon = (\sqrt{2})[(R/l) \sin^{-1} (l/R) - 1] \qquad (12.8)$$

12.3.2 Geomembranes

The bending resistance of a geomembrane is negligible. Strains in the membrane are due to stretching that occurs either when the membrane spans a void or when the membrane conforms to the settlement profile of the landfill. For example, suppose a membrane spans a void with a length $2l$ and is loaded by overlying materials imposing a stress p. Then if the resulting deflection, δ, is less then l, it is reasonable to assume that the shape of the geomembrane is that of a spheroid (Koerner et al., 1990). In this case, Eqs. 12.6 and 12.8 may be used to estimate the radius of curvature and tensile strain in the membrane. The stress in the membrane is calculated assuming an internal pressure, p, generated by the overburden above the void.

$$\sigma_t = pR/2t' \tag{12.9}$$

where σ_t is the tensile stress in the geomembrane and t' is the corrected thickness of the geomembrane after deformation:

$$t' = t \cdot l/(1.414R \sin^{-1} l/R) \tag{12.10}$$

For a large void, the pressure p is equal to the unit weight of the overburden times the thickness. For a small void, the pressure imposed on the membrane is reduced due to arching. The reduced pressure is computed the same as that on a pipe (see Chapter 11) and p in Eq. 12.10 is replaced by p'. For a cylindrical depression with width $2a$, p' is estimated as

$$p' = \gamma(2a)(1 - \exp - 2ku'H/2a)/2ku' \tag{12.11}$$

where γ is the average unit weight of the overburden above the membrane, H is the thickness of the overburden, k is the lateral earth pressure coefficient (the ratio of horizontal to vertical earth pressure), and u' is the coefficient of friction of the overburden above the geomembrane (which equals the tangent of the friction angle). For a rectangular void $2a$ wide and $2b$ long, arching in the third dimension is considered. p' is estimated as

$$p' = \gamma(1/F_1)(1 - \exp - ku'HF_1)/ku' \tag{12.12}$$

where

$$F_1 = (1/a + 1/b)$$

The maximum values of ku' for typical soils (EPA, 1983) are 0.19 for granular material without cohesion, 0.165 for sand and gravel, 0.13 for clay, and 0.11 for saturated clay. No data are available for refuse. Assuming a friction angle of 32° and an active lateral earth pressure coefficient of 0.3, the parameter ku' is close to 0.19.

Geomembranes used for liners or caps can withstand only a small void, which can be generated by corrosion of drums, differential settlements, or the geomembrane not conforming to the settlement of the subgrade. In the latter case, a geogrid (Figure 12.3) can be used to help span a bigger void. For uniform strains, the stiffer and stronger the geogrid is, the lesser the stresses in the membrane will be.

Figure 12.3
Cap reinforcement

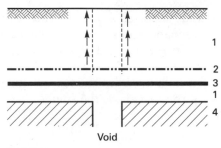

Void

1. Soil cover
2. Geogrid
3. Geomembrane
1. Soil cover/venting median
4. Refuse

(a) Arching in a cap structure above a void

Example 12.1

A clay cap is 2 ft thick. If the tensile strain at failure is 0.5%, determine the maximum tolerable settlement in the middle of a 100-ft-wide depression.

Solution: Using $2l = 100$ ft and Eq. 12.4, we have

$$\varepsilon_f = 0.005$$

$$\delta/l = 0.60 \times 0.005 \times 25(1 + 3.36 \times 0.02 \times 0.02) = 0.075$$

$$\delta = 0.074 \times 50 = 3.7 \text{ ft}$$

From Eq. 12.5 (neutral axis at the bottom), we obtain

$$\delta/l = 0.30 \times 0.005 \times 25(1 + 13.44 \times 0.02 \times 0.02) = 0.038$$

$$\delta = 0.036 \times 50 = 1.8 \text{ ft}$$

From Eq. 12.7,

$$0.005 = 1.414 \times 2 \times /2R$$

$$R = 282.8 \text{ ft}$$

Finally, from Eq. 12.6,

$$282.8 = 2500/2\delta$$

$$\delta = 4.42 \text{ ft}$$

Example 12.2

Consider a 60-mil HDPE cover, an allowable strain of 2.5%, and an allowable stress of 1500 psi (10.3 mPa). The geomembrane is overlain by soil 3-ft (0.91-m) thick having a unit weight of 120 pcf (18.9 kN/m³). What distance can the liner span without failure?

Solution:

$$p = 120 \times 3 = 360 \text{ psf} = 2.5 \text{ psi}$$

We will use a reduced thickness of 0.04 in. (to be verified later) for t'. From Eq. 12.9,

$$1500 = 2.5 \times R/2 \times 0.04$$

$$R = 48 \text{ in.}$$

From Eq. 12.8,

$$0.025 = 1.414(R/l \sin^{-1} l/R - 1)$$

$$0.0176 = R/l \sin^{-1} l/R - 1$$

Solving by trial and error, we get $l/R = 0.32$ and $l = 15.36$ in. The width of the depression is $2l = 30.72$ in. or $l = 15.36$. For $l = 15.36$ in., $R = 48$ in., and the critical deflection is estimated from Eq. 12.6.

$$48 = 15.36^2/2\delta + \delta/2$$

$$\delta = 2.52 \text{ in.}$$

We now calculate the reduced thickness t' from Eq. 12.10.

$$t' = 0.06 \times 15.36/1.414 \times 48 \sin^{-1}(15.36/48) = 0.042 \text{ in.}$$

which is close to the assumed value of 0.04.

If the membrane is supported but complies with landfill settlements, then from Eq. 12.8, the limiting R/l for 0.025 strain is 3.15. Eq. 12.6 may be rewritten as

$$\delta/l = R/l - \sqrt{((R/L)^2 - 1)}$$

for $R/l = 3.15$, $\delta/l = 0.16$. Thus the liner can withstand a settlement equal to 8% of the span length.

12.3.3 Geosynthetic Clay Liner (GCL)

Details on the GCL are given in Chapter 10. A GCL is a thin layer (about 6 mm thick) of bentonite sandwiched between two geotextiles or glued to a geomembrane. At placement time, the bentonite will be dry and permeable to gas. After exposure to water and hydration, the bentonite barrier in intact condition is virtually impermeable to both gas and water. Because of the reinforcing effect of both the geotextile and the geomembrane and the relatively high tensile failure strain for bentonite, the GCL is expected to be more resistant to cap settlement. Weiss et al. (1995) and La Gatta et al. (1997) reported laboratory measured tensile strains of 1 to more than 10% with no compromise in the hydraulic performance of the GCL.

The major limitation in the use of GCL is on slopes because the shear strength of hydrated bentonite is low (typically less than 10°). Daniel (1993) reported tests that showed that geotextile-encased, adhesive-bonded GCL will retain its low permeability when wetted after a dry cycle. Bentonite swells when wetted after shrinking due to drying. If low permeability to gas is important, the GCL has to remain hydrated.

GCLs have several advantages. For example, GCLs are relatively easy to install. They can be rolled using light equipment and sheets can be overlapped. Also, GCLs require much less quality control effort than do geomembranes and clay liners.

12.4   COVER EFFICIENCY

Defects in either the clay cap or geomembrane will lead to increased leakage. For a clay cap with a fully penetrating crack of width b, the flow through the crack can be modeled as flow through parallel plates (Albertson and Simons, 1964):

$$q = \gamma_w b^3 i/12v \qquad (12.13)$$

where b is the crack width, q is the flow per unit length of the crack, v is the dynamic viscosity of water (2.1×10^{-5} psf-s), and i is the average hydraulic gradient. Assuming $\gamma_w/v = 30 (\text{ft/s})^{-1} \times 10^5$ then

$$q = 2.5b^3 i \times 10^5 (\text{ft}^3/\text{s/ft}) \qquad (12.14)$$

If the head loss across the depth of the crack is assumed to be the same as in the intact liner, then

$$k_e l_c = k_1 l_c + 2.5b^3 \times 10^5 \qquad (12.15)$$

$$k_e = k_1 + 2.5(b^3/l_c) \times 10^5 \qquad (12.16)$$

where k_e is the effective hydraulic conductivity of the cap (ft/s), k_1 is the hydraulic conductivity of the intact cap (ft/s), l_c is the crack spacing (ft), and b is the crack width (ft). If the crack is filled with soil having conductivity $k_s > k_1$, then

$$k_e = k_1 + k_s(b/l_c) \qquad (12.17)$$

The cap efficiency may be defined as

$$E = 1 - e/P \qquad (12.18)$$

where E is the cover efficiency, e is the percolation rate through the cover, and P is the precipitation.

Figure 12.4
Cover efficiency versus
cover hydraulic
conductivity
(Oweis and Biswas, 1993)

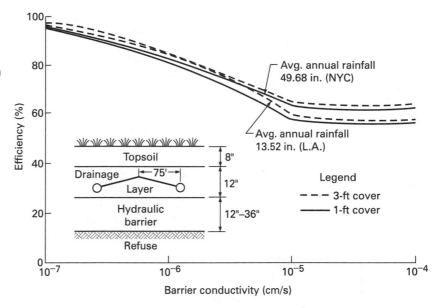

Figure 12.4, as well as similar analyses presented by Schroeder and Peyton (1990), suggests that the efficiency of a clay cap drops substantially as the hydraulic conductivity of the barrier layer approaches 10^{-5} cm/s. For a geomembrane cap, the effectiveness depends on the number and size of defects and the contact condition with the underlying material. Formulas are given in Chapter 10 for estimating the radius of flow beneath a defect. The surface area of the flow over which leakage occurs is computed from this radius. The flow through the geomembrane is calculated as (Schroeder and Peyton, 1990)

$$Q_p = (LF)k_1(h + t)/t \qquad (12.19a)$$

For $h \ll t$,

$$Q_p = k_1(LF) \qquad (12.19b)$$

where Q_p is the vertical percolation rate per unit area of the geomembrane, k_1 is the hydraulic conductivity of the soil underneath the geomembrane, h is the head above the geomembrane, and t is the thickness of the geomembrane. LF is the leakage fraction, which is the total area of the saturated flow through the subsoil beneath all the geomembrane defects divided by the horizontal area covered by the geomembrane. For example, if there is one defect per acre and the radius of the wetted area under the defect is 5 ft, then the leakage fraction LF is 1.8×10^{-3}.

In Eq. 12.19b, it is assumed that the hydraulic head beneath the geomembrane is small compared to the soil layer underneath with a hydraulic conductivity k_1. In Eq. 12.19a, the assumption is that there is no head loss as water seeps through and underneath the flaw in the geomembrane; thus Eq.

Figure 12.5
Effect of leakage
fraction on system
performance
(reproduced from
Schroeder and Peyton,
1990, by permission)

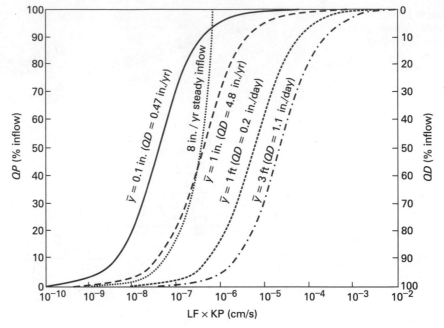

K_p = Hydraulic conductivity of liner
QP = Percolation rate/unit area
QD = Drainage per unit area of liner

12.19a is conservative for leakage prediction. It is clear from Eq. 12.19b that the effective hydraulic conductivity for a geomembrane cap is the leakage fraction times the conductivity of the underlying soil.

Figure 12.5 shows the results of analyses by Schroeder and Peyton (1990) for a geomembrane underlain by a native soil 3 m thick. The figure is based on a drainage layer above the geomembrane having a hydraulic conductivity of 10^{-2} cm/s and a drainage length of 23 m. If the soil underneath the liner is sandy with a hydraulic conductivity of 10^{-3} cm/s and the leakage fraction is 10^{-4}, the membrane could be very effective for relatively large heads. However, the leakage rate Q_p will be high for small heads above the membrane. The formulas presented in Chapter 10 for the radius of a wetted area may be used to estimate the leakage fraction. It is not possible to know the number of defects (holes, pinholes) in a geomembrane after construction. Potential sources of such defects are discussed in Chapter 10.

12.5   EROSION PROTECTION

As previously indicated, RCRA caps are intended for landfills with gentle slopes (not exceeding 5%) where gully development is not severe and sheet flow assumptions are reasonable. Figure 12.6 shows erosion of a cap after a

Figure 12.6
Failure of cap by erosion after a heavy storm

heavy storm. The damage is usually severe and the cost of repair is substantial. Overland flow on a slope usually starts as a sheet flow but in a short distance flow concentrates to form rills and then gullies that erode the cap structure and carry sediments downslope. The expected nonuniform settlements and obstructions on a slope (such as fences, pipes, or other objects) contribute to flow concentration.

The distance at which sheet flow concentrates cannot be analytically predicted; it depends on the steepness of the slope, the type of cover protection, the smoothness of the grading, and the nature of any objects on or protruding from the cover. For covers on steep slopes with poor grass protection, the distance may be less than 50 ft (based on visual observations). A maximum distance of 300 ft has been suggested to distinguish sheet flow from shallow to channelized flow (SCS, 1986). Abt et al. (1987) performed laboratory experiments and found that for a riprap with a D_{50} of 2.2 in., the channelized flow rate is 2.24 to 3.33 times the sheet flow rate. For finer grained material like fine poorly graded sand, this ratio may be higher.

The objective of erosion protection measures is twofold. First, for a given slope, erosion protection measures should limit the distance of overland flow. This requires the construction of swales, diversion berms, downslope chutes, and other hydraulic structures. Second, these measures should provide erosion-resistant materials and vegetation that resists erosion without root penetration that may damage the barrier component of the cap.

Erosion protection measures usually have been based on permissible velocity concepts. Tables 12.3 through 12.6 (Lane, 1955 as quoted by Nelson et al., 1986) can be used to assess the limiting velocities for covers after correcting for depth of flow and flow concentration. The values in these tables are consistent

Table 12.3
Maximum permissible
velocities in erodible
channels

Channel material	Water transporting colloidal silts, v (ft/s)
Fine sand, colloidal	2.50
Sandy loam, noncolloidal	2.50
Silty loam, noncolloidal	3.00
Alluvial silts, noncolloidal	3.50
Firm loam	3.50
Volcanic ash	3.50
Stiff clay, colloidal	5.00
Alluvial silts, colloidal	5.00
Shales and hardpans	6.00
Fine gravel	5.00
Graded loam to cobbles, noncolloidal	5.00
Graded silts to cobble, colloidal	5.50
Coarse gravel, noncolloidal	6.00
Cobbles and shingles	5.50

Source: Lane, 1955

Table 12.4
Maximum allowable
velocities in sand-
based material

Material	Velocity (ft/s)
Very light sand of quicksand character	0.75 to 1.00
Very light loose sand	1.00 to 1.50
Coarse sand to light sandy soil	1.50 to 2.00
Sandy soil	2.00 to 2.50
Sandy loam	2.50 to 2.75
Average loam, alluvial soil, volcanic ash	2.75 to 3.00
Firm loam, clay loam	3.00 to 3.75
Stiff clay soil, gravel soil	4.00 to 5.00
Coarse gravel, cobbles, and shingles	5.00 to 6.00
Conglomerate, cemented gravel, soft slate, tough hardpan, soft sedimentary rock	6.00 to 8.00

Source: Lane, 1955

Table 12.5
Limiting velocities in
cohesive materials

Principal cohesive material	COMPACTNESS OF BED			
	Loose velocity (ft/s)	Fairly compact velocity (ft/s)	Compact velocity (ft/s)	Very compact velocity (ft/s)
Sandy clay	1.48	2.95	4.26	5.90
Heavy clayey soils	1.31	2.79	4.10	5.58
Clays	1.15	2.62	3.94	5.41
Lean clayey soils	1.05	2.30	3.44	4.43

Source: Lane, 1955

Table 12.6
Maximum permissible
velocities in feet per
second (fps) for
channels lined
with uniform stands
of various well-
maintained grass
covers

Cover	Slope range (%)	MAXIMUM PERMISSIBLE VELOCITIES[a]	
		Erosion-resistant soils	Easily-eroded soils
Bermuda grass	0–5	8	6
	5–10	7	5
	Over 10	6	4
Buffalo grass	0–5	7	5
Kentucky bluegrass	5–10	6	4
Smooth brome	Over 10	5	3
Blue grama[b]	0–5	5	4
Grass mixture[b]	5–10	4	3
Lespedeza sericea			
Weeping lovegrass			
Yellow bluestem[c]	0–5	3.5	2.5
Kudzu			
Alfalfa			
Crabgrass			
Common lespedeza[c,d]	0–5	3.5	2.5
Sudan grass[d]			

[a] Use velocities over 5 fps only where good covers and proper maintenance can be obtained.
[b] Do not use on slopes steeper than 10%.
[c] Use on slopes steeper than 5% is not recommended.
[d] Annuals are used on mild slopes or as temporary protection until permanent covers are established.
Source: SCS, 1984

with the U.S. Corps of Engineers' suggested permissible mean velocities for channels (COE, 1991). It should be noted that the values given are for channels. For sheet flow on a cap, the depth of flow is usually less than a foot and the average velocity is close to the maximum. Therefore, the channel values in Tables 12.3 through 12.6 should be corrected as per Table 12.7 (USNRC, 1990).

12.5.1 *Soil Loss*

The factors contributing to erosion are empirically expressed in the universal soil loss equation (USLE) developed in the 1930s for evaluating soil conservation practices. The modified universal soil loss equation (MUSLE) was developed in 1978 by the Utah Water Research Laboratory for estimating soil losses in connection with highway construction. The equation is used for estimating average soil losses but not gully development.

The MUSLE is defined as (Nelson et al., 1986)

$$A = RK(LS)(VM) \tag{12.20}$$

Table 12.7
Correction factors for
permissible velocity

Depth of flow (ft)	Correction factor
≥3	1.0
1.9	0.9
1.0	0.8
0.65	0.7
0.4	0.6
≤0.25	0.5

Source: USNRC, 1990

where A is the loss per unit area per year (tn/acre/yr) and R is the rainfall factor. The rainfall factor (Lutton et al., 1979) is

$$EI/100$$

where E is the total kinetic energy of a given storm ($E = 916 + 331 \log i$) (ft-tn/acre-in.), i is the rainfall intensity (in./hr), and I is the maximum 30-min rainfall intensity (in./hr). The E for an individual storm can be obtained by dividing the storm into individual increments with uniform intensity and summing the incremental Es.

The rainfall factor R for specific areas of the United States can be obtained from the local Soil Conservation Service (SCS) office. For average annual soil loss, R can be estimated from Figure 12.7 (Lutten et al., 1979). The topographic factor, LS, is

$$LS = (L/72.6)^m(65s^2 + 450s + 650)/(s^2 + 10,000) \qquad \textbf{(12.21)}$$

where
L = slope length (ft),
s = slope steepness (%)
m = exponent
$= 0.2$ for $s < 1$
$= 0.3$ for $1 < s < 3$
$= 0.4$ for $3 < s < 5$
$= 0.5$ for $5 < s < 10$
$= 0.6$ for $s > 10$

The erosion control factor, VM, accounts for erosion control measures at a particular site and may vary from 1.0 to 0.01, as explained in Table 12.8 (Nelson et al., 1986). The soil erodibility factor, K, depends on the soil composition. Table 12.9 (Lutton et al., 1979) shows approximate values of K for USDA textural soil classification (see Chapter 2).

K is the average soil loss in tn/acre per unit of R for a given soil on a "unit plot." A unit plot is 72.6 ft long with a 9% slope, has continuous fallow, and is tilled parallel to the land slope. Nomograph empirical solutions of K are available (NAVFAC DM-7.1). Such solutions require grain size data, the permeability classification of the soil and organic matter, and the soil structure classification.

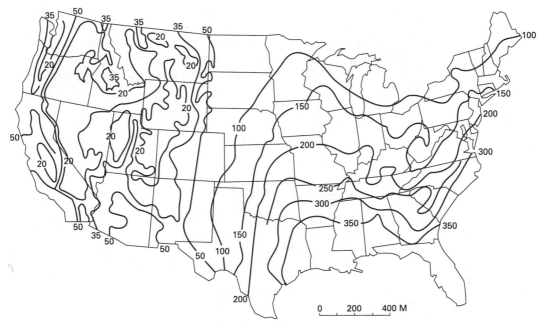

Figure 12.7 Average annual rainfall — erosivity factor R

Table 12.8
Typical VM factor values

Condition	VM factor
Bare soil condition	
Freshly disked, 6–8 in.	1.0
After one rain	0.89
Loose, 12 in. thick	
Smooth	0.9
Rough	0.8
Compacted bulldozer scraped up and down	1.3
Same except roots raked	1.2
Compacted bulldozer scraped across slope	1.2
Rough irregular tracked in all directions	0.9
Seed and fertilize fresh	0.9
Same after 6 months	0.54
Compacted fill	1.24–1.71
Saw dust, 2 in. deep disked in	0.61
Dust binder	
605 gal/acre	1.05
1210 gal/acre	0.29–0.78
Hydromulch (wood fiber slurry), fresh	
1000 lb/acre	0.05
1400 lb/acre	0.01–0.02
Seedings	
Temporary, 0–60 days	0.4
After 60 days	0.05
Permanent, 0–60 days	0.4
2–12 months	0.05
After 12 months	0.01
Excelsior blanket with plastic net	0.04–0.1

Example 12.3

A landfill in south New Jersey is designed to have a cover with a slope of 5% of a top plateau extending from a central ridge (high point) for a distance of 300 ft. Beyond this distance, the cover slopes down to the toe at a grade of $1V$ on $4H$. The upper cover component is loamy sand with 2% organic content. Grass is the only means of erosion control. Determine the expected soil loss from sheet flow.

Solution: From Figure 12.7, $R = 200$. From Table 12.9, $K = 0.1$. From Eq. 12.21:

LS (top plateau), $m = 0.4$

$LS = (300/72.6)^{0.4}(65 \times 25 + 450 \times 5 + 650)/(25 + 10{,}000) = 0.794$

LS (side slope), $m = 0.6$

$LS = (500/72.6)^{0.6}(65 \times 625 + 450 \times 25 + 650)/(625 + 10{,}000) = 15.73$

To determine the soil loss, we begin by using Eq. 12.20 for the top plateau:

$$A = 200(0.1)(0.79)(VM) = 15.8(VM)$$

From Table 12.8, the VM factors are 0.4, for grass seedings less than 2 months old, 0.05 for those 2 to 12 months old, and 0.01 for those over 12

Table 12.9
Approximate values of factor K for USDA textural classification

	ORGANIC MATTER CONTENT		
Texture class	<0.5% K	2% K	4% K
Sand	0.05	0.03	0.02
Fine sand	0.16	0.14	0.10
Very fine sand	0.42	0.36	0.28
Loamy sand	0.12	0.10	0.08
Loamy fine sand	0.24	0.20	0.16
Loamy very fine sand	0.44	0.38	0.30
Sandy loam	0.27	0.24	0.19
Fine sandy loam	0.35	0.30	0.24
Very fine sandy loam	0.47	0.41	0.33
Loam	0.38	0.34	0.29
Silt loam	0.48	0.42	0.33
Silt	0.60	0.52	0.42
Sandy clay loam	0.27	0.25	0.21
Clay loam	0.28	0.25	0.21
Silty clay loam	0.37	0.32	0.26
Sandy clay	0.14	0.13	0.12
Silty clay	0.25	0.23	0.19
Clay		0.13–0.29	

The values shown are estimated averages of broad ranges of specific-soil values. When a texture is near the borderline of two texture classes, use the average of the two K values.
Source: Lutten et al., 1979

months old. Thus,

$A = 15.8 \times 0.4 = 6.32$ tn/acre/yr for areas with seedings 0–60 days,

$A = 15.8 \times 0.05 = 0.79$ for areas with seedings 2–12 months old, and

$A = 15.8 \times 0.01 = 0.16$ tn/acre/yr for seedings over 12 months old

Assuming that runoff from the plateau is collected, we calculate the soil loss for the side slope as

$A = 200(0.1)(15.73)(VM) = 315(VM)$

$= 315 \times 0.4 = 126$ tn/acre/yr for recently planted areas (<2 months old)

$= 315 \times 0.05 = 15.75$ tn/acre/yr for areas with seedings 2–12 months old

$= 315 \times 0.01 = 3.15$ tn/acre/yr for mature grass cover 12 months and older.

The above estimates for the side slopes are higher than the usually accepted criterion of 2 tn/acre/yr.

12.6   EROSION ANALYSIS (PERMISSIBLE VELOCITY)

The velocity of sheet flow over a landfill slope can be estimated by solving the Manning formula expressed as

$$V = 1.486 R^{2/3} s^{1/2}/n \tag{12.22}$$

where V is the average velocity of a specified cross section, R is the hydraulic radius (=area/wetted perimeter), s is the slope of the channel bottom (length/length), and n is the surface roughness coefficient. Typical values of n for landfill covers are 0.02 or 0.025. Table 12.10 (SCS, 1986) may be used for sheet flow.

Eq. 12.22 accounts for the average velocity in an open channel. Velocity reaches a maximum near the free surface and decreases with depth. In cap design, the depth of flow is usually a few inches and the base flow velocity tending to erode the slope is nearly equal to the maximum velocity. For a strip a unit length wide, the velocity, V, is

$$V = C_1 Q/d \tag{12.23}$$

where d is the depth of flow, Q is the flow volume computed by the methods of Section 12.8, and C_1 is the flow concentration factor. A value of 3 for C_1 may be used if differential settlements are minimal and uniform grading is accomplished during construction. Limited data are available on flow concentration; engineers must use their judgment based on local conditions.

Table 12.10
Roughness coefficients
(Manning's *n*) for sheet
flow

Surface description	n^a
Smooth surfaces (concrete, asphalt, gravel, or bare soil)	0.011
Fallow (no residue)	0.05
Cultivated soils	
Residue cover $\leq 20\%$	0.06
Residue cover $> 20\%$	0.17
Grass	
Short grass prairie	0.15
Dense grasses^b	0.24
Bermuda grass	0.41
Range (natural)	0.13
Woods^c	
Light underbrush	0.40
Dense underbrush	0.80

^{a} The *n* values are a composite of information compiled by Engman (1986).
^{b} Includes species such as weeping lovegrass, bluegrass, buffalo grass, blue grama grass, and native grass mixtures.
^{c} When selecting *n*, consider cover to a height of about 0.1 ft. This is the only part of the plant cover that will obstruct sheet flow.
Source: SCS, 1986

For shallow flow over the landfill slope, the hydraulic radius (area/wetted perimeter) is close to flow depth *d*. Eqs. 12.22 and 12.23 can be solved for *d* as

$$d = (C_1 nQ/1.486S^{0.5})^{3/5} \qquad \textbf{(12.24)}$$

Based on the computed depth of flow in Eq. 12.24, the velocity is calculated from Eq. 12.23. The computed velocity is compared to the permissible velocity, V_p, which is estimated as

$$V_p = C_2 V_u \qquad \textbf{(12.25)}$$

where V_u is the uncorrected permissible velocity (Tables 12.3 to 12.6) and C_2 is the depth correction factor (Table 12.7). If the calculated actual flow velocity is larger than the permissible velocity, either a cover protection or a design for lesser velocities (by fattening slopes and/or intermediate collection swales) is needed.

Example 12.4

Consider a landfill cover with a maximum flow distance to a collection swale of 300 ft along a 4*H* on 1*V* slope. The flow quantity per unit width (from the methods in Section 12.8) is 0.1 cubic ft/s (cfs). The upper 2 ft of the landfill is sandy silt. Determine the flow velocity and analyze the potential for erosion.

Solution: If we assume a bare cover, the Manning n from Table 12.10 is 0.011. For the concentration factor, we assume 3.0. From Eq. 12.24, we obtain

$$d = (3 \times 0.1 \times 0.011/1.486(0.5))^{3/5} = 0.04 \text{ ft}$$

From Eq. 12.23, we have

$$V = 0.1 \times 3/(1 \times 0.04) = 7.5 \text{ ft/s}$$

The uncorrected permissible velocity (from Table 12.3) is 2.0 ft/s. The corrected factor from Table 12.7 is 0.5. Thus, the permissible velocity (from Eq. 12.25) is 1.0 ft/s, indicating that the cover will erode for the design storm indicated.

If we assume a gravel cover, the uncorrected permissible velocity (from Table 12.3) is 6 ft/s. The permissible velocity is now $3.0 < 5.45$ ft/s. Thus a gravel layer will also erode. Redesign or riprap protection should be considered.

12.7   EROSION ANALYSIS (PERMISSIBLE SHEAR STRESS METHOD)

In this method of erosion analysis, the effective tractive shear stress applied to the soil slope is such that it is less than a permissible shear stress. In estimating the tractive shear stress, the resistance of vegetation and detached soil particles at the soil–water interface can be empirically considered (Tempe et al., 1987). The vegetal impedance is expressed as a function of the product of the hydraulic radius, R, and the velocity, V (i.e., RV). For shallow flow over a slope, the hydraulic radius is the flow depth, d, and hence RV is equal to the flow, q, per unit strip width. The relationship takes the form:

$$n_r = \exp\left(C_I(0.0133(\ln q)^2 - 0.0954 \ln q + 0.297) - 4.16\right) \qquad \textbf{(12.26)}$$

where

n_r = Manning's computed flow resistance coefficient
C_I = empirical parameter that depends on vegetation condition
    (see Table 12.11 from Tempe et al., 1987 quoted by USDOE, 1989).
$q = dV$ ft^3/s
d = depth of flow (ft)
V = flow velocity (ft/s)

The resistance caused by the soil grain roughness is expressed as (Lane, 1955 quoted by Tempe et al., 1987)

$$n_s = (d_{75})^{1/6}/39 \qquad \textbf{(12.27)}$$

Table 12.11
Empirical vegetal
parameters for good
uniform stands of each
cover

Cover	C_f	C_I[a]	Reference stem density (stems/ft^2)[b]
Bermuda grass, 12-inch height	0.9	10.00	500
Weeping lovegrass			500
Buffalo grass			400
Kentucky bluegrass	0.87	7.64	350
Blue grama			350
Grass–legume mixture	0.75	5.60	200
Weeping lovegrass	0.75	5.60	350
Bermuda grass, 6-inch height	0.75	5.60	350
Yellow bluestem	0.75	5.60	350
Alfalfa[c]	0.5	4.44	350
Lespedeza sericea, 2-inch height[c]	0.5	4.44	300
Common lespedeza	0.5	4.44	150
Sudan grass	0.5	2.88	50
Bermuda grass, burned stubble	0.5	2.88	50

[a] If vegetation is not uniformly distributed over the areas present, C_I and C_f will be set equal to zero. In other words, the cover will be designed as if it were bare soil only.

[b] Multiply the stem densities given by 1/3, 2/3, 1, 4/3, and 5/3, for poor, fair, good, very good, and excellent covers, respectively. The equivalent adjustment to C_f remains a matter of engineering judgment until more data are obtained or a more analytical model is developed. A reasonable, but arbitrary, approach is to reduce the cover factor by 20% for fair stands and 50% for poor stands. Values of C_f for untested covers may be estimated by recognizing that the cover factor is dominated by density and uniformity of cover near the soil's surface. Thus, the sod-forming grasses near the top of the table exhibit higher C_f values than the bunch grasses and annuals near the bottom.

[c] For the legumes tested, the effective stem count for resistance (given) is approximately five times the actual stem count very close to the bed. Similar adjustment may be needed for other unusually large-stemmed, branching, and/or woody vegetation.

where n_s is the Manning coefficient associated with soil particles of a size capable of being detached by the flow at the stability-limiting condition s (soil grain roughness) and d_{75} is the particle diameter (in.), for which 75% of the material is finer by weight.

The above relationship is applicable for d_{75} larger than 0.05 in. (1.27 mm). For particles finer than 0.05 in., the influence of the soil grain roughness is assumed constant at 0.0156 and is included in the empirical development of n_r.

The combined influence of vegetation and roughness is expressed (Tempe et al., 1987) as

$$n = (n_r^2 - (0.0156)^2 + n_s^2)^{1/2} \tag{12.28}$$

where n is the Manning coefficient for the channel under specified flow conditions.

Table 12.11, which shows empirical vegetal parameters, includes the parameter C_f. This parameter characterizes the energy dissipation due to vegetation in the immediate vicinity of the soil–water boundary. If the vegetal and soil friction impedance are considered, the tractive shear stress is computed (Tempe et al., 1987) as

$$\tau_e = \gamma_w dS(1 - C_f)(n_s/n)^2 \tag{12.29}$$

where τ_e is the effective tractive shear stress on the soil, γ_w is the unit weight of water, and S is the slope of the energy line (assumed to be equal to the cover slope).

The reduction in tractive shear stress due to grass cover cannot be applied without considerable judgment and experience, especially if the grass cover is not uniform in density or if some parts of the cover are better vegetated than other areas. In most cases where the vegetative cover has not matured, no credit is given unless the design calls for sodded cover. In this case the tractive shear stress is expressed as

$$\tau_e = \gamma_w dS \tag{12.30}$$

The allowable shear stress may be estimated from the empirical relationships developed by Tempe et al. (1987) and presented in Table 12.12.

Example 12.5

For the landfill described in Example 12.4, assume a sandy loam cover classified as SM material with a plasticity index of 10 and a void ratio of 0.7. Determine the permissible shear and assess the erosion potential.

Solution: From Table 12.12,
$$C_{e1}^2 = (1.42 - 0.61 \times 0.7)^2 = 0.987$$
$$\tau_{ab} = (1.07 \times 100 + 7.15 \times 10 + 11.9)10^{-4} = 0.019 \text{ lb/ft}^2 \text{ (psf)}$$
$$\tau_a = 0.987(0.019) = 0.0187 \text{ psf}$$
The effective tractive shear force from Example 12.4 is $d = 0.04$ ft. For a bare cover, $C_f = 0.0$ and $\tau_e = 0.04 \times 62.5 \times 0.25 = 0.625$ psf $\gg 0.019$. The cover will erode.

For a Bermuda grass cover in fair to poor condition, we reduce the values of C_1 and C_f by 50%. Thus, from Table 12.11, $C_1 = 5$ and $C_f = 0.45$. To determine q, we assume a concentration factor of 3.
$$q = 0.1 \text{ cfs} \times 3 = 0.3 \text{ cfs}$$
$$n_r = \exp(5(\ln(0.3)^2 \times 0.0133 - (0.09543 \ln 0.3) + 0.297) - 4.16) = 0.134$$
$$n_s = 0.0156$$
$$n = \text{Manning coefficient (from Eq. 12.28)} = 0.134$$
The depth of flow (from Eq. 12.24) is

$$d = (3 \times 0.1 \times 0.134/(1.486 \times 0.5))^{3/5} = 0.17 \text{ ft}$$

Table 12.12
Allowable effective
tractive shear stress

Soil type	Plasticity index, PI	τ_a (psf)
GW, GP, SW, SP		
$d_{75} < 0.05$ in.	<10	0.02
$d_{75} \geq 0.05$ in.	<10	$0.4 d_{75}$
SM	$10 \leq PI \leq 20$	$C_{e1}^2 (1.07(PI)^2 + 7.15\ PI$ $+ 11.9)10^{-4}$
	>20	$C_{e1}^2 (0.058)$
GM, SC	$10 \leq PI \leq 20$	$C_{e1}^2 (1.07(PI)^2 + 14.3\ PI$ $+ 47.7)(10^{-4})$
	> 20	$C_{e1}^2 (0.076)$
GC	$10 \leq PI \leq 20$	$C_{e1}^2 (0.0477(PI)^2 + 2.86\ PI$ $+ 42.9)(10^{-3})$
	>20	$C_{e1}^2 (0.119)$
CH		$0.0966\ (C_{e2})^2$
CL	$10 \leq PI \leq 20$	$(C_{e3})^2 (1.07(PI)^2 + 14.3\ PI$ $+ 47.7)(10^{-4})$
	>20	$(C_{e2})^2(0.076)$
MH	$10 \leq PI \leq 20$	$(C_{e2})^2(0.0477(PI)^2 + 1.43\ PI$ $+ 10.7)(10^{-3})$
	>20	$(C_{e2})^3(0.058)$
ML	$10 \leq PI \leq 20$	$(C_{e3})^2(1.07)(PI)^2 + 7.15\ PI$ $+ 11.9)(10^{-4})$
	>20	$(C_{e3})^2(0.058)$

$C_{e1} = 1.42 - 0.61\ e,\ C_{e2} = 1.38 - 0.373\ e,\ C_{e3} = 1.48 - 0.57\ e,\ e = $ void ratio

The tractive shear stress (from Eq. 12.29) is

$$\tau_e = 62.5(0.17)(0.25)(1 - 0.45)(0.0156/0.134)^2 = 0.02 \text{ psf}$$

which is close to 0.0187 psf and considered acceptable.

For a gravel layer or a stone mulch with $d_{75} = 3$ in., the allowable shear (from Table 12.12) $= 0.4 \times 3.0 = 1.2$ psf.

$$n_s = 3.0^{1/6}/39 = 0.031$$
$$C_1 = 0.0$$
$$n_r = \exp - 4.16 = 0.0156$$
$$n = 0.031$$

The depth of flow is

$$d = (0.1 \times 3 \times 0.031/1.486 \times 0.5)^{3/5} = 0.073 \text{ ft}$$

The tractive shear stress is

$$\tau_e = 62.5 \times 0.073 \times 0.25 = 1.14 \text{ psf} < 1.2$$

which is acceptable.

12.8 FLOW ESTIMATES

To assess the potential for erosion using either the permissible velocity or shear stress methods, the flow quantity q must be estimated. To do this, the Soil Conservation Service (SCS) curve number (CN) method (SCS, 1986) may be used. The CN method requires the following data and procedure.

  a. Determine the soil hydrologic group and land use conditions.

  b. Based on information from step (a), determine the curve number (CN). The methods in Chapter 11 can be used to assign a CN number for varied cover and slope conditions.

  c. Determine the potential maximum retention (S) after runoff begins.

$$S(\text{in.}) = (1000/CN) - 10 \qquad (12.31)$$

  d. Select design storm and determine rainfall, P (in.). Figure 12.8 shows a 25-year, 24-hour duration storm for the eastern and midwestern United States.

  e. Determine the runoff Q (in.)

$$Q = (P - I_a)^2/(P + 4I_a) \qquad (12.32)$$

where I_a is the loss before runoff (initial abstraction) which equals $0.2S$.

  f. Determine the ratio I_a/P, using 0.1 as the minimum ratio.

  g. Determine the time of concentration, t_c. Use a minimum value of 6 minutes.

In step (g), t_c is defined as the time required for water to reach the point in question from the most hydraulically distant point in the basin. For sheet flow less than 300 ft,

$$t_c(\text{hr}) = 0.007(nL)^{0.8}/(P_2^{0.5}s^{0.4}) \qquad (12.33)$$

where

  t_c = time of concentration (hr)
  n = Manning coefficient (Table 12.10)
  s = land slope (ft/ft)
  L = flow length (ft)
  P_2 = 2-year, 24-hour rainfall (in.).

Because Figure 12.8 shows a 25-year, 24-hour storm, the values on the figure are about 1.7 to 2 times that of a 2-year storm. Thus, using Figure 12.8 will underestimate t_c and consequently provide a more conservative estimate of unit flow.

For $L > 300$ ft, t_c for shallow concentrated sheet flow should be used and is expressed as

$$t_c = L/16.1345s^{0.5} \qquad (12.34)$$

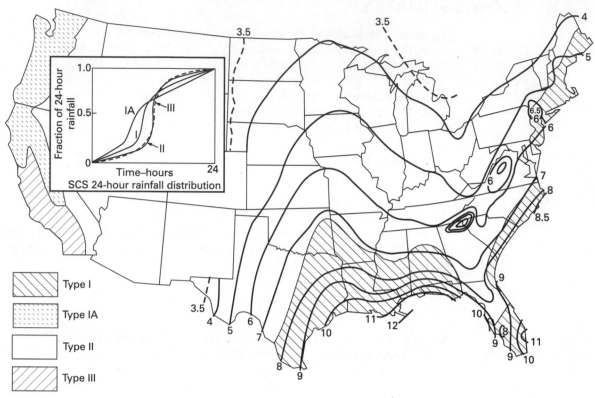

Figure 12.8   25-year, 24-hour storm

The HELP model (Schroeder et al., 1994) defines the time of concentration based on the Manning formula as follows:

$$t_c = (1.5/(P_s - I_i))(L^2/s)^{1/3}(1.49/n)^{-2/3} \qquad \text{(12.35)}$$

where t_c is the concentration time (min), P_s is the steady-state rainfall (in./hr), and I_i is the infiltration rate (in./hr). For channel flow, the time of concentration is simply the length of the channel divided by the velocity:

$$t_c(\text{hr}) = L/V \qquad \text{(12.36)}$$

The velocity is not known at the outset. The velocity used in the calculations should be checked against what is actually designed for based on the Manning formula and the erosion protection measures employed. In the SCS manual (SCS, 1986), natural channels are considered and assumed full. In this way, all the parameters for the Manning formula are known and the velocity, v, is computed.

h. Determine the type of storm distribution (see Figure 12.8).

i. From information in steps (g) and (h), determine the unit peak discharge, q_u, in units of cubic feet per square mile, per inch of runoff, Q.

$$\log q_u = C_0 + C_1 \log T_c + C_2(\log T_c)^2 \qquad (12.37)$$

where t (min) = 0.1 hr; t_c (max) = 10 hr; and C_0, C_1, and C_2 are the coefficients from Table 12.13.

j. Calculate the peak discharge, q_p, as

$$q_p = q_u Q A_m \qquad (12.38)$$

where A_m is the drainage area in mi^2.

Example 12.6

The landfill diagrammed in Figure 12.9 has a circular shape for ease in calculation. The hypothetical landfill is assumed to be located in south New Jersey. Consider a 25-yr, 24-hr storm, and an SCS CN number without slope correction. Estimate the flow volume for:
  a. Assessing the erosion potential of the cap
  b. Flow quantity for sizing swale ab
  c. Flow quantity for sizing swale cd
  d. Flow quantity for sizing chute bd
Assume swales with cross sections and erosion protection such that the flow velocity is 5 ft/s. Assume also bare soil with $CN = 91$ and a Manning coefficient of 0.011.

Table 12.13 Coefficients for Eq. 12.37

Rainfall type	I_a/P	C_0	C_1	C_2	Rainfall type	I_a/P	C_0	C_1	C_2
I	0.10	2.30550	−0.51429	−0.11750	II	0.10	2.55323	−0.61512	−0.16403
	0.20	2.23537	−0.50387	−0.08929		0.30	2.46532	−0.62257	−0.11657
	0.25	2.18219	−0.48488	−0.06589		0.35	2.41896	−0.61594	−0.08820
	0.30	2.10624	−0.45695	−0.02835		0.40	2.36409	−0.59857	−0.05621
	0.35	2.00303	−0.40769	0.01983		0.45	2.29238	−0.57005	−0.02281
	0.40	1.87733	−0.32274	0.05754		0.50	2.20282	−0.51599	−0.01259
	0.45	1.76312	−0.15644	0.00453					
	0.50	1.67889	−0.06930	0.0	III	0.10	2.47317	−0.51848	−0.17083
						0.30	2.39628	−0.51202	−0.13245
IA	0.10	2.03250	−0.31583	−0.13748		0.35	2.35477	−0.49735	−0.11985
	0.20	1.91978	−0.28215	−0.07020		0.40	2.30726	−0.46541	−0.11094
	0.25	1.83842	−0.25543	−0.02597		0.45	2.24876	−0.41314	−0.11508
	0.30	1.72657	−0.19826	0.02633		0.50	2.17772	−0.36803	−0.09525
	0.50	1.63417	−0.09100	0.0					

Figure 12.9

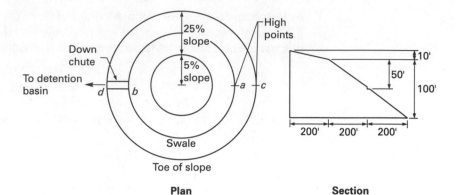

Plan   **Section**

Solution: From Figure 12.8, $P_{25} = 6$ in. and $P_2 = 3.5$ in. (est).
$$S = (1000/91) - 10 = 0.989$$
$$Q = (6 - 0.2(0.989))^2)/(6 + 0.8 (0.989) = 4.96 \text{ in.}$$
For the longest flow path, the cover is $(100 + 40,000)^{0.5}$
$+ (2500 + 40,000)^{0.5} = 200.25 + 206.15$.

$$t_{c_1} = 0.007(0.011 \times 200.25)^{0.8}/3.5^{0.5} \times 0.05^{0.4} = 0.023 \text{ hr}$$

For swale a, Eq. 12.33 is used for the first 300 ft (200.25 and 99.75) and Eq. 12.34 is used for the remaining distance (106.4 ft).

$$t_{c_2} = 0.007(0.011 \times 99.75)^{0.8}/(3.5^{0.5} \times 0.25^{0.4}$$
$$+ (106.4)/(3600)(16.1345 \times (0.25)^{0.5}$$
$$= 0.007 + 0.004 = 0.011$$
hr $t_c = 0.011 + 0.023 = 0.034$

We will use 0.1 hr.
$$I_a = 0.2 \, s = 0.2(0.989) = 0.2$$
$$I_a/p = 0.2/6 = 0.033$$
We will use 0.1. For the rainfall distribution we will use Type 3.

$$\log q_u = 2.47317 - 0.51848 \log 0.1 - 0.17083(\log 0.1)^2 = 2.82082$$

$$q_u = 661.94 \text{ cfs/mi}^2/\text{in. runoff}$$

Area of cover per unit (ft) width $= (200.25 + 206.15) = 406.4 \text{ ft}^2 = 0.0095$ acres (ft^2/42,560) $= 1.49 \times 10^{-5}$ mi^2 (acres/640)
$$q_p = 661.94 \times 1.49 \times 10^{-5} \times 4.96 = 0.049 \text{ cfs}$$
For swale ab, the longest flow path is $200.25 + 206.15 + 3.14(400)$.

$$t_c = 0.034 + 3.14 \times 400/5 \times 3600 = 0.034 + 0.069 = 0.103 \text{ hr}$$

$$\log q_u = 2.47317 - 0.51848 \log 0.103 - 0.17083(\log 0.103)^2 = 2.81$$

$$q_u = 658.55$$

The surface area is

$$(1/2(200 \times 200.25) \times 3.14)/42,560 + (0.5(200 + 400))$$

$$(206.15)(3.14))/42,560 = 1.477 + 4.56 = 6.04 \text{ acres} = 0.0094 \text{ mi}^2$$

$$q_p = 658.55 \times 0.0094 \times 4.96 = 30.7 \text{ cfs}$$

For swale cd, the longest flow path is $206.15 + 3.14 \times 600$.

$$t_c = 0.007(0.011 \times 206.5)^{0.8}/(3.5^{0.5} \times 0.25^{0.4})$$

$$+ 3.14 \times 600/5 \times 3600 = 0.0126 + 0.105 = 0.1176 \text{ hr}$$

$$\log q_u = 2.47317 - 0.51848 \log 0.1176 - 0.17083(\log 0.1176)^2$$

$$= 2.807$$

$$q_u = 641.9$$

The surface area is $0.5(3.14)(400 + 600)(206.15) = 7.6 \text{ acres} = 0.012 \text{ mi}^2$.

$$q_p = 641.9 \times 0.012 \times 4.96 = 38.2 \text{ cfs}$$

The flow for chute bd is $30.7 \times 2 = 61.4$ cfs. The flow channel is $61.4 + 2 \times 38.2 = 137.8$ cfs. The flow from the two swales were combined because the time of concentrations are not too different.

12.9   EROSION PROTECTION METHODS

12.9.1 *The Stephenson Method*

If the calculated velocities or tractive shear for landfills are too excessive, some form of erosion protection will be needed. This may include flattening the slopes, construction of drainage swales to reduce the flow distance, or placement of protective covers that may include riprap protection.

In designing riprap protection, several methods may be used. The Stephenson method was found to be appropriate (Abt et al., 1987) for sheet flow protection for slopes steeper than 10%. Using the Stephenson method, the D_{50} size is expressed as

$$D = \left\langle \frac{q \tan \theta^{7/6} n^{1/6}}{C g^{1/2}[(1 - n)(G_s - 1) \cos \theta(\tan \phi - \tan \theta)]^{5/3}} \right\rangle^{2/3} \qquad \textbf{(12.39)}$$

where

D = representative rock diameter D_{50} at which rock movement is expected for a unit discharge, q

q = low rate per unit width

n = rock porosity

g = acceleration of gravity

G_s = specific gravity of rock

θ = slope angle from horizontal

ϕ = angle of friction

C = empirical factor

= 0.22 for gravel and pebbles

= 0.27 for crushed granite

To ensure stability, the representative size D is multiplied by Olivier's constant (1.2 for gravel and 1.8 for crushed stone) (Abt et al., 1987). The resulting median rock diameter is D_{50}. Rock stone should be well graded, rock thickness should be at least twice the size of D_{50}, and a filter layer should underlie the riprap.

12.9.2 *The* USCOE *Method*

The toe or an intermediate level of a landfill is usually surrounded by a channel (swale) that diverts the flow to an outlet structure. The toe of the landfill and the bottom and sides of the channel are usually most heavily subjected to erosive forces. The design of protective riprap for diversion channels is made using the Corps of Engineers equation (COE, 1991). The riprap stone size is expressed in terms of D_{30}. After the D_{30} is determined, a gradation size is constructed to limit D_{85}/D_{15} in the range between 1.4 and 2.2. Table 12.14 shows standardized gradations having narrow range for D_{85}/D_{15} of 1.4 to 2.2.

The equation is

$$D_{30} = S_f C_s C_v C_T \, d((\gamma_w/(\gamma_s - \gamma_w))^{1/2} V/(K_1 gd)^{1/2})^{2.5} \qquad (12.40)$$

In this equation D_{30} is the riprap size, 30% of which is finer by weight, and S_f is the safety factor. We use 1.1 as a basic safety factor but this number may be

Table 12.14
Riprap standard sizes

D_{100} (max) (in.)	D_{30} (min) (ft)	D_{90} (min) (ft)
12	0.48	0.7
15	0.61	0.88
18	0.73	1.06
21	0.85	1.23
24	0.97	1.4
27	1.1	1.59
30	1.22	1.77
33	1.34	1.94
36	1.46	2.11
42	1.7	2.47
48	1.95	2.82
54	2.19	3.17

adjusted to account for unknowns in estimating the unit weights, freeze–thaw conditions, impact forces, and so on. We double the thickness if the riprap is placed under water.

C_s is 0.3 for crushed rock and 0.36 for rounded rock. The thickness of the riprap is $1.0D_{100}$ or $1.5D_{50}$, whichever is greater.

$$D_{85}/D_{15} = 1.7 - 5.2$$

For twice the required thickness, C_s may be reduced by 15%, 30%, and 46% for D_{85}/D_{15} of 1.7, 2.5, and 5.2, respectively.

C_v is 1.0 for a straight channel and 1.25 for the end of the dikes and for downstream of the concrete channels. For bends, this parameter increases by up to 22%, depending on the radius of the bend (sharp bends have a low radius).

C_T is the thickness coefficient and is 1.0 for a thickness equivalent to D_{100} max. d is the local depth of flow, defined to correspond to the depth at which a characteristic velocity is defined. γ_w is the unit weight of water, γ_s is the unit weight of stone, and g is the gravitational constant. V is the local average velocity. In a channel bend, the characteristic velocity (V_{ss}) for side slopes should be used. This velocity is the depth-averaged local velocity over the slope at a point 20% of the slope length from the toe of the slope. For trapezoidal channels and sharp (low radius) bends, it is about 47% higher than the local average velocity.

Finally, K_1 is the side slope correction coefficient factor where
$K_1 = 1.0$ for $3.5H$ on $1V$ or flatter
    $= 0.98$ for $3H$ on $1V$
    $= 0.95$ for $2.5H$ on $1V$
    $= 0.87$ for $2H$ on $1V$
    $= 0.7$ for $1.5H$ on $1V$

Eq. (12.40) can be used with either SI (metric) or English units of measure. The equation is applicable for side slopes of $1V$ on $1.5H$ or flatter. In most cases the velocity is not known and the Manning formula is needed to compute the velocity. The parameter n in the Manning formula is computed according to the COE method in terms of D_{90} for velocity and stone size calculations as follows:

$$n = 0.034(D_{90}(\text{min}))^{1/6} \tag{12.41}$$

Because D_{90} is not known to start with, the process of calculating D_{30} is iterative. D_{90} is selected for a trial gradation (e.g., from Table 12.14), and D_{30} is computed and compared with D_{30} in the trial gradation. If needed, another trial is performed.

12.9.3 Tractive Shear Stress Method

A quicker estimate for the mean grain size D_{50} can be made using the tractive shear stress method (USNRC, 1990). The design shear stress is computed based on

$$\tau_0 = \gamma_w \, y \cdot S \qquad\qquad (12.42)$$

where τ_0 is the design shear stress (lb/ft^2), y is the depth of flow (ft), S is the slope (dimensionless), and

$$D_{50} = \tau_0/0.04(\gamma_s - \gamma_w) \qquad\qquad (12.43)$$

In the above expression, the γ variables are in lb/ft^3 and D_{50} is in ft. The depth of flow y may be computed from the Manning formula. The Manning coefficient n for steep slopes is expressed as

$$n = 0.0456(D_{50} \cdot S)^{0.159} \qquad\qquad (12.44)$$

Here D_{50} is expressed in in.

The riprap layer should be underlain by a base material with a thickness 1/2 the riprap thickness but not less than 6 to 9 in. The base blanket should vary in size from 3/16 in. to an upper limit of 3 to 3.5 in. and should satisfy the filter criterion for category 4 (see Chapter 11).

The filter material is usually underlain by another filter (usually a geotextile) in intimate contact with the base of the channel or cover. In areas where energy dissipation occurs, such as the toe of a slope, the riprap thickness should be at least twice the thickness of the riprap on the slope and should extend laterally at least three times the thickness. Although there are no set rules for estimating the geometry at the toe, useful guidance is given in COE (1991). Procedures for estimating the depth of scour for culverts, given in USDOT (1983), are useful because the riprap protection must extend to or beyond the zone of scour.

The relation between the size and the weight of the stone can be described (COE, 1991) as

$$D_\% = (6W_\%/3.14\gamma_s)^{1/3} \qquad\qquad (12.45)$$

where $D_\%$ is the equivalent-volume spherical stone diameter (ft) and $W_\%$ is the weight of an individual stone having a diameter of $D_\%$.

Example 12.7

Determine the riprap protection size for a trapezoidal channel. Hydraulic computations indicate that the velocity is 8 ft/s and flow depth is 3 ft. Assume that quarry stone is available, the unit weight is 165 pcf, and the side slope is 2H on 1V. Use a safety factor of 1.2.

Solution: Use the COE method, and Eq. 12.40 with

$S_f = 1.2$

$C_s = 0.3$

$C_v = C_T = 1$

$$K_1 = 0.87$$

$$D_{30} = 1.2 \times 0.3 \times 3\{[62.5/(165 - 62.5)]^{0.5} \times [8/(0.87 \times 32.2 \times 3)^{0.5}]\}^{2.5}$$

$$= 0.41 \text{ ft}$$

Using a standard size of 0.48 ft results in $D_{100} = 12$ in. and $D_{90} = 0.7$ ft.

Example 12.8

A 20-ft-wide chute is to be constructed on a slope of $5H$ on $1V$. The side slope of the channel is $2H$ on $1V$ and the design discharge is 100 cfs. Determine the size of riprap protection. Assume a rock unit weight of 165 pcf.

Solution

Method 1 (USNRC, 1990)

Assuming $D_{50} = 2$ ft, from Eq. 12.44, we have

$$n = 0.0456(24 \times 0.2)^{0.159} = 0.058$$

We solve the Manning equation for depth d.
   Area $= (20 + 2d)d$, $R = (20 + 2d)d/(20 + 4.48d)$
   $Q = 1.486 \ AR^{2/3}S^{1/2}/n$
For $d = 0.6$ ft,
   $A = 21.2 \times 0.6 = 12.72$, $R = 12.72/(20 + 2 \times 2.24 \times 0.6) = 0.56$
From Eq. 12.22
   $Q = VA$
   $Q = 1.486 \times 12.72 \times 0.56^{2/3} \times 0.2^{0.5}/0.058 = 99$ cfs
From Eq. 12.42
   $\tau_0 = 0.6 \times 0.2 \times 62.5 = 7.5$ psf
From Eq. 12.43
   $D_{50} = 7.5/0.04(165 - 62.5) = 1.83$ ft
The calculated size is close to the assumed size and no second iteration is necessary.

Method 2 (COE method)

If we assume $D_{90} = 3.17$ ft, then from Table 12.14, the assumed $D_{30} = 2.19$ ft.

$$n = 0.036(D_{90})^{0.166} = 0.044$$

From iterative solution of the Manning equation, $d = 0.5$ ft
   $S_f = 1.1$, $C_s = 0.3$, $C_v = C_T = 1.0$, $K = 0.87$
   $A = (20 + 1)0.5 = 10.5$ ft^2
   $V = 100/10.5 = 9.52$ fps
   $D_{30} = 1.1 \times 0.3 \times 0.5\{[62.5/(165 - 62.5)]^{0.5}$
         $\times [9.52/(0.87 \times 32.2 \times 0.5)^{0.5}]\}^{2.5}$
      $= 0.91$ ft
For $D_{90} = 1.59$, $D_{30} = 1.10$ (from Table 12.14).

$$n = 0.036(1.59)^{0.166} = 0.039$$

From the trial solution of the Manning formula: $d = 0.48$ ft.
$$V = 100/(20 + 2 \times 0.48)0.48 = 9.94 \text{ fps}$$
$$d_{30} = 1.1 \times 0.3 \times 0.48\{[62.5/(165 - 62.5)]^{0.5}$$
$$\times [9.94/(0.87 \times 32.2 \times 0.48)^{0.5}]\}^{2.5} = 1.03$$
We choose $D_{30} = 1.1$ ft. From Table 12.14, $D_{100} = 27$ in. $= 2.25$ ft. The above are minimum sizes; actual sizes should be equal or larger.

Method 3 (*Stephenson method*)
Maximum flow per unit width $= 100/20 = 5$ cfs/ft
Rock porosity $n_p = 0.4$ (assumed)
Rock specific gravity $= 2.64$ (assumed)
Slope angle $= 11.31°$
$C = 0.27$ (crushed rock)
Rock friction angle $= 40°$ (assumed)

$$D_{50} = (5 \times (0.2)^{7/6}0.4^{1/6}/\{0.27 \times 5.67[0.6 \times 1.64 \times 0.98$$
$$\times (\tan 40 - \tan 11.3)]^{5/3}\})^{2/3} = 1.0 \text{ ft}$$

If the calculated D_{50} is multiplied by Oliver's constant of 1.8 for stone, the representative size D_{50} is 1.8 ft, which is consistent with other methods. However, it should be noted that the method was developed for sheet flow (Abt et al., 1987) over and through rock fill on steep slopes. The method does not account for uplift on stone due to emerging flow. Alternative methods should therefore be used for toe and channel bank protection.

12.10   USE OF FABRICS

In recent years, fabrics have been increasingly used in erosion protection construction. Fabrics can be classified as (Simons, Li & Associates, 1988):

1. Two-dimensional woven meshes and fabrics using natural or synthetic (polymer) fiber, through which grass is allowed to grow.

2. Three-dimensional open synthetic mats that are filled with top soil and seeded.

3. Three-dimensional filled synthetic mats or grid confinement systems filled with rock or asphalt. Table 12.15 (Simons, Li & Associates, 1988) provides some guidance for assessing the suitability of some geotextiles.

Erosion control fabrics allow the establishment of vegetation. If open weave nets are used, seed is sown into the open fabric and grass grows through it and eventually covers it. The soil may also be fertilized and seeded before installing the open mat. A major problem (besides the high cost) is the potential for cover erosion from underneath the mat. A filter layer should therefore underlie the mat and be in intimate contact with the base soil to be protected. If a geotextile is used as a filter and gaps exist between the geotextile and the base soil, erosion from underneath could still occur. In most cases a soil layer

Table 12.15
Permissible shear
stress for selected
geotextiles and other
lining material

Lining type	Manning's n (depth = 0–0.5 ft)	Permissible unit shear stress (psf)
Woven paper net	0.016	0.15
Jute net	0.028	0.45
Single fiberglass	0.028	0.6
Double fiberglass		0.85
Straw with net	0.065	1.45
Curled wood mat	0.066	1.55
Synthetic mat	0.036	2.0
6-in. Gabion on SM–SC soil		35
4-in. Geoweb* on SM–SC soil		10
Gravel		
$D_{50} = 1$ in.		0.4
$= 2$ in.		0.8
Rock		
$D_{50} = 6$ in.		2.5
$= 2$ in.		5.0

* Gravel filled, wired to Tenax netting to prevent boiling of gravel
Source: Simons, Li and Associates 1988

above the geotextile is necessary to achieve the intimate contact. This in turn will produce still higher costs of installation. If the cap is not well drained, flow normal to the slope may occur and may cause boiling of the surface soil and uplifting of the fabric mat.

SUMMARY

Waste disposal facilities require daily cover during operation, thicker intermediate cover if operation is halted for a limited time, and final cover at closure. The regulatory reason for the cover is to minimize water infiltration and thus minimize leachate. Another major reason is to control gas emissions.

A fact known for many years is that cracks that develop as a result of settlements will make failure of a clay cap almost certain. Geomembranes are also sensitive in this way but to larger settlements. This chapter has presented methods for assessing the stresses produced by settlements and their effects on the hydraulic conductivity of the cap.

Material used for liners is typically used for caps. The survival of such materials as caps is heavily impacted by large landfill settlements and other factors. The use of such materials is usually limited to side slopes no steeper than $1V$ on $3H$; for some geomembranes and GCLs, slopes steeper than 8 degrees may not be possible.

In most cases, cap failure is characterized by erosion. The universal soil loss equation is used to estimate soil losses due to sheet flow. Erosion of caps is rarely a sheet-flow problem and flow concentration is certain. Analytical methods are presented in this chapter for estimating the potential for erosion and ways to mitigate the problem.

NOTATIONS

ε	bending strain	G	shear modulus (F/L^2)	Q	static moment of cross-sectional area from the neutral axis to the extreme fiber (L^3)
ε	strain	H	thickness of the overburden (L)		
δ	deflection at the center (L)				
γ	average unit weight of the overburden above the membrane (F/L^3)	i	average hydraulic gradient		
		i	rainfall intensity (in./hr)	Q	flow volume (L^3/T)
		I	moment of inertia about the neutral axis (L^4)	R	radius of curvature (L)
v	dynamic viscosity of water $(2.1 \times 10^{-5}$ psf-s)			R	rainfall factor
		I_i	infiltration rate (in./hr)	R	hydraulic radius (L)
τ_e	effective tractive shear stress on the soil (F/L^2)	k	lateral earth pressure coefficient (ratio of horizontal to vertical earth pressure)	S	slope of the energy line
				s (%)	slope
γ_w	unit weight of water (F/L^3)			S	slope (dimensionless)
τ_0	design shear stress (F/L^2)			s	slope of the channel bottom
σ_t	tensile stress in the geomembrane (F/L^2)	k_e	effective hydraulic conductivity (L/T)	S_f	safety factor
				t'	corrected thickness of the geomembrane after deformation (L)
A	loss per unit area per year (tn/acre/yr)	k_1	hydraulic conductivity of intact cap (L/T)		
A_m	drainage area (mi^2)	l	half-span length (L)	t_c	concentration time (min) (T)
b	crack width (ft)	L	slope length (ft)	u'	coefficient of friction of the overburden above the geomembrane
C	distance from the neutral axis to the extreme fiber (L)	l_c	crack spacing (ft)		
		LS	topographic factor		
C_1	flow concentration factor	n	surface roughness coefficient	V	average velocity of a specified cross section (L/T)
C_2	depth correction factor	d	depth of flow (L)		
C_I	empirical parameter	n_r	Manning's computed flow resistance coefficient	V_p	permissible velocity (L/T)
e	percolation rate through the cover (L/T)			V_u	uncorrected permissible velocity (L/T)
		P	precipitation (L/T)		
E	cover efficiency	P_s	steady-state rainfall (in./hr)	y	depth of flow (ft) (d)
E	total kinetic energy of a given storm (ft-ton/acre-in.)	q	flow per unit length of crack (L^2/T)		
E	Young's modulus (F/L^2)				

PROBLEMS

12.1 A 200-acre waste pond in east Texas is covered with sandy loam with no organic content. The cover has an average slope of 5%.

(a) Determine the annual soil loss for both a bare soil cover and a vegetated cover.

(b) If a 2-ft thick cover is not maintained, in approximately how many years will the waste be exposed?

12.2 A landfill design calls for a square footprint of 120 acres. The landfill maximum height is 100 ft.

(a) Determine the steepest side slope that will still limit soil loss to 2 tn/acre/yr.

(b) The number of swales required to fully develop the full permitted height of the landfill. Assume that the area of the plateau is 5 acres and nonerodible. Also assume temporary seedings and $K = 250$.

(c) Determine three options for the maximum length of the side slope and cover types that will result in soil loss less than 2 tn/acre/yr.

(d) For a nonorganic loamy sand cover, what cover protection is needed for the designs in part (c) to be valid?

12.3 Consider a landfill in eastern Wyoming with an area distribution as follows: 40 acres at 3% slope and slope length of 400 ft, 80 acres at $3H$ on $1V$ slope and a slope length of 300 ft. Determine the annual soil loss for the following conditions.

(a) Bare inorganic silty clay cover.

(b) Top soil of sandy loam, 2% organic content, and grass over 1 year old.

12.4 A landfill slope is 900 ft long at 15% slope without any intermediate swales. The calculated runoff per unit width of the slope is 0.4 cfs.

Using both the permissible velocity method and the permissible shear stress method determine whether the cover will erode for covers of

(a) fine sand, $D_{75} = 0.1$ mm

(b) coarse sand, $D_{75} = 3$ mm

(c) gravel mulch, $D_{75} = 30$ mm

(d) silty clay, plasticity index $= 19$, void ratio $= 0.7$

12.5 (a) Repeat Problem 12.4 assuming good Bermuda grass cover.

(b) For covers with the potential erosion determined in Problem 12.4, determine the maximum length of slope needed to avoid erosion.

12.6 If the 900 ft slope in Problem 12.4 is not to be interrupted by diversion channels, use both the Stephenson formula and the shear stress method to select a riprap and filter design to avoid erosion. Assume that the unit weight of rock is 165 pcf.

12.7 An unprotected trapezoidal swale is proposed for a landfill cover. To accommodate landfill settlement, the bottom slope has to be at least 1%. The computed flow volume is 50 cfs. Determine the size of the swale to limit the shear stress to 0.08 psf.

12.8 For the swale in Problem 12.7, determine the size if there will be a good Bermuda grass channel lining.

12.9 A 25-ft-wide channel on a 20% landfill slope will carry a flow of 150 cfs. Determine the protection requirements using

    (a) the COE method.

    (b) the Stephenson method.

    (c) the shear stress method.

12.10 For the required protection in Problem 12.9, recommend at least two alternatives to the riprap protection.

13 Foundation and Slope Stability

13.1 FOUNDATION STABILITY

The stability of landfills can be assessed based on the usual limit equilibrium methods commonly used in geotechnical engineering (Oweis et al., 1985; Oweis, 1993). Computer programs are available (e.g., STABL) that use limit equilibrium techniques such as the general wedge method, modified Bishop, etc.

The DM-7 wedge method (NAVFAC DM-7.3, 1982) illustrated in Figure 13.1 is adapted to hand calculations. The waste may be treated as an engineering material and characterized by cohesion and friction. The cohesion may be interpreted as an apparent cohesion of a reinforced soil mass. In limit equilibrium analysis, deformation is not an input parameter. The failure envelope for refuse is likely to be nonlinear although data on the mechanical properties of refuse is scarce.

Figure 13.2 illustrates an example of a stability analysis using a simple sliding wedge analysis. One of the potential failure planes is taken along the geomembrane–clay liner interface. In this case,

$$\alpha_1 = 45 + \phi_1/2$$

$$\alpha_2 = 45 + \phi_2/2$$

where α_1 is the angle between the failure plane in layer 1 and the horizontal, α_2 is the angle between the failure plane in layer 2 and the horizontal, and ϕ_1 and ϕ_2 are the friction angles in layers 1 and 2, respectively.

The angle α_3 is the slope of the liner along the base and β_1 is the slope of the liner as it meets the surface. The clay–geomembrane interface friction angle is assumed to be 12°. The analysis is sensitive to the liner slope and interface friction. If a leachate mound exists (as in the case of old landfills), the analysis is sensitive to the level of the mound.

The potential failure planes described above are shallow planes. In typical situations where weak soil layers are present, a deep-seated failure mechanism should be investigated (as illustrated in Figure 13.2). In the formulations explained in Figure 13.1, the interwedge forces are assumed to be horizontal.

The stability problem may be analyzed as a bearing capacity problem, as illustrated in Figure 13.3. Figure 13.3 depicts an embankment with half-width b supported on a clay layer having an undrained shear strength increasing with depth. A layer of reinforcement within the fill over the clay provides some

Figure 13.1 Stability analysis: wedge method

Definition of Terms

P_a = Resultant horizontal force for an active or central wedge along potential sliding surface a b c d e

P_β = Resultant horizontal force for a passive wedge along potential sliding surface e f g

W = Total weight of soil and water in wedge above potential sliding surface

R = Result of normal and tangential forces on potential sliding surface considering friction angle of matrial

P_w = Resultant force due to pore water pressure on potential sliding surface calculated as

$$P_w = \frac{hw_i + hw_{ii}}{2}\,(L)(\gamma_w)$$

ϕ = Friction angle of layer along potential sliding surface

c = Cohesion of layer along potential sliding surface

L = Length of potential sliding surface across wedge

h_w = Depth below phreatic surface at boundary of wedge

γ_w = Unit weight of water

Procedures

1. Except for central wedge where a is dictated by stratigraphy use

    $$a = 45° + \frac{\phi}{2}, \quad \beta = 45° - \frac{\phi}{2}$$

    for estimating failure surface.

2. Solve for P_a and P_β for each wedge in terms of the safety factor (F_s) using the equations shown below. The safety factor is applied to soil strength values (tan ϕ and c). Mobilized strength parameters are therefore considered as $\phi_m = \tan^{-1}(\tan \phi / F_s)$ and $c_m = c/F_s$.

    $$P_a = [W - c_m L \sin a - P_w \cos a]\tan[a - \phi_m] - [c_m L \cos a - P_w \sin a]$$

    $$P_\beta = [W + c_m L \sin \beta - P_w \cos \beta][\tan(\beta + \phi_m)] + [c_m L \cos \beta + P_w \sin \beta]$$

    in which the following expansions are to be used.

    $$\tan(a - \phi_m) = \frac{\tan a - \dfrac{\tan \phi}{F_s}}{1 + \tan a \dfrac{\tan \phi}{F_s}} \qquad \tan(\beta \neq \phi_m) = \frac{\tan \beta + \dfrac{\tan \phi}{F_s}}{1 - \tan \beta \dfrac{\tan \phi}{F_s}}$$

3. For equilibrium $\Sigma P_a = \Sigma P_\beta$, sum P_a and P_β forces in terms of F_s, select a trial F_s, calculate ΣP_a and ΣP_β. If $\Sigma P_a \neq \Sigma P_\beta$, repeat. Plot P_a and P_β versus F_s with sufficient trials to establish the point of intersection (i.e., $\Sigma P_a = \Sigma P_\beta$), which is the correct safety factor.

4. Depending on stratigraphy and soil strength, the centre wedge may act to maintain or upset equilibrium.

5. The safety factor for several potential sliding surfaces may have to be computed to find the minimum safety factor for the given stratigraphy.

Figure 13.1 (*Continued*)

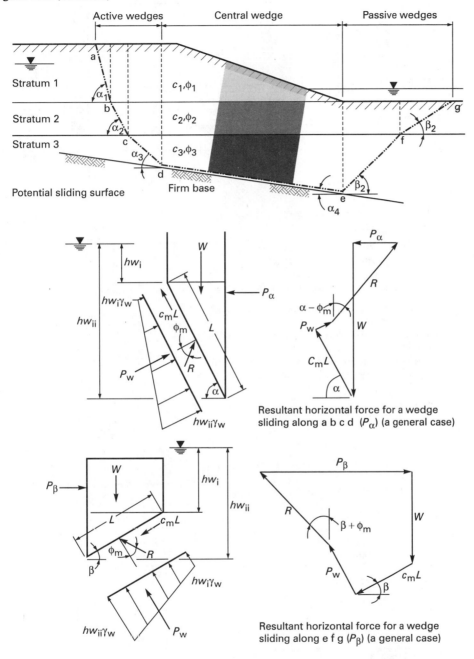

Resultant horizontal force for a wedge sliding along a b c d (P_α) (a general case)

Resultant horizontal force for a wedge sliding along e f g (P_β) (a general case)

resisting force T against the lateral pressure force P (active or at-rest pressure) from the fill. The failure mechanism shown in Figure 13.3 is assumed (Michalowski, 1993). Considering side slopes at angle 0, the average imposed load q_{av} is expressed as

Figure 13.2
Example of stability analysis

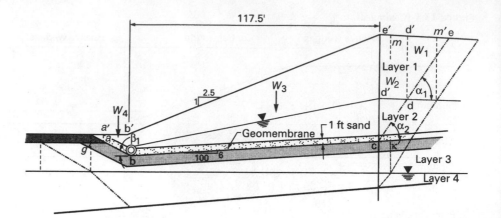

Shallow Potential Failure Surface

Refuse and soil properties

Layer 1, refuse $\phi = 26°$, $c = 100$ lb/ft², $\gamma = 35$ lb/ft³
Layer 2, refuse $\phi = 26°$, $c = 100$ lb/ft², $\gamma = 70$ lb/ft³
Layer 3, sand $\phi = 30°$, $c = 0$, $\gamma = 120$ lb/ft³
Layer 4, soft clay $\phi = 0$, $c = 500$ lb/ft², $\gamma = 100$ lb/ft³

Geomembrane: interface friction angle with clay liner = 12°
Clay liner: $\phi' = 30°$, $c' = 0$, $\gamma = 120$ lb/ft³
Consider the one ft sand layer above the membrane as part
   of layer 2.
Consider potential failure wedge along the geomembrane a b c d e
   $a'a' = 5'$, $a'b' = 10'$, $bb' = 10'$, $d'e' = 30'$, $d'c' = 20'$

$\alpha_1 = 45 + \phi/2 = 58°$

$W_1 = (30)^2/2 \tan 32 (0.035) = 9.8$ K, $P_{w1} = 0$

$\alpha_2 = 58°$

$W_2 = (20 \tan 32)(30)(0.035) + \dfrac{20^2}{2} \times 0.07 \times \tan 32$

$\quad = 13.1$ K $+ 8.8$ K $= 21.9$ K

$P_{w2} = \dfrac{20}{\sin 58} \times \dfrac{20 \times 0.0625}{2} = 14.7$ K

$W_3 = \left(\dfrac{30+10}{2}\right)(117.5)(0.035) + \left(\dfrac{20+0}{2}\right)(117.5)(0.07)$

$\quad = 82.3 + 82.25 = 164.55$ K

$\alpha_3 = \tan^{-1}(0.06) = 3.43$ α

$P_{w3} = 0$ no leakage through membrane

$\quad = \dfrac{20(0.0625)(117.5)}{2 \cos 3.43} = 73$ K   with leakage

$\beta_1 = \tan^{-1} 1/2 = 26.56$

$W_4 = \dfrac{5+10}{2}(10)(0.035) = 2.6^K$, $P_w = 0$

$P_{\alpha 1} = \left[9.8 - \dfrac{0.1}{F_s}(30)\right]\left[\dfrac{\tan 58 - \dfrac{\tan 26}{F_s}}{1 + \dfrac{\tan 58 \tan 26}{F_s}}\right]$

$\quad - \left[\dfrac{0.1}{F_s}\dfrac{30}{\sin 58}\cos 58\right]$

$\quad = \left(9.8 - \dfrac{3}{F_s}\right)\left(\dfrac{1.6F_s - 0.49}{F_s + 0.78}\right) - \dfrac{1.88}{F_s}$

$P_{\alpha 2} = \left[21.9 - \left(\dfrac{0.1}{F_s}\right)0.20 - 14.8 \cos 58\right]\left[\dfrac{1.6F_s - 0.49}{F_s + 0.78}\right]$

$\quad - \left[\dfrac{0.1(20)}{F_s \tan 58} - 14.8 \sin 58\right]$

$\quad = \left(21.9 - \dfrac{2}{F_s} - 7.8\right)\left(\dfrac{1.6F_s - 0.49}{F_s + 0.78}\right)$

$\quad - \dfrac{1.25}{F_s} + 12.6$

$P_{\alpha 3} = $ (impervious liner)

$\quad = [164.55]\left[\dfrac{\tan 3.43 - \dfrac{\tan 12}{F_s}}{1 + \dfrac{\tan 3.43 \tan 12}{F_s}}\right]$

$\quad = 164.55\left[\dfrac{0.06F_s - 0.212}{F_s + 0.013}\right]$

$P_{\beta 1} = 2.6\left[\dfrac{\tan 26.56 + \dfrac{\tan 12}{F_s}}{1 - \dfrac{\tan 26.56 \tan 12}{F_s}}\right]$

$P_{\beta 1} = 2.6\left[\dfrac{0.5F_s + 0.212}{F_s - 0.106}\right]$

F_s	$P_{\alpha 1}$	$P_{\alpha 2}$	$P_{\alpha 3}$	ΣP_α	ΣP_β
1.1	3.02	19.68	−21.5	1.2	1.99
1.2	3.73	20.54	−19.0	5.27	1.93

$$F_s \simeq 1.12$$

Figure 13.3
Bearing capacity of an
embankment on a clay
layer

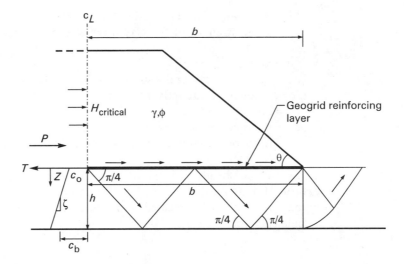

$$q_{av} = \gamma H[1 - H/(2b \tan \theta)] \tag{13.1a}$$

$$H = b \tan \theta \{1 - \sqrt{[1 - (2q_{av}/b\gamma \tan \theta)]}\} \tag{13.1b}$$

where γ is the average unit weight of the refuse embankment, q_{av} is the average stress imposed by the fill on the clay layer, and P is the lateral thrust from the landfill.

Alternately, the computational width b may be taken at mid-height of the embankment. In this case,

$$H = q_{av}/\gamma \tag{13.1c}$$

Horizontal thrust from the landfill is introduced as a uniform shear stress, τ_{av}, where

$$\tau_{av} = (P - T)/b \tag{13.2}$$

where H is the height of the refuse embankment and T is the resisting force at the base of the fill generated by reinforcement.

The limit load estimated from Michalowski (1993) based on the failure mechanism in Figure 13.3 is

$$\frac{q_f}{c_0} = 3 + \frac{\pi}{2} + \frac{1}{2}\frac{b}{h}(\kappa - \chi) + \frac{\xi b}{c_0}\left[\frac{h}{b}(1 + \sqrt{2}) + \frac{\kappa}{2}\right], \quad \frac{b}{h} = 2, 4, 6 \tag{13.3a}$$

where
q_f = average stress on the clay layer over the width b causing failure (limit vertical stress)

$h =$ thickness of the clay layer

$c =$ shear strength of the clay layer at depth z below the top of the clay layer

$c = c_0 + \xi z$

$c_0 =$ shear strength at the surface of the clay layer

$\xi =$ gradient of strength with depth

$\kappa = c_b/c$

$c_b =$ shear strength at the base interface. For rough base, $c_b = c_0 + \xi h$.

$\chi =$ ratio of the applied shear stress at the surface to the shear strength (τ_{av}/c_0)

Because the height H is unknown to start with, this ratio must be assumed first or estimated initially from Eq. 13.3b and corrected after critical H is computed. The process is repeated until the assumed value becomes close to the calculated value.

$$\chi = \frac{\gamma H^2 K_0 - 2T}{2bc_0} \qquad -1 \le \chi \le 1 \qquad \text{(13.3b)}$$

where b is the half-width of the landfill embankment.

The analysis in Eq. 13.3 presumes that the value of P is given and no factor of safety is applied to the refuse parameters in computing P. It is recommended, therefore, that a conservative estimate of P be made, such as using the at-rest lateral pressure coefficients such that

$$P = K_0 \gamma H^2/2 \qquad \text{(13.4)}$$

where $K_0 = 1 - \sin \phi$ and ϕ is the friction angle of the embankment material.

Example 13.1

Consider the situation depicted in Figure 13.4. Determine the critical height, H, of the landfill using both the Eq. 13.3 and the DM-7 wedge method.

Solution: *Eq. 13.3 method*
  $K_0 = 1 - \sin 30 = 0.5$
  $c_0 = c = 600$ psf

Figure 13.4
Data for Example 13.1

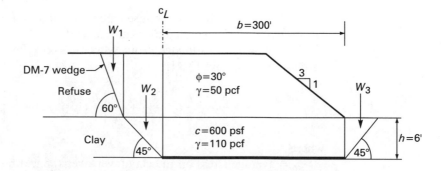

$$\xi = 0$$
$$\kappa = 1 \text{ (rough base)}$$
$$b/h = 50$$
Assume $\chi = 0.3$
$$(q_f/c) = 4.57 + 0.5 \times 50 \times 0.7 = 22.07 \quad \text{(for rough base)}$$
$$q_f = 13{,}242 \text{ psf}$$

Applying Eq. 13.1b and substituting q_f for q_{av}, we can compute the critical height. For this case, the function under the square root is negative. Thus the height H is limited by the maximum value $b \tan \theta = 100$ ft. There is no need to correct χ because the height of the fill would still be limited by b and the slope angle. This method (bearing capacity failure mode) yields results that may be unconservative with respect to other modes of failure, such as slope failure or spreading. It is therefore recommended that other potential failure modes (Figure 13.5) be investigated.

DM-7 method

Considering a factor of safety of 1.0 in applying the DM-7 wedge method, we have

$$W_1 = (H^2/2) \tan (45 - \phi/2)50 = 14.43H^2$$
$$P_{a_1} = 14.43H^2(\tan 60 - \tan 30)/(1 + \tan 60 \times \tan 30) = 8.33H^2$$
$$W_2 = 6 \times H \times 50 + \text{weight of clay wedge}$$
$$P_{a_2} = W_2 - 2 \times 600 \times 6 = W_2 - 7200$$
$$P_{a_3} = -600 \times 3 \times H = -1800H$$
$$P_{\beta_1} = 2 \times 6 \times 600 + \text{weight of clay wedge}$$
$$\Sigma P_\alpha = \Sigma P_\beta$$
$$8.33H^2 + 300H - 1800H - 7200 = 7200$$
$$H^2 - 180.1H - 1728.7 = 0$$
$$H = 189.2 \text{ ft}$$

Figure 13.5
Potential modes of failure

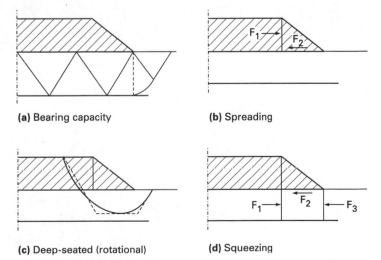

(a) Bearing capacity

(b) Spreading

(c) Deep-seated (rotational)

(d) Squeezing

In this case the critical height of the slope is limited by the half-width b and slope angle and H cannot exceed 100 ft.

13.2 SIDE SLOPE AND STABILITY

The stability of covers along the side of a landfill can be assessed using the wedge method described in Section 13.1. Figure 13.6 shows a cover of constant thickness where seepage occurs parallel to the slope within the cover. The more general case of variable thickness and a seepage surface not parallel to the slope could be analyzed by the wedge method as well (Oweis, 1993). The cover is kept from sliding by reinforcement providing a resisting tensile force T. To develop T, the reinforcing element (geogrid or geotextile) is anchored at the top of the slope.

Considering the limit equilibrium of the central wedge and Figure 13.1, the following equation can be solved for the factor of safety (knowing T) or for the required T if the factor of safety is assigned. (The latter case is less time consuming.)

$$P_a + P_c = P_p \tag{13.5}$$

where

$$P_a = (W_a - c_m L_a \sin \alpha - P_{w_a} \cos \alpha)\tan(\alpha - \phi_m)$$
$$\quad - (c_m L_a \cos \alpha - P_{w_a} \sin \alpha)$$
$$P_c = (W_c - c_{im} L_c \sin i - T \sin i - P_{w_c} \cos i)\tan(i - \phi_{im})$$
$$\quad - ((c_{im} L_c + T) \cos i - P_{w_c} \sin i)$$
$$P_p = (W_p + c_m L_p \sin \beta - P_{w_p} \cos \beta)(\tan \beta + \phi_m)$$
$$\quad + (c_m L_p \cos \beta + P_{w_p} \sin \beta)$$

Figure 13.6
Cap wedge analysis

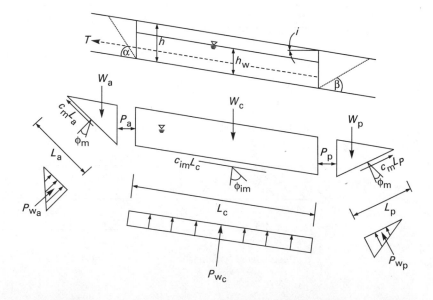

It is tacitly assumed that the unit weight of the cover above the water table is the same as below it.

$$W_a = [(\gamma h^2)/2] \cos i \cos \alpha/\sin(\alpha - i)$$
$$W_c = \gamma h L_c \cos i$$
$$W_p = (\gamma h^2/2) (\cos i \cos \beta/\sin(i + \beta))$$
$$P_{w_a} = \gamma_w h_w (\cos i)^3/2 \sin(\alpha - i)$$
$$P_{w_c} = \gamma_w h_w L_c \cos i^2$$
$$P_{w_p} = (\gamma_w h_w^2/2) \cos i^3/\sin(\beta + i)$$
$$F = \text{factor of safety}$$
$$\tan \phi_m = \tan \phi/F$$
$$c_m = c/F$$
$$\tan \phi_{im} = \tan \phi_i/F$$
$$c_{im} = c_i/F$$

The angle α is chosen to maximize P_a and the angle β is chosen to minimize P_p. This is accomplished by choosing the angles as follows:

$$\tan \alpha = \tan \phi + \sqrt{[1 + \tan \phi^2 - (\tan i/\sin \phi \cos \phi)]}$$

$$\tan \beta = -\tan \phi + \sqrt{[1 + \tan \phi^2 - (\tan i/\sin \phi \cos \phi)]} \qquad \textbf{(13.6)}$$

For a long slope where P_a and P_β are relatively small compared to the weight of the cap, the factor of safety can be defined by Eq. 13.7 for an infinite slope with seepage parallel to the slope (Figure 13.7a). Assume that the soil above the seepage line is also saturated.

$$F = \frac{1}{1-r} \left[\left(1 - \frac{h_w}{h} \frac{\gamma_w}{\gamma} \right) \frac{\tan \phi_i}{\tan i} + \frac{c}{\gamma h} \left(\frac{1}{\sin i \cos i} \right) \right] \qquad \textbf{(13.7)}$$

where
$$r = T/(\gamma h L \cos i \sin i) = T/W_c \sin i$$
$$W_c = \text{weight of the cap}$$

If the seepage line is not parallel to the slope but intersects the slope at an angle i' (Figure 13.7b), the factor of safety is defined as

$$F = \frac{\tan \phi_i}{\tan i} - \tan \phi_i \frac{\gamma_w}{\gamma} \frac{\cos i'}{(\sin i) \cos (i - i')} + \frac{c_i}{\gamma h} \frac{1}{\sin i \cos i} \qquad \textbf{(13.8)}$$

If the seepage is perpendicular to the slope (Figure 13.7c), the critical gradient can be estimated based on (Rhee and Bezuijen, 1992)

$$i_{cr} = -(1 - n)\left(\frac{\gamma_s - \gamma_w}{\gamma_w} \right) \frac{\sin (\phi - i)}{\sin \phi} \qquad \textbf{(13.9)}$$

where γ_s is the unit weight of the solid particles. The parameter $(\gamma_s - \gamma_w)/\gamma_w$ is typically about 1.65 and n is typically 0.4.

The above analysis also may be applied to a cap with multilayer reinforce-

Figure 13.7
Stability analysis—
infinite slope

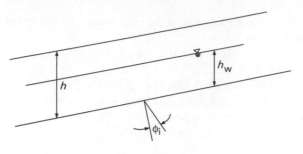

(a) Seepage parallel to slope

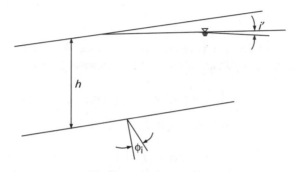

(b) Seepage intersecting slope

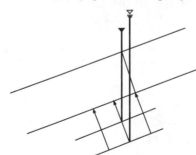

(c) Seepage perpendicular to slope

ment, as shown in Figure 13.8. Figure 13.8 shows a cap with three plastic components. The soil cover is underlain by a geotextile filter to help avoid clogging of the underlying geonet. The geonet is underlain by a geotextile that overlies a geomembrane. All the plastic elements are anchored at the top of the slope. The destabilizing force is the weight of the cap times the sine of slope angle ($W \sin i$). The tension in the element is the difference between the drag on top of the element ($W_c \cos i \tan \delta u$ [u means upper] for element 1) and the frictional resistance at the bottom of the element. The shear (or drag) at the top of the element is limited by $W_c \sin i$ or $W_c \cos i \tan \delta_{1u}$, whichever is smaller.

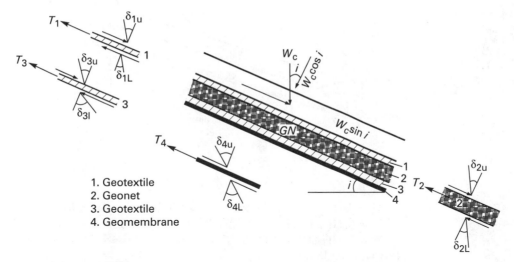

Figure 13.8 Multilayer cap structure

Consider for example that the interface friction angle δ_{1u} is 32° and δ_{11} (*l* means lower) is 14°. Then for a slope angle of 14° or less, the frictional resistance at the base of the element is larger or equal to the drag force at the top. Hence there is no tension in the membrane. As the slope angle increases but remains equal to or less than 32°, the drag force is equal to $W_c \sin i$ (which for *i* less than 32° is less than $W \cos i \tan 32$); hence the drag force controls. The resisting force is $W_c \cos i \tan \delta_1$ and the force T_1 is the difference. In general,

$$(T/W_c \sin i) = 1 - (\tan \delta_1/\tan i) \qquad \text{for } i \text{ less } \delta_1 \qquad \textbf{(13.10)}$$

As the slope increases in steepness to 32° or more, the drag force is now controlled by δ_u because $W \sin i$ becomes larger than $W \cos i \tan \delta_u$. In general,

$$(T/W_c \sin i) = (\tan \delta_u - \tan \delta_1)/\tan i \qquad \text{for } i \text{ larger than } \delta_u \qquad \textbf{(13.11)}$$

Example 13.2

A cover consists of 4 ft of sand over a geomembrane, as illustrated in Figure 13.9. The geomembrane is in contact with a drainage net below. The drainage net is in contact with another geomembrane supported on sand subgrade. The friction angle of the sand subgrade and sand fill is 32°. The weakest interface is between the drainage net and the geomembrane — the interface friction is assumed to be 19°. The unit weight of sand is 120 pcf. If the slope angle is 21.8°, determine the factor of safety.

Solution:

$$\alpha = 45 + 32/2 = 61°$$
$$\tan \beta = -\tan 32 + (1 + \tan^2 32 - (\tan 21.8/\sin 32 \cos 32))^{0.5}$$
$$\beta = 4.75°$$

Figure 13.9
Stability analysis of a
cover

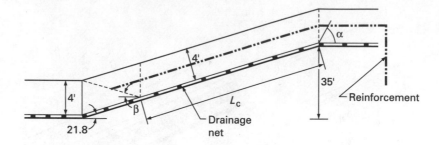

First we calculate P_a.

$W_a = 4.31 \times 4.31 \times 120 \cot 61/2 = 620$ lb/ft $= 0.31$ tn/ft

$P_a = 0.31(\tan 61F - \tan 32)/(F + \tan 61 \tan 32)$
   $= 0.31(1.804F - 0.625)/(F + 1.126)$

Next we calculate P_c.

$L_c = (35/\sin 21.8) - (4.3 \cos 21.8/\sin (21.8 + 4.75))$
   $\times \cos 4.75/\cos 21.8$
   $= 94.25 - (4/0.447)(1.07) = 84.67$ ft

$W_c = 84.67 \times 4 \times 120 = 40,641.6$ lb/ft $= 20.3$ tn/ft

$P_c = 20.3(F \tan 21.8 - \tan 19)/(F + \tan 21.8 \tan 19)$
   $= 20.3(0.4F - 0.34)/(Fs + 0.14)$

Then we calculate P_p.

$W = 120(4.3)^2/2(\cos 21.8 \cos 4.75)/(\sin 21.8 + 4.75)$
   $= 1109.4 \times 2.07 = 2297$ lb/ft $= 1.15$ tn/ft

32) $P_p = 1.15(F \tan 4.75 + \tan 32)/(F - \tan 4.75 \tan$
   $= 1.15(0.083F + 0.625)/(F - 0.052)$

Now we solve $P_a + P_c = P_p$.
With $F = 1.0$, we have

$$LHS = 0.17 + 1.068 = 1.238, \ RHS = 0.858$$

With $F = 0.95$, we have

$$LHS = 0.162 + 0.75 = 0.912, \ RHS = 0.901$$

Thus, we use $F = 0.94$.

Example 13.3

For the situation described in Example 13.2, determine the reinforcing force, T, required to develop a factor of safety of 1.3.

Solution:

$P_a = 0.31(1.804 \times 1.3 - 0.625)/(1.3 + 1.126) = 0.22$ tn/ft

$P_c = (20.3 - T \sin 21.8)(0.4 \times 1.3 - 0.34)/(1.3 + 0.14) - T \cos 21.8$
   $= 2.54 - 0.046T - 0.93T = 2.54 - 0.98T$

$P_p = 1.15(0.083 \times 1.3 + 0.625)/1.3 - 0.052 = 0.67$ tn/ft

$$P_a + P_c = P_p$$
$$0.22 + 2.54 - 0.98T = 0.67$$
$$T = 2.13 \text{ tn/ft width}$$

Example 13.4

Consider a cap with three plastic components, as illustrated in Figure 13.10. Determine the force in each of the reinforcing layers as a function of the weight above it and the slope angle (Long et al., 1994).

Solution: For the geotextile layer, no tension in the geotextile will develop until the slope angle increases beyond 13.5°. For slope angles higher than 13.5° but less than 35.3°,

$$T_1/W_c \sin i = 1 - (\tan 13.5/\tan i) = 1 - (0.24/\tan i)$$

where T_1 is the tension in the geotextile layer. For a 300-ft slope of $1V$ on $3H$, an average cap thickness of 3 ft perpendicular to the slope, and a unit weight of 120 pcf,

$$W_c = 300 \times 3 \times 120 = 108,000 \text{ lb/ft}$$
$$i = 18.3°$$
$$T_1 = 108,000 \sin 18.3(1 - 0.73) = 9156 \text{ lb/ft}$$

For slopes greater than 35.3°, the drag on the fabric is limited by tan 35.3. Hence,

$$T_1/W_c \sin i = (\tan 35.3 - \tan 13.5)/\tan i = 0.47/\tan i$$

If the length of the slope is the same (300 ft) and the slope angle i is $1V$ on $1.5H$, then $i = 33.7°$ and

$$T_1 = 108,000 \sin 33.7 \times 0.24/0.666 = 21,594 \text{ lb/ft}$$

For the geonet (element 2), no tension would develop according to the limit approach until the slope angle is beyond 7.5°, but the drag on top of the

Figure 13.10
Example 13.4

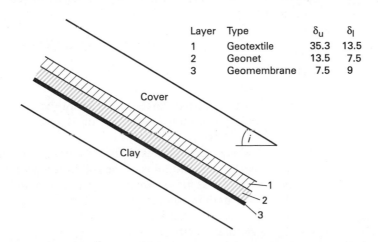

Layer	Type	δ_u	δ_l
1	Geotextile	35.3	13.5
2	Geonet	13.5	7.5
3	Geomembrane	7.5	9

Cover

Clay

layer is limited by tan 13.5°. Hence for the same slope length, $7.5° < i < 13.5°$, and

$$T_2/W_c \sin i = 1 - \tan 7.5/\tan i$$

For $1V$ on $5H$ slope ($i = 11.3°$),

$$T_2 = 108,000 \sin 11.3(1 - 0.13/0.2) = 7407 \text{ lb/ft}$$

For the same slope length but $1V$ on $1.5H$ slope angle,

$$T_2 = (108,000 \sin 33.7)(\tan 13.5 - \tan 7.5)/\tan 33.7 = 9983 \text{ lb/ft}$$

For the geomembrane, note that δ_u is smaller than δ_1. Therefore, it is not possible to develop tension because the drag on the top of the membrane is controlled by tan 7.5, which is less than the resistance at the base controlled by tan 9 and adhesion.

Example 13.3 reveals some problems with the limit analysis (Long et al., 1994). In the range of slope angles of 7.5° to 13.5°, the geotextile is free of tension whereas the geonet is stressed. Also, the analysis does not account for any load transfer between various elements.

13.3   STRAIN COMPATIBILITY

To overcome some of the limitations of the limit method described above for the analysis of cap reinforcement, Long et al. (1993) attempted to model the problem using a strain compatibility approach. In the analysis of Section 13.2, the stiffness of the reinforcing layer is presumed much larger than the soil to be reinforced. If the soil layer and the reinforcing layer are treated as components of a composite column (Long et al., 1993) and if the column is fixed at both ends (Figure 13.11), then for displacement compatibility the stretch in the reinforcing layer over a length L_t must be equal to the compression in the soil over a length L'.

$$fL_t^2/2E_t A_t = f(L')^2/2E_c A_c \qquad (13.12)$$

where

f $= F/L$

F = unbalanced force that cannot be resisted by interface friction ($= W_c \cos i(\tan i - \tan \delta_1)$ or $W_c \cos i(\tan \delta_u - \tan \delta_1)$, whichever is less)

E_t = Young's modulus of the reinforcing layer in tension ($E_c = 0$)

E_c = Young's modulus of the soil in compression ($E_t = 0$)

A_c, A_t = cross-sectional areas in compression and tension, respectively

L' = length in compression

L_t = length in tension

L_c = total length

Figure 13.11
Simple composite column

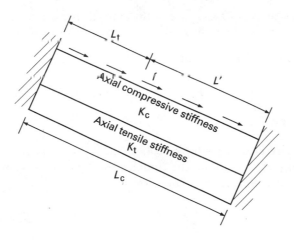

Noting that

$$L' = L_c - L_t \tag{13.13}$$

from Eqs. 13.12 and 13.13, we have

$$L_t = L_c/(1 - (K_c/K_t)) \tag{13.14}$$

where
$$K_c = E_c A_c$$
$$K_t = E_t A_t$$

Multiplying both sides of Eq. 13.14 by the uniformly distributed traction f, the tensile force in the reinforcing layer is expressed as

$$T_{gs} = T/(1 + (K_c/K_t)^{0.5}) \tag{13.15}$$

where T_{gs} is the tensile force in the reinforcing (geosynthetic) layer ($L_t f$) and T is the tensile force computed from Eq. 13.10 or Eq. 13.11.

If the cover contains more than one tension-resisting component, the tensile load is distributed by assuming that strains are identical in all components. In this case the load is distributed in proportion to their respective stiffness (Long et al., 1994):

$$T_{gsi} = T_{gs}(E_{ti} A_i/K_t) \tag{13.16}$$

where T_{gsi} is the tensile force in the geosynthetic component i and K_t is the total stiffness of the geosynthetic components in tension.

Example 13.5

For the situation described in Example 13.10, assume that the soil cover stiffness is 90 tn/ft, the geomembrane stiffness is 6.6 tn/ft, and the geotextile and geonet stiffnesses are both 1.4 tn/ft. Assess the forces in the soil and the tension members using a slope angle i of 18.26° ($1V$ on $3H$).

Solution: The lowest interface friction is 7.5°. Thus

$$F = W \cos 18.26(\tan 18.26 - \tan 7.5)$$

$$= 108,000 \times 0.95(0.198) = 20,315 \text{ lb/ft}$$

As described in Example 13.4, since $\delta_u < \delta_l$ the geomembrane will not be subjected to tension. The combined stiffness of the geotextile and geonet is 2.8 tn/ft. With $K_c = 90$ tn/ft, we apply Eq. 13.15:

$$T_{gs} = 20,315/(1 + \sqrt{90/2.8}) = 3050 \text{ lb/ft}$$

The load carried by the soil cap $= 20,315 - 3050 = 17,265$ lb/ft. Because the stiffness of the geonet and geotextile is identical, each member will carry 1525 lb/ft.

If the surface of the reinforcing layer in Example 13.5 is considered to act in unison with the cover (no slippage), then the problem could be treated as that of a beam on an elastic foundation (Long et al., 1993). In this case the spring constant k_s characterizes the shear stiffness (shear stress times area divided by displacement) and this shear spring characterizes the resistance at the interface.

Assuming that the reinforcing layer is fixed at the top of a slope (displacement $= 0$) and that there is no stress at the downslope end, it can be shown (Long et al., 1993) that the following equation describes the mechanics of the problem.

$$EA d^2 z/dx^2 + f' - k_s z = 0 \qquad \textbf{(13.17)}$$

and

$$z = C_1 \exp - \lambda x + C_2 \exp \lambda x + f'/k_s \qquad \textbf{(13.18)}$$

where
  $C_1 = -(f'/k_s)/(1 + \exp - 2\lambda L_c)$
  $k_s =$ spring constant in shear on shear stiffness
  $C_2 = C_1 \exp(-2\lambda L_c)$
  $f' =$ applied shear per unit length
  $\lambda = (k_s/EA)^{0.5}$
  $x =$ distance measured from the top of the slope
  $z =$ displacement
  $EA =$ stiffness of the geosynthetic
  $L_c =$ length over which the shear is acting

The tension T is found by differentiating Eq. 13.18 for strain and multiplying the result by EA. Thus

$$T = EA(C_2 \lambda \exp \lambda x - C_1 \lambda \exp - \lambda x) \qquad (13.19)$$

In many cases (see Figure 13.11), the ultimate interface friction is achieved at very small strains. In this case, the product $k_s z$ is constant and the displacement reduces to the familiar expression

$$z = [(f' - f_s)/EA](xL - x^2/2) \qquad (13.20)$$

where f_s is the interface friction value.

Example 13.6

If the slope for Example 13.3 is assumed to be very long (infinite), a geogrid with axial stiffness, EA, of 40 tn/ft is used for reinforcement. The sand fill has an axial stiffness of 25 tn/ft. Determine the total deflection if the shear stiffness, k_s, is 40 tn/ft^2.

Solution

$$W_c = 20.3 \text{ tn}, \ L_c = 84.67 \text{ ft (from Example 13.2)}$$
$$f' = W_c \sin i/84.67 = 20.3 \sin 21.8/84.67 = 0.09 \text{ tn/ft}^2$$
$$f'/k_s = 0.09/40 = 0.00225$$
$$\lambda = (40/40)^{0.5} = 1$$
$$C_1 = -0.00225/(1 + \exp - 2 \times 1 \times 84.67) = -0.00225$$
$$C_2 = -0.00225 \exp - 2 \times 1 \times 84.67 = 0.0$$
$$z(x = L) = -0.00225 \exp - 84.67 + 0.00225 = 0.0025 \text{ ft}$$

If full mobilization of the interface shear is assumed, then

$$f = f' - f_s$$
$$f_s = W_c \cos i \tan 19/84.67 = 20.3 \cos 21.9 \tan 19/84.67 = 0.077$$
$$f = 0.09 - 0.077 = 0.013$$
$$z = (1/EA)fL^2/2 = (1/40)(0.013)(84.67)^2/2 = 1.16 \text{ ft}$$

Because interface shear stiffness is linear only over a very small displacement, a linear analysis would not be representative for a long slope. If the above equations are to be used, it will have to be in an iterative mode for an equivalent linear model: A value of k_s is selected and strains are computed from differentiation of Eq. 13.18. The calculated strains are compared to the strain used to compute k_s (from relationships such as in Figure 13.12) and new improved values of strain and k_s are selected until an agreement between the successive iterations is achieved.

13.4   ANCHOR TRENCH

Geosynthetics are usually fixed at the top of the slope in liner construction. Cap reinforcements require that the anchor develop its ultimate resistance without pulling out of the anchor trench. The actual configuration depends on the site conditions. Considering the configuration in Figure 13.13a, the maximum tension developed in the vertical reinforcement in the trench is equal to the lateral earth pressure on both sides times the interface friction

Figure 13.12
Characteristics of
interfaces

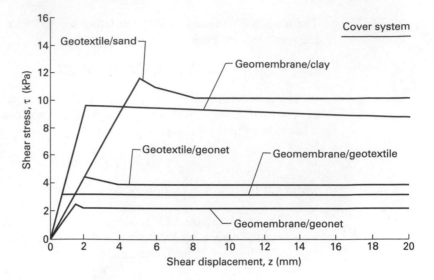

(a) Shear stress versus shear displacement at interface

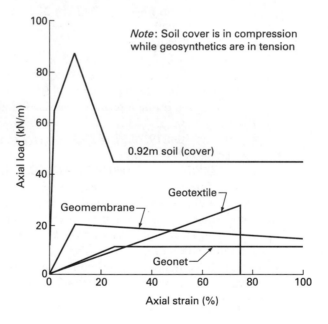

(b) Axial load versus strain relationship for structural components

angle. An active state of earth pressure is assumed on one side and an at-rest condition on the resisting side. The assumption of an at-rest condition is reasonably conservative. The embedment is usually estimated considering a factor of safety on the calculated T.

Figure 13.13
Anchor trenches

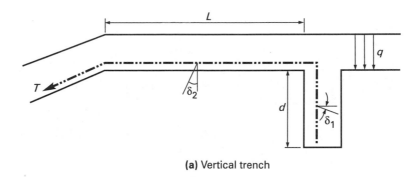

(a) Vertical trench

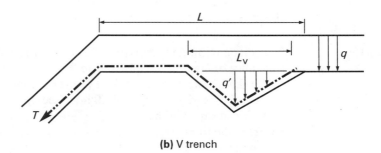

(b) V trench

Based on these assumptions, the following equation describing the equilibrium of forces may be used.

$$\tan \delta_1 (0.5\gamma d^2 + qd)(k_0 + k_a) + qL \tan \delta_2 + T \sin i \tan \delta_2 = T \cos i F_s \quad (13.21)$$

where
F_s = factor of safety
δ_1 = interface friction angle between the geosynthetic and the soil in the trench
δ_2 = interface friction between the geosynthetic and the cover soil
q = effective stress on the top of the trench
d = depth of embedment
$k_0 = 1 - \sin \phi$
$k_a = \tan^2(45 - \phi/2)$
ϕ = friction angle of the embedment soil
If the vertically embedded geosynthetic is replaced by a concrete wall, then the full passive resistance in front of the wall is assumed to develop and Eq. 13.21 takes the form

$$(k_p - k_a)(0.5\gamma d^2 + qd) + qL \tan \delta_2 + T \sin i \tan \delta_2 = T \cos i F \quad (13.22)$$

where

$$k_p = \tan^2(45 + \phi/2)$$

If the anchor is developed as a V anchor (Figure 13.13b), equilibrium requires that

$$\tan \delta_2(q(L - L_v + L_v/\cos i') + q'L_v/2 \cos i')$$
$$+ T \sin i \tan \delta_2 = T \cos iF \qquad \textbf{(13.23)}$$

where i' is the slope of the V trench, q' is the maximum stress at the base of the V trench due to the effective stress of the soil within the V portion, and q is the effective stress due to the fill above the top of the V trench.

Example 13.7

Consider a reinforcing geosynthetic designed to stabilize a cover. The ultimate resistance of the geotextile is 2000 lb/ft. Assume that 4 ft of sand overlies the top of the trench. The unit weight of the overburden sand and trench backfill is 120 pcf. The friction angle at all interfaces is 28°. The friction angle of the overburden sand and backfill is 32°. The slope angle is 21.8° and the trench is 6 ft from the top of slope. Determine the depth of an anchor trench to develop this load with a factor of safety of 1.2.

Solution:
$$k_a = \tan^2(45 - 16) = 0.31$$
$$k_0 = 1 - \sin 32 = 0.47$$
$$k_a + k_0 = 0.78$$

$$\tan 28(0.5 \times 120d^2 + 4 \times 120 \times d)(0.78) + 480 \times 6 \tan 28$$
$$+ 2000 \sin 21.8 \tan 28 = 2000 \cos 21.8 \times 1.2$$

$$d^2 + 8d - 74 = 0$$
$$d = (-8 + (64 + 48.6)^{0.5})/2 = 1.31 \text{ ft}$$

13.5 SEISMIC LOADING

RCRA Subtitle D regulations require that new landfills be designed to resist a minimum horizontal ground acceleration having at least a 90% probability of not exceeding in 250 years. Figure 13.14 is a seismic risk map cited in EPA (1993). Although it is not the intent of this book to be a source on seismic design, the following criteria should be observed.

a. The accelerations in Figure 13.14 are applicable for firm sites such as rock sites, glacial till, or rocklike materials. In general, the site should fit type S_1 as described in various seismic codes: rocklike material characterized by a shear wave velocity greater than 2500 ft/s, or stiff or dense soil conditions where soil depth is less than 200 ft. The accelerations in Figure 13.14 therefore may apply to rock outcrops or the surface of dense soil. For other profiles,

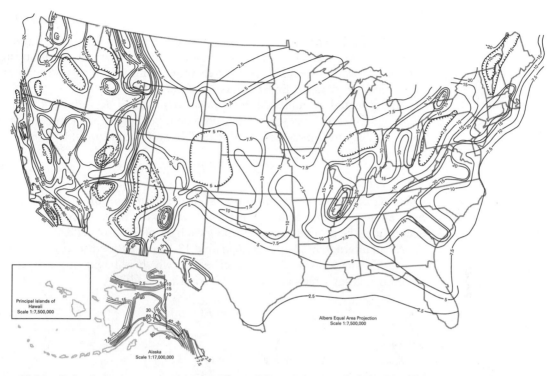

Figure 13.14 Seismic impact zones (areas with a 10% or greater probability that the maximum horizontal acceleration will exceed 0.10 g in 250 years)

response analyses may be needed using available computer codes such as SHAKE.

The soil profile could also be modeled as a shear beam. If the response spectrum of the rock motion is known, the acceleration at the surface may be approximated using the following relationships that were based on many runs on a wide range of soil profiles using SHAKE (Oweis, 1978).

$$a_s = 0.32(Sa_1)^{0.61}(4 + (Sa_2/Sa_1)^2)^{0.5} \qquad \textbf{(13.24)}$$

where a_s is the surface acceleration (in g's). Sa_1 and Sa_2 are the absolute spectral accelerations of input motion (in g's) at the base of the profile corresponding to mode i having a period of T_i. Sa_1 and Sa_2 correspond to the first and second modes, respectively.

The first mode period is usually referred to as the site period T_s and corresponds to the fundamental period of vibrations of the profile. The second mode period (for which Sa_2 is read from the response spectrum) is 0.3 T_s to 0.5 T_s. T_s is a function of the shear strains induced by ground shaking and stress–strain relationships of the profile. For small strains and a uniform layer, the site period is (Dobry et al., 1976).

$$(T_s)\text{min} = 4H/(V_s)\text{max} \tag{13.25}$$

where H is the thickness of the layer above where the base ground motion is specified. (V_s)max is the shear wave velocity of the layer at small strains as determined from crosshole shear wave velocity measurements. (T_s)min is the site period based on the linear behavior of the layer.

Under actual ground shaking, the shear modulus degrades and the wave velocity decreases. Thus the site period would be higher than estimated by Eq. 13.25. In this case, an increased value of T_s is estimated (Oweis, 1978) as

$$T_s = 1.512(a_I)^{0.146}(T_p)^{0.126}((T_s)\text{min})^{0.874} \tag{13.26}$$

where T_p is the predominant period(s) of the input motion at the base of the profile and corresponds to the period at which the peak spectral acceleration occurs. a_I is the ground acceleration at the base of the profile (%g).

As an alternative to Eq. 13.26, Table 13.1 may be used to estimate the site period T_s. The following simpler criterion, which is based on early SEAOC recommendations, may also be used (Griffith et al., 1994).

$$T_s = R(T_s)\text{min} \tag{13.26a}$$

where
$$R = \text{correction factor for seismicity}$$
$$= 1.25 \text{ for low seismicity } (a_I < 0.15 \text{ g})$$
$$= 1.5 \text{ for high seismicity } (a_I > 0.15 \text{ g})$$

Table 13.1   Estimate of site period T_s

	$M = 6.5 \pm 0.25$ SOIL PROFILE				$M = 7.25 \pm 0.25$ SOIL PROFILE				$M = 8.0 \pm 0.25$ SOIL PROFILE			
g	B	C	D	E	B	C	D	E	B	C	D	E
0.1	0.32	0.45	0.46	0.44	0.41	0.53	0.56	0.56	0.51	0.69	0.71	0.71
0.2	0.37	0.44	0.49	0.64	0.42	0.53	0.55	0.74	0.47	0.61	0.65	0.85
0.3	0.35	0.43	0.50	0.73	0.38	0.51	0.55	0.76	0.48	0.64	0.65	0.98
0.4	0.39	0.47	0.50	0.87	0.42	0.56	0.59	0.93	0.46	0.62	0.66	1.04
0.5	0.37	0.46	0.50	—	0.42	0.53	0.62	—	0.45	0.59	0.70	—
0.6	0.35	0.44	0.50	—	0.43	0.54	0.64	—	0.46	0.60	0.76	—
0.7	—	—	—	—	0.50	0.66	0.76	—	0.54	0.71	0.80	—

Profile B: Rock with $2500 < v_s \le 5000$ ft/s (760 m/s $< v_s \le 1500$ m/s).
Profile C: Very dense soil and soft rock with shear wave velocity 1200 ft/s $< v_s \le 2500$ ft/s (360 m/s $< v_s < 760$ m/s) or with a weighted average SPT $N > 50$, or $S_u \ge 2000$ psf (100 kPa) in the upper 100 ft (30 m) of the soil profile. N is not to exceed 100 for any layer. S_u is not to exceed 5000 psf (25.0 kPa) for any layer.
Profile D: Stiff soil with shear wave velocity 600 ft/s $< v_s \le 1200$ ft/s (180 m/s $< v_s \le 360$ m/s) or with either SPT $15 \le N \le 50$, or 1000 psf $\le S_u \le 2000$ psf (50 kPa $\le S_u \le 100$ kPa) in the upper 100 ft. S_u is not to exceed 5000 psf (250 kPa) and N is not to exceed 100 for any layer.
Profile E: $v_s < 600$ ft/s (180 m/s) or any profile with more than 10 ft (3 m) but less than 25 ft (8 m) of soft clay ($PI > 20$, $w \le 40\%$, $S_u < 500$ psf (25 kPa).
Profile F: Soils requiring site-specific evaluation. Peat and highly organic clay, highly plastic clay ($PI > 75$) over 25 ft (8 m). Thick, soft to medium clays more than 120 ft (36 m) thick, soil subject to failure during a seismic event.
Source: After ATC-32, 1996

b. The foundation of the landfill must have an adequate factor of safety against foundation failure due to liquefaction. During earthquake shaking, excess pore water pressures develop in saturated sands and silts. With the reduction in effective stress, the strength of such soils is reduced to a near-fluid condition. Techniques are available for liquefaction analysis (e.g., NAVFAC DM-7.3, 1982; Seed et al., 1983).

c. If the landfill is supported on firm ground, accelerations from Figure 13.14 apply to the base of the landfill. The average acceleration within the landfill may be different and a response analysis may be needed. In general, ground motion generated by a local shock is usually rich in high frequencies and the landfill will cause attenuation of ground motion. Thus the use of the surface acceleration at the base will be conservative and perhaps very conservative. Reported measured shear wave velocities in California landfills ranged from 103 to 900 ft/s (Singh and Murphy, 1990). In general, the most important parameter in characterizing the response of the landfill is the fundamental period.

As an example of the application of these criteria, consider a landfill 400 ft high. The period (from Eq. 13.25) is 2.46 s if the average wave velocity is 650 ft/s. This is well beyond the predominant period of a local or even a distant event and attenuation of ground motion would be expected. Considering for example a base acceleration of firm ground of 0.1 g and using the response spectrum in Figure 13.15, the following estimates are made:

$T_p = 0.32$ s (period at highest peak in the spectrum)

$T_s = 1.512(10)^{0.146} \times (0.32)^{0.126} \times (2.46)^{0.874} = 4$ s

From Figure 13.15, the first-mode spectral acceleration is essentially nil and according to Eq. 13.24, accelerations at the surface would be nil. If the thickness of the landfill is only 100 ft, the minimum period would be 0.615 s ($4 \times 100/650$). The compatible strain period from Eq. 13.26 is 1.2 s, and the second-mode period is 0.3 to 0.5 times the first-mode period (0.36 to 0.6 s). From Figure 13.15, Sa_1 at $T = 1.2$ is 0.03 g (amplification ratio of 0.3×0.1 g). The spectral magnification ratio in the period range of 0.36 to 0.6 s is about

Figure 13.15
Example of a response spectrum

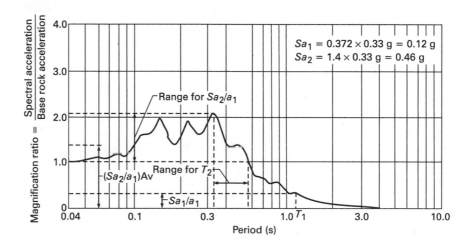

1.3, leading to $Sa_2 = 0.13$ g. From Eq. 13.24, the surface acceleration would be about 0.18 g. Therefore, it is clear that landfills could attenuate or amplify the base motion depending on the characteristics of the landfill and the frequency content of the ground motion. In our example we used a wave velocity value of 650 ft/s.

The effect of earthquake loading on the stability may be assessed using the equations in Figure 13.1 or Eq. 13.4 with an added destabilizing force. In this case, P_α and P_β in Figure 13.1 should be replaced by

$$P'_\alpha = P_\alpha + Wa_g$$
$$P'_\beta = P_\beta - Wa_g$$

where a_g is the average acceleration (in g's) within the postulated sliding wedge and W is the weight of the wedge associated with P_α or P_β.

For a cap on a long slope and an acceleration a_g parallel to the slope, the safety factor is computed by using r' instead of r in Eq. 13.7. The parameter r' is defined as

$$r' = r - a_g/\sin i \qquad \textbf{(13.27)}$$

Example 13.8

Consider a landfill similar to that in Example 13.1 with a thickness of 100 ft. If the side slope is $1V$ over $3H$, determine the factor of safety if the average acceleration is 0.05 g. Assume a unit weight of clay $= 110$ pcf.

Solution

tn $W_1 = 14.43H^2 = 144,300$ lb $= 72.15$

$P_{a_1} = 0.05 \times 72.15 + 72.15\ (F \tan 60 - \tan 30)/(F + \tan 60 \tan 30)$
$\quad = 3.6 + 72.15(1.73F - 0.58)/(F + 1)$

$W_2 = 300H + (6 \times 6) \times 110/2 = 31,980$ lb $= 16$ tn

$P_{a_2} = 16 - (2 \times 600 \times 6/F)/2000 + 16 \times 0.05 = 16.8 - 3.6/F$

$W_3 = 100(300)(50)/2 + 300 \times 6 \times 110 = 750,000 + 198,000 = 948,000$ lb
$\quad = 474$ tn

$P_{a_3} = -600 \times 3 \times 100/2000 \times F + 474 \times 0.05 = -90/F + 23.7$

$W_\beta = (6 \times 6 \times 110/2) = 495$ lb $= 1980 = 1$ tn

$P_\beta = 1 + 2 \times 600 \times 6/2000F - 0.05 = 0.95 + 3.6/F$

To solve $\Sigma P_\alpha = \Sigma P_\beta$, we first try $F_s = 1.4$.

$$LHS = 59 + 14.2 - 40.6 = 32.6 \qquad RHS = 3.52$$

For $F = 1.0$, we have

$$LHS = 45.4 + 13.2 - 66.3 = -7.7 \qquad RHS = 4.55$$

For $F = 1.1$, we have

$$LHS = 49.05 + 13.5 - 58 = 4.55 \qquad RHS = 4.22$$

Thus we use $F = 1.1$.

The above types of analyses can provide a value of ground acceleration that produces a factor of safety of 1.0. Considering a square acceleration pulse of duration t, the acceleration and velocity diagrams of an assumed rigid block are plotted in Figure 13.16. The force resisting motion at the bottom of the block is NW, where W is the weight of the block and N is some dimensionless friction coefficient that in this case can be thought of as resisting acceleration (because for accelerations less than N, no movement is possible). The net velocity diagram is shown as a hatched zone in Figure 13.16.

The area of the diagram is the displacement experienced by the rigid block. Thus,

$$u = (V/2)(t - t_0) = (V/2)(V/Ng - V/Ag) = (V^2/2gN)(1 - N/A) \qquad \textbf{(13.28)}$$

where u is the displacement, N is the acceleration (in g's) that will result in a factor of safety of 1.0, and V is the maximum velocity of the earthquake motion. This expression is for a single pulse on a horizontal surface. When the ground motion reverses direction, the block will move in the opposite direction (back and forth) by the same amount if the resistance N and the acceleration amplitude A are the same.

If the block is at a slope sliding surface, movements will be downhill and each time N is exceeded, the block will move farther. The number of effective pulses may be assumed to be equivalent to A/N. If we multiply Eq. 13.28 by A/N, the resulting formula for downhill movement is

$$u = (V^2/2gN)(A/N - 1) \qquad \textbf{(13.29)}$$

The above developments are by Nemark (1965) and are easy to apply. Other procedures that have been proposed or used for dams include those from Makdisi and Seed (1978), Franklin and Chang (1977), and Yegian et al.

Figure 13.16
Acceleration
displacement
relationships,
Nemark
method

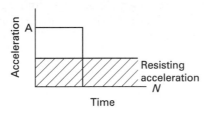

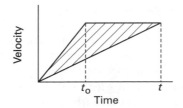

(1991). Franklin and Chang (1977) used the Nemark method in conjunction with actual earthquake records.

Example 13.9

For the landfill described in Example 13.8, determine the resistance factor N.

Solution:

N = acceleration ion g's required to produce a factor of safety of 1.0

$W_1 = 72.15$

$P_{a_1} = 72.15N + 72.15(0.58) = 41.85 + 72.15N$

$W_2 = 16$ tn

$P_{a_2} = 16N + 16 - 3.6 = 12.4 + 16N$

$W_3 = 474$ tn

$P_{a_3} = 474N - 90$

$W_p = 1$ tn

$P_\beta = 1 + 3.6 - N = 4.6 - N$

$\Sigma P_\alpha = \Sigma P_\beta$

$563.15N - 40.3 = 0$

$N = 0.07$ g

Example 13.10

Seismological investigation and site response analysis indicated that a landfill will be subjected to an average acceleration of 0.3 g and velocity of 14 in./s. Determine the expected movements.

Solution: Sliding will occur along the clay base in a back and forth motion. If we assume that the clay strength remains the same, then from Eq. 13.28 we have

$$u = (14 \times 14/2 \times 32.2 \times 12 \times 0.07)(1 - 0.07/0.3) = 2.8 \text{ in.}$$

$$u = 11.3 \text{ in.}$$

The allowable seismic movements in landfill design is not as yet regulated and is a matter of engineering judgment. A limit of 6 in. has been used in practice but actual limits must be site-specific. For cap designs where static factors of safety are usually low, a limit of 6 in. is difficult to achieve unless the cap is reinforced.

SUMMARY

The stability of landfill caps, slopes, and foundations is an important design element. Potential failure can occur by sliding along the base, slope failure, or a deep-seated failure involving the soil supporting the landfill. The liner may also squeeze laterally under stresses imposed by the landfill. Limit equilibrium methods can be used to assess slope or deep-seated failures. Caps are usually constructed of soil in combination with geotextiles, geomembranes, or geosynthetic clay liners (GCLs). The interface friction between various components is usually low and careful analysis is needed that considers the long-term behavior of the material and field conditions, including proper drainage of the slope. Simplified procedures have been presented to assess the seismic response of a landfill. Such procedures are no substitute for a detailed site-response analysis. The procedures, however, help define potential design problems.

NOTATIONS

ϕ friction angle

ϕ_m mobilized friction angle

ϕ_i interface friction angle-soil cap

δ interface friction angle-soil and geosynthetic cap

θ slope angle of embankment

α angle between potential failure plane and horizontal-active side

β angle between potential failure plane and horizontal-passive side

γ unit weight of embankment (F/L^3)

γ_{unit} weight of water (F/L^3)

γ_s unit weight of solid particles (F/L^3)

τ_{av} average shear stress imposed on the top of the clay layer (F/L^2)

ξ gradient of strength with depth (M/L^2T^2)

χ τ_{av}/c_0

κ c_b/c ratio of interface strength to the undrained shear strength at the base of the clay layer ($=1.0$ for rough base)

λ $(EA/k_s)^{0.5}$

a_I interface acceleration (%g)

a_s surface acceleration (g)

A acceleration (g)

A_c area of compression per unit length (L^2)

A_t area of tension per unit length (L^2)

b half-width of embankment

c undrained shear strength (F/L^2)

c_b undrained shear strength at the base interface of the clay layer (F/L^2)

c_0 undrained shear strength at the top of clay layer (F/L^2)

c_m mobilized undrained shear strength

d depth of anchor trench (L)

EA stiffness of a geosynthetic

E_t Young's modulus of reinforcing layer (F/L^2)

E_c Young's modulus of soil in compression

A_c cross-sectional area in compression (L^2)

A_t cross-sectional area in tension (L^2)

H height of embankment (L)

h thickness of the clay layer (L)

h thickness of cap (LL)

F factor of safety

f_s interface friction per unit length (F/L)

f' applied shear per unit length

h_w depth of water column (L)

i slope of a cap

k_0 coefficient of lateral earth pressure at rest

P horizontal thrust per unit length (F)

P_w hydrostatic force per unit length acting on potential failure plane (F)

P_α Active force per unit length-wedge defined by angle α (F)

P_β Passive force per unit length-wedge defined by angle β (F)

K_c $E_c A_c$

K_t $E_t A_t$

k_s spring constant in shear (F/L^2)

L length of potential failure plane (L)

L_c	length of central wedge (L)	Sa_1	spectral acceleration — first mode (g)	T_p	predominant period of input motion at the base of the soil profile(s)
L_c	length of cap in compression (L)	Sa_2	spectral acceleration — second mode (g)	T_{gs}	tensile force in per unit length reinforcing layer
L_t	length of cap in tension	T	resisting force per unit length at the base of the fill generated by reinforcement (F)	u	displacement (L)
L_v	width of v trench			V	velocity (L/T)
N	yield acceleration (g)			V_s	shear wave velocity (L/T)
q	effective stress on top of the trench	T	tensile force per unit length in a cap reinforcement (F)	W	weight per unit length of slope (F)
q_{av}	average stress imposed by the fill on the clay layer (F/L²)	T_s	site fundamental period (s)	W_c	weight of cap per unit length (F)
q_f	average vertical stress causing failure of the clay layer (F/L²)	$(T_s)_{min}$	site period corresponding to small strain shear wave velocity (s)		

PROBLEMS

13.1 If the slope in Example 13.2 is a long slope, determine the factor of safety.

13.2 Repeat Problem 13.1 assuming that the interface strength includes a cohesion intercept of 100 psf, half of the sand fill is saturated, and seepage is parallel to the slope.

13.3 If seepage surface in Problem 13.2 intersects the slope at an angle of 15°, determine the factor of safety.

13.4 In Example 13.2, if the seepage from the subgrade is perpendicular to the slope, determine the gradient that will cause failure in the 4 ft sand cover.

13.5 Repeat Example 13.3 assuming an infinite slope.

13.6 For Example 13.2, assume a geogrid reinforcement with 40 tn/ft and soil stiffness of 25 tn/ft. Determine the required tensile resistance for a safety factor of 1.3.

13.7 For example 13.7, assume a V trench with side slope of 21.8°. The nearest edge of the V to the slope is 2 ft and the farthest is 6 ft. Determine the resistance to slippage.

13.8 For Problem 13.5, determine the average acceleration to produce a factor of safety = 1.0. Assume the limit load in reinforcement = 2.6 tn/ft. If the average acceleration is 0.3 g and the velocity is 14 in./s, estimate the cap movement.

13.9 For the landfill geometry shown in Figure 13.17, determine the critical height H.
   a. Try a failure surface through the refuse.
   b. Try a failure surface through the stiff clay using both the bearing capacity method and the wedge method.

13.10 Repeat Problem 13.9 assuming a reinforcing geogrid at the base of the refuse with an allowable tension of 1500 lb/ft.

Figure 13.17

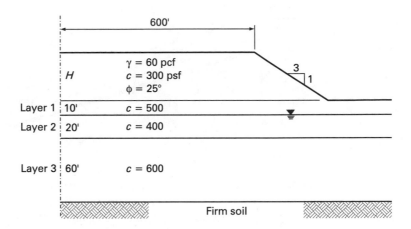

13.11 Repeat Problem 13.9 assuming that layers 2 and 3 are sand with a friction angle of 30° and a unit weight of 120 pcf.

13.12 Repeat Problem 13.9 assuming that the landfill and foundation are subjected to an average ground acceleration of 0.3 g and velocity of 15 in./s. For these conditions estimate the expected slope movement.

13.13 For the cap shown in Figure 13.18, calculate the factor of safety against a slope failure for the following conditions.
   a. Accounting for the toe resistance.
   b. Assuming an infinite slope.

13.14 Repeat Problem 13.13 assuming a seepage surface at the ground surface parallel to the slope and intersecting the slope.

13.15 For Problem 13.13, determine the critical hydraulic gradient if the seepage flow lines are perpendicular to the surface.

13.16 For the slope shown in Figure 13.19, determine the required tension in the geogrid for a factor of safety of 1.2. Perform the calculation with and without toe restraint.

13.17 For Problem 13.12, determine the factor of safety if the slope is subjected to 0.2 g acceleration parallel to the slope. If the velocity is 15 in./s, determine the slope movement.

13.18 Repeat Problem 13.17 if the slope is subjected to 0.15 g ground acceleration.

13.19 For Problem 13.13, calculate the stretch in the reinforcing element. Assume an axial stiffness of 40 tn/ft and a full mobilization of the interface shear resistance.

Figure 13.18

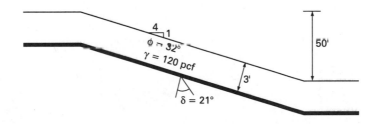

Figure 13.19

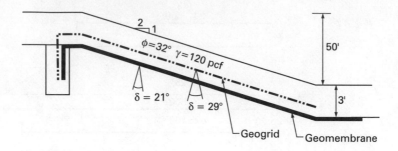

13.20 A landfill is 400 ft high and supported on dense glacial till. The shear wave velocity measurements indicated an average shear wave velocity of 800 ft/s. Using the response spectrum of Figure 13.15 for the ground motion at the base of the landfill, determine the average acceleration through the landfill.

14 Gas Management

14.1 INTRODUCTION

Gas generation in landfills impacts both nearby communities and construction on landfills in both negative and positive ways. On the negative side, the gas usually contains toxic chemicals that have an adverse health impact. Also, nontoxic explosive gases (e.g., methane) can travel large distances outside landfills and present the risk of explosion or fire in buildings (EPA, 1975). Explosion hazards are usually expressed in terms of the lower explosive limit (LEL), which is the lowest percentage of volume of a mixture of explosive gases that will propagate a flame in air at 25°C and atmospheric pressure. For methane, the LEL is about 5%. On the positive side, methane in gas can be extracted and processed for energy use if commercially feasible.

Also, minimal gas generation could mean the landfill is old and has gone through the decomposition process leaving very few solids to further decompose. These landfills will be subjected to less long-term creep or secondary compression than will newer landfills and so can be used as building sites.

14.2 ANALYTICAL ESTIMATES OF GAS GENERATION

As described in Chapter 11, gas results from decomposition of waste. Barlaz and Ham (1993) estimated that 73.4% of the methane production is due to cellulose ($C_6H_{10}O_5$), which constitutes 51.2% (by dry weight) of municipal refuse. Another 17.1% of the methane production is attributed to hemicellulose ($C_5H_8O_4$), which constitutes about 11.9% (by dry weight) of municipal refuse. The remaining 9.5% of the total methane production is produced by protein, starch, and soluble sugar, which together make up about 5.05% (by dry weight) of municipal refuse. Lignin, which makes up about 15.2% (by dry weight) of municipal refuse, has no methane potential. Tchobanoglous et al. (1993) provided the guidelines shown in Table 14.1 on the biodegradable fraction of municipal waste.

The methane potential generation may be assessed based on the following mass balance equation (Barlaz and Ham, 1993):

$$C_nH_aO_bN_c + \left(n - \frac{a}{4} - \frac{b}{2} + \frac{3c}{4}\right)H_2O \rightarrow \left(\frac{n}{2} - \frac{a}{8} + \frac{b}{4} + \frac{3c}{8}\right)CO_2$$

$$+ \left(\frac{n}{2} + \frac{a}{8} - \frac{b}{4} - \frac{3c}{8}\right)CH_4 + CNH_3 \quad \textbf{(14.1)}$$

Table 14.1

Composition	Percentage by weight	Biodegradable fraction (%)
Food waste	9	82
Newspaper	21	22
Office paper	21	82
Cardboard	6	47
Yard waste	20	72

From Equation 14.1, Barlaz and Ham (1993) calculated a theoretical methane generation of 415 and 424 liters of methane at standard pressure and temperature for each kilogram of cellulose and hemicellulose, respectively. These estimates can be derived from Eq. 14.1 as follows.

$$C_6H_{10}O_5 + H_2O \rightarrow 3CO_2 + 3CH_4$$

$$(162) \quad (18) \quad (132) \quad (48)$$

The methane generated for 1 kg of cellulose is (48/162) (1 kg) = 0.296 kg. The methane density at 1 atm and 0°C = 0.7167 g/l (see Table 14.2). Thus, the methane generation = 296/0.7167 = 413 l/kg cellulose at the pressure and temperature indicated.

$$C_5H_8O_4 + H_2O = 2.5CO_2 + 2.5CH_4$$

$$(132) \quad (18) \quad (110) \quad (40)$$

The methane generation for 1 kg of hemicellulose is (40/132) (1 kg) = 0.303 kg. At 0°C and 1 atm (methane specific weight is 0.7167), the production is 422.8 l/kg of hemicellulose.

If we assume that cellulose constitutes 51.2% of refuse by dry weight and hemicellulose constitutes 11.9% by dry weight, then a kilogram of dry refuse is expected to generate (0.512 × 413 + 0.119 × 422.8), or about 262 l of methane at 1 atm and 0°C. Assuming an ideal gas, the above calculations may be made at other temperatures and pressures using Eqs. 14.2 and 14.3.

$$C = P/RT \tag{14.2}$$

where C is the molar density (mol/m³), R is the universal gas constant (8.313 m³ Pa/mol.k), and T is the absolute temperature (Kelvin or K).

$$\rho = \rho_0(P/P_0)(T_0/T) \tag{14.3}$$

where ρ is the gas density at pressure P and temperature T (K) and ρ_0 is the gas density at standard pressure P_0 and standard temperature T_0 (K). (Note: 1 K = 0°C + 273.)

Many factors affect gas generation. As indicated in Eq. 14.1, water is essential. Thus successful efforts to "seal" landfills by caps would lead to drastic reduction in gas generation. A continued significant generation of gas is an indication of continued availability of moisture.

As described above (Barlaz and Ham, 1993), cellulose and hemicellulose account for about 90% of methane generation and constitute about 63.1% by dry weight of municipal refuse. The remaining methane-generation constituents account for about 5% of refuse by dry weight. If we assume a complete conversion to gas, a landfill would shrink because of material loss (see Chapter 4). Because decomposition occurs during the filling operation and is affected by many variables (moisture, toxic content, pH, etc.), it is not possible to accurately predict the gas generation rate with time based on Eq. 14.1.

Landfill gas production cannot go on forever. Mathematical models have attempted to predict the decay in gas production over time. A decay rate is typically in the form

$$C_2/C_1 = \exp - \zeta(t_2 - t_1) \tag{14.4}$$

where C_1 is the amount of methane-producing substrata at time t_1 (yr), C_2 is the amount of methane-producing substrata at time t_2 (yr), and ζ is the decay constant. The decay constant is sensitive to many factors including waste composition and moisture conditions. In the absence of site-specific data, a value of $0.05 \ y^{-1}$ is recommended (USEPA, 1996).

Based on initial predictions from Eq. 14.1 (or some other suitable technique) and a relationship such as Eq. 14.4, a gas production curve may be developed for a particular landfill. The total methane production for each year is calculated by adding up the gas produced from each ton of refuse from layers of different ages.

14.3 EMPIRICAL ESTIMATES OF GAS GENERATION

As previously mentioned, gas production in landfills is sensitive to many variables and an accurate prediction of gas generation over time is often difficult. Pumping tests are often used to develop estimates but such tests are of short duration (hours or days). Also, the gas pumped could often be drawn from storage so the pumping rate of a well may not be indicative of the landfill's gas production.

Figure 14.1 is an example of empirical estimates of gas generation. Under favorable conditions, methane generation is 0.2 ft^3/lb/yr. If the refuse is saturated and the moisture content is 0.2 on a dry weight basis, the methane production would be 0.11 ft^3/lb of dry refuse per year (7.1 l/kg/yr). Considering an average landfill life of 30 years, the production per kilogram of dry weight would be 213 l/kg, which is consistent with the analytical prediction cited above.

Figure 14.1
Gas production rates
(Washington State
Department of Ecology)

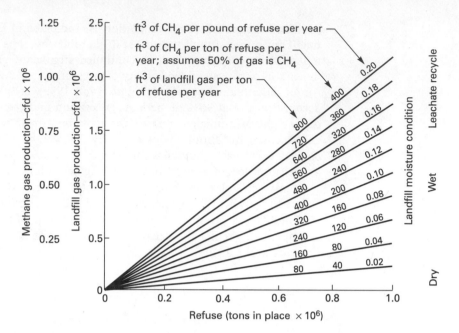

As illustrated in Figure 14.1, the methane generation could differ by as much as 1 order of magnitude, depending on moisture conditions. Other factors to consider when assessing gas production include (Washington State Department of Ecology) site size, average depth of fill, refuse composition, refuse disposal rates, and quantity of in-place refuse. Topographic surveys and hydrogeologic information are useful sources.

Another source of information is the EPA. In its effort to characterize global methane emissions, the EPA collected data from several gas recovery projects (EPA, 1993). The methane generation rates were estimated by dividing the data for methane recovered by the gas collection efficiency. (In the absence of site data, a gas collection efficiency of 0.75 was assumed.) The data were evaluated for arid regions (less than 25 in./yr rainfall) and nonarid regions. The statistical relationships for nonarid regions take the form

$$CH_4(m^3/min) = 7.66W \qquad \text{std. dev.} = 0.46 \qquad \textbf{(14.5)}$$

For arid regions the relationships are

$$CH_4(m^3/min) = 5.87W \qquad \text{std. dev.} = 0.82 \qquad \textbf{(14.6)}$$

In both equations, W is the welled waste in 10^6 megagram. The landfills from which data were used were about 30 years old.

Example 14.1

Estimate the gas production in $ft^3/yr/lb$ of refuse in an old landfill in the northeastern United States, where the average rainfall is over 25 in./yr.

Solution: From Eq. 14.5, the average production is

$$CH_4 = 7.66 \times 10^{-9} \text{ m}^3/\text{min/kg}$$
$$= 7.66 \times (0.525 \times 10^6 \text{ min/yr})(3.28 \text{ ft/m})^3$$
$$\times 10^{-9}/(2.204 \text{ lb/kg})$$
$$= 0.065 \text{ ft}^3/\text{yr/lb}$$

If we consider the maximum as the average plus one standard deviation, the production would be $0.065 \times (7.66 + 0.46)/7.66 = 0.068$. From Figure 14.1, a rate of 0.08 is reasonable for an average wet condition.

Example 14.2

If the generation rate of 0.065 computed in Example 14.1 is for a 30-year-old landfill, what was the generation rate 10 years earlier? What will be the generation 20 years later? Assume $\xi = 0.04$.

Solution: The generation rate 10 years earlier is C_1.

$$0.064/C_1 = \exp - 0.04 \times 10 = 0.67$$
$$C_1 = 0.095$$

The generation rate 20 years later is C_2.

$$C_2/0.064 = \exp - 0.04 \times 20 = 0.449$$
$$C_2 = 0.03$$

As mentioned earlier, short-term (several hours to few days) gas extraction by pumping can provide some indication of gas generation. However, up to 50% of the pumped gas could come from storage depending on the volume of voids. Thus the test results may not reflect the landfill's actual gas generation rate. However, the test will establish the pumping rate to preclude significant air intrusion from the atmosphere. The well is usually pumped at different rates with pumped gas composition determined at each rate. A long-term pumping rate can then be selected.

14.4 INVESTIGATIONS FOR GAS

The presence of gas in a landfill usually can be detected during the drilling of a borehole using field instruments that monitor for gas (including methane) (see Chapter 7). If a gas sample from a shallow depth is desired, commercial hole makers with a slap hammer are available in lengths up to 7 ft. At least one hour before sampling, a probe is inserted into the hole, which is then closed with a stopper on the upper end. This allows gas lighter than air (such as methane) to displace the air that may have entered the hole. The gases are withdrawn with a suction pump into a portable field instrument. An apparatus calibrated for methane would indicate the percent LEL of methane.

The combustible gas indicator (CGI) detects methane as well as other combustible gases (such as those from hydrocarbons). In and around a sanitary landfill, methane is known to be the primary combustible gas but this should be verified by testing gas samples. The measurements are usually in terms of % LEL. Errors may result if the measurements are conducted in a low oxygen or high carbon dioxide environment or when the relative humidity is above 90%.

If methane is present in low concentrations, the FID (flame ionization detector) and PID (photoionization detector) may be used. FID detects methane and other compounds (petroleum hydrocarbons (PHC) and volatile organic compounds (VOCs)); PID detects PHC and VOCs. If both instruments are used, the presence of methane in small concentrations (parts per million) can be detected.

Where gas sampling is a part of an overall geotechnical program, a vacuum probe (as shown in Figure 14.2) could be used to extract a gas sample. The probe consists of a stainless steel rod with a drive tip and a drive head. It has machine-cut, very fine slots connecting to an axial tube 3/8 in. in diameter. This tube in turn connects to a 3/8-in. teflon tubing that extends to the ground surface through a 5-ft rod. The probe is advanced by driving with a standard 140-lb hammer. To sample the gas, the lines are evacuated by pumping for a few minutes. The gas is injected directly into the detection device to measure the % LEL or concentration of organic vapors. Gas samples for laboratory analysis are injected into gas bags provided by the testing laboratory.

In designing structures on or in the vicinity of a landfill, the design requires an assessment of the gas flux emanating from the ground surface. The most direct and accurate procedure to accomplish this is to determine gas pressures and concentration in the field, determine material properties with respect to the diffusion coefficient and permeability, determine concentration and pressure gradients, and calculate gas discharge, as explained in the following sections.

An alternative is to measure the flux actually emanating from the ground surface using a surface flux chamber (API, 1985), which is installed on the ground surface. Clean dry air is added to the chamber at a known rate and

Figure 14.2
Vacuum probe

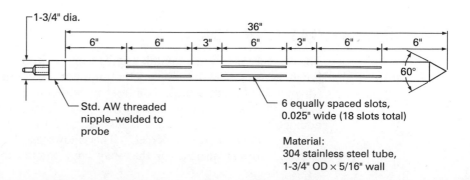

the concentration of VOCs (mostly methane in landfills) of discharge from the chamber is measured. The flux of gas emanating from the surface can then be calculated. This method does not apply if the methane exists in low concentrations because the dry clean air injected may cause a methane concentration below the detection limit of the instrument in the field.

14.5   GAS MIGRATION

14.5.1 *Migration Under a Pressure Gradient*

Like fluids, gas can migrate under a pressure gradient, a concentration gradient (by molecular diffusion), or both. Under a pressure gradient (assuming the gas to be a viscous fluid and that Darcy's law is valid), for one-dimensional flow

$$q = -\left(\frac{K}{\mu}\right)\frac{dp}{dx} \qquad \textbf{(14.7)}$$

where q is the specific discharge (Darcy velocity) (L/T), μ is the gas viscosity (M/LT), and K is the intrinsic permeability (L^2).

The intrinsic permeability K is a function of the average pore size. The average pore radius r and permeability K may be estimated for sands and gravels (Sherard et al., 1984) as

$$r = 0.1D_{15} \qquad \textbf{(14.8a)}$$

$$K = 0.125r^2 \qquad \textbf{(14.8b)}$$

where D_{15} is the particle size at which 15% of the soil is finer by weight. The intrinsic permeability is often expressed in darcies: One darcy is approximately 10^{-8} cm^2.

Table 14.2 provides specific weights and viscosities for gases found in sanitary landfills.

Example 14.3   If D_{15} from the grain size curve is 0.05 mm, estimate the gas permeability.

Solution:

$$K = 0.125(0.1 \times 0.005)^2 = 3.125 \times 10^{-8}\text{cm}^2 = 3.125 \text{ darcies}$$

The above relationships may be used to solve simple problems with practical significance. Consider for example one-dimensional flow of gas into a slot (Figure 14.3). The law of conservation of mass states that the difference in the mass of gas entering a control volume and that of gas exiting the control

Table 14.2
Viscosity and specific
weights of gases found
in sanitary landfills

	Temperature (°C)	Viscosity (μP)	SPECIFIC WEIGHT (0°C, 1 atm)	
			g/l	lb/ft³
Air	0	170.8	1.2928	0.0808
	18	182.7		
	54	190.4		
Carbon dioxide	0	139	1.9768	0.1235
	20	148		
	40	157		
Hydrogen sulfide	0		1.5392	0.0961
	17	124.1		
Nitrogen	0		1.2507	0.0782
	10.9	170.7		
	27.4	178.1		
Oxygen	0	189	1.4289	0.0892
	19.1	201.8		
Carbon monoxide	0	166	1.2501	0.0781
	21.7	175.3		
Ammonia	0	91.8	0.7708	0.0482
	20	98.2		
Hydrogen	0	83.5	0.0898	0.0052
	20.7	87.6		
	28.1	89.2		
Methane	0	102.6	0.7167	0.0448
	20	108.7		

1 μP $= 10^{-6}$ poise. 1 poise $= 1$ dyn·s/cm$^2 = 0.0672$ lb/(s·ft) $= 1$ g/(s·cm) $= 0.00209$
(lbf·s)/ft^2.
Source: Tchobanoglous et al., 1977; Weast and Melvin, 1980

volume is equal to the change of mass (dM/dt) with time in the control
volume. Assuming a steady state $(dM/dt = 0)$ in a homogeneous and isotropic
medium, from Figure 14.3 we have

$$-\frac{d}{dx}(\rho q_x) = 0 \qquad (14.9)$$

Substituting Eqs. 14.3 and 14.7 into Eq. 14.9, we obtain the steady-state flow
equation for one-dimensional flow in an isotropic homogeneous medium
under isothermal conditions:

$$\frac{d^2(P^2)}{dx^2} = 0 \qquad (14.10)$$

The solution to Eq. 14.10 is subject to the boundary conditions $P = P_1$ at
$x = 0$ and $P = P_2$ at $x = L$:

$$P^2 = P_1{}^2 + \left(\frac{P_2{}^2 - P_1{}^2}{L}\right)x \qquad (14.11)$$

Figure 14.3
Gas flow into a trench

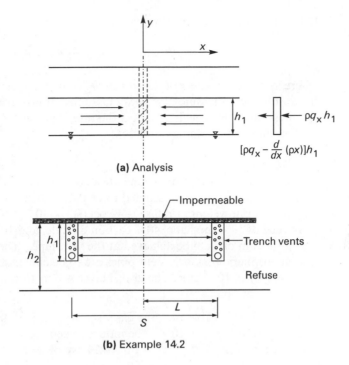

(a) Analysis

(b) Example 14.2

From Eqs. 14.11 and 14.7, the specific discharge in terms of the distance x and the boundary pressures can be expressed as

$$q_x = -\frac{K}{2\mu L} \frac{P_2^2 - P_1^2}{\sqrt{P_1^2 + \frac{P_2^2 - P_1^2}{L} x}} \qquad (14.12)$$

Example 14.4

Consider trench vents beneath an impervious slab of a structure (Figure 14.3). If the rate of gas generation is q_g (L^3 per unit of time per unit volume of refuse) and for the dimensions shown in Figure 14.3, determine the rate of gas withdrawal at the trench.

Solution: From Eq. 14.12 ($x = 0$),

$Q = qh_1$
$Q = -h_1(K/2L\mu P_1)[(P_2)^2 - (P_1)^2]$
$L = S/2$
$Q = (S/2)h_2 q_g$

$$S^2 = -2 \frac{h_1 K}{h_2 q_g \mu P_1} (P_2^2 - P_1^2) \qquad (14.13)$$

If P_2 differs little from P_1, then

$$S^2 = (4h_1 K/h_2\, q_g\, \mu)(P_2 - P_1) \qquad\qquad \textbf{(14.14)}$$

where h_1 is the thickness of the trench, h_2 is the thickness of the refuse, P_2 is the absolute gas pressure at $x = S/2$ (assumed atmospheric), P_1 is the absolute gas pressure at the trench, and q_g is the rate of gas generation (volume of gas/unit time/unit volume of refuse).

Example 14.5

A structure is supported on a concrete slab on the top of 20 ft of old refuse. The gas generation rate is assumed to be 0.04 ft^3/yr/lb (0.008 cm^3/s/kN) of refuse. The unit weight of refuse is 40 pcf (6.3 × 10^{-6} kN/cm^3). The gas is composed of 50% methane, 40% carbon dioxide, and 10% air. It is proposed that a vacuum of 2 cm be achieved at the vent pipes. The gas in the landfill is at atmospheric pressure. Vent pipes are to be embedded in stone with a permeability of 10^{-6} cm^2. The gravel layer is 6 in. thick. Determine the spacing of vent pipes.

Solution: From Table 14.2, the gas viscosity is $0.5 \times 108.7 + 0.45 \times 148 + 0.05 \times 170.8 = 129.5 \times 10^{-6}$. The induced vacuum $= 2$ cm $= 2 \times 980 = 1960$ g/cm/s$^2 = P_2 - P_1$. From Eq. 14.14, we obtain

$$S^2 = 4(0.5 \times 10^{-6} \times 1960/20 \times 129.5 \times 10^{-6} \times 0.008 \times 6.3 \times 10^{-6})$$

$$= 30 \times 10^6 \text{ cm}^2$$

$$S = 5480 \text{ cm} = 54.8 \text{ m}$$

If we use the more accurate Eq. 14.13, we have

$$P_2 = P_{atm} = 1000 \times 980 = 0.98 \times 10^6 \text{g/cm/s}$$

$$P_1 = (1000 - 2)980 = 0.97804 \times 10^6$$

$$(P_2{}^2 - P_1{}^2/P_1) = 0.0038 \times 10^{12}/0.97804 \times 10^6 = 0.00392 \times 10^6$$

$$S^2 = 2 \times 0.5 \times 10^{-6} \times 0.00392 \times 10^6/20 \times 129.5 \times 10^{-6} \times 0.008 \times 6.3 \times 10^{-6}$$

$$= 29.877 \times 10^6$$

$$S = 546.6 \text{ cm} = 54.66 \text{ m}$$

For the above examples, the given rate of gas generation per unit weight was converted to rate per unit volume using the given unit weight of refuse.

As the induced vacuum becomes larger, the difference in computed spacing derived from Eqs. 14.12 and 14.13 becomes larger. Eq. 14.13 is given in a different form in a manual by Washington State (Washington State Department of Ecology).

14.5.2 *Migration by Diffusion*

Gas will flow from one point to the other if there is a difference in concentration. Under one-dimensional flow, the gas flux J is described as

$$J = -DdC_g/dx \tag{14.15}$$

where J is the quantity of gas diffusing per unit of time through a unit area perpendicular to the concentration gradient, C_g is the gas concentration by volume, and D is the diffusion coefficient that depends on the gas and flow medium.

The area available for gas flow is characterized by the effective porosity n', which is expressed in terms of the porosity n and the degree of saturation S as

$$n' = n(1 - S) \tag{14.16}$$

The coefficient D for unobstructed free air diffusion of some gases are given in published tables (e.g., Weast and Melvin, 1980). The diffusion coefficient in soils or waste depends on the effective porosity and tortuosity, which results in an increase in the length of the flow path. Considering an unobstructed diffusion coefficient D_0, the soil gas diffusion coefficient D may be estimated based on the Millington-Quirk model in column experiments (Silka and Jordan, 1993):

$$D = D_0(a)^{10/3}/n^2 \tag{14.17}$$

where n is the porosity and a is the volumetric air content of the soil.

Example 14.6

Consider a free, unobstructed diffusion coefficient of methane of 21 ft^2/day at 20°C. Determine the diffusion coefficient in refuse with a total porosity of 0.4 and air content of 0.1.

Solution:

$$D = 21(0.1)^{10/3}/0.16 = 0.061 \text{ ft}^2/\text{day}$$

The diffusion coefficient is usually given at standard temperatures. Conversion to other temperatures may be accomplished using the following formula (Farmer et al., 1980).

$$(D)_{T_2} = (D)_{T_1}(T_2/T_1)^{0.5} \tag{14.18}$$

where $(D)_{T_1}$ is the diffusion coefficient at temperature T_1 (Kelvin) and $(D)_{T_2}$ is the diffusion coefficient at temperature T_2 (Kelvin).

Example 14.7

For the problem in Example 14.6, determine the diffusion coefficient at 30°C.

Solution:

$$D = 0.061(303/293)^{0.5} = 0.062 \text{ ft}^2/\text{day} \quad \underline{\hspace{2cm}}$$

If the diffusion coefficient of gas 1 is known, the diffusion coefficient for gas 2 may be determined as follows (Farmer et al., 1980).

$$D_2 = D_1(M_1/M_2)^{0.5} \tag{14.19}$$

where D_1 is the diffusion coefficient for gas 1, M_1 is the molecular weight for gas 1, M_2 is the molecular weight for gas 2, and D_2 is the diffusion coefficient for gas 2.

Example 14.8

For the problem in Example 14.6, determine the diffusion coefficient if the gas is carbon dioxide.

Solution:
M_1 for $CH_4 = 16$
M_2 for $CO_2 = 44$

$$D_2 = 0.062(16/44)^{0.5} = 0.037 \text{ ft}^2/\text{day} \quad \underline{\hspace{2cm}}$$

14.6   CONTROL OF GAS BY PUMPING

Gas pumping wells are often used to control migration of gas off-site. By virtue of the ability to intersect a deep layer of refuse, pumping wells offer an advantage over the usual trenches for gas venting around the landfill. The design parameters often needed are the radius of influence and the gas permeability of the subsurface materials. Pumping tests are usually conducted where the well screen is a few feet above the water or leachate level. The screen length is set at about 2/3 of the unsaturated zone to help reduce air intrusion. Pumping is conducted in increments. When gas sampling indicates dilution by air, the pumping rate is set at a lower rate.

Figure 14.4 is an idealization of Hantush's (1960) analysis for a ground-water pumping well in a leaky aquifer. The "gasfer" in this case is layer 2. Layer 1 is leaky (has a small permeability) as is layer 3. When the well is pumped, gas is drawn from the three layers. The flow is radial in layer 2 and vertical (recharging layer 2) in layers 1 and 3. The formulation of the problem follows Jacob (1940) and Hantush (1960), assuming ideal gas under an isothermal condition and making use of Eq. 14.3.

Figure 14.4
Gas flow into a well

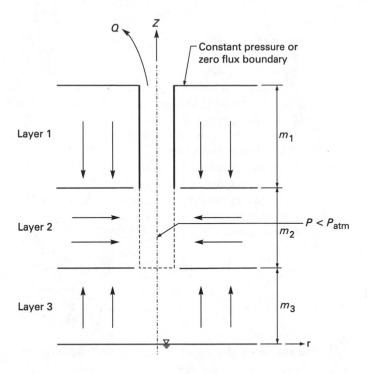

A gas pumping well has a construction similar to a water pumping well. A screen is surrounded by high permeability, washed gravel or stone. Above the screened portion, the space between the well casing and the soil (or refuse) around it is sealed. A vacuum is applied at the well head, which will create a pressure gradient toward the well. In some applications, the surface around the well is covered with a geomembrane or low permeability soil. This will minimize air intrusion (i.e., short circuiting). In the absence of such a cover, the well screen is placed over the lower portion of the well.

14.6.1 *Flow Mechanics for Large Vacuum*

For the system shown in Figure 14.4 and following Jacob (1940) and Hantush (1960), the differential equation describing the transient flow of compressible gas flow into the well takes the form

$$\frac{1}{r}\frac{\partial \phi_r}{\partial r} + \frac{\partial^2 \phi_r}{\partial r^2} + \frac{K_1}{T_r}\frac{\partial \phi_1}{\partial z} - \frac{K_3}{T_r}\frac{\partial \phi_3}{\partial z} = \frac{s_g}{T_r}\frac{\partial \phi_r}{\partial t} \qquad \textbf{(14.20)}$$

where

$$\phi_r = (P_0)^2 - (P_r)^2$$
$$\phi_1 = (P_0)^2 - (P_1)^2$$
$$\phi_3 = (P_0)^2 - (P_3)^2$$

P_0 = ambient absolute pressure (atmospheric pressure prior to pumping) (assumed the same in layers 1, 2, and 3) (M/LT^2)
P_1 = absolute pressure in layer 1
P_r = absolute pressure in layer 2
P_3 = absolute pressure in layer 3
K_1 = gas permeability in layer 1
K_3 = gas permeability in layer 3
T_r = gas transmissivity in layer 2 = $K_r m_2$ (L3)
K_2 = gas permeability in layer 2
m_2 = thickness of layer 2   (L)

$$S_g = (P_0/P_{av})S_g' = \text{gas storage coefficient in layer 2,} \quad (LT) \tag{14.21}$$

$$S_g' = \mu m_2\, n_2'\left(\frac{1}{E_2\, n_2'} + \frac{1}{P_0}\right) \tag{14.21a}$$

The storage coefficients in layers 1 and 3 are computed as

$$S_1 = (P_0/P_{av})S_1' \tag{14.22}$$

$$S_1' = \mu m_1 n_1'\left(\frac{1}{E_1 n_1'} + \frac{1}{P_0}\right) \tag{14.22a}$$

$$S_3 = (P_0/P_{av})S_3' \tag{14.23}$$

$$S_3' = \mu m_3\, n_3'\left(\frac{1}{E_3\, n_3'} + \frac{1}{P_0}\right)$$

where

E_1, E_2, and E_3 = Young's modulus of layers 1, 2, and 3, respectively
μ = gas viscosity (assumed the same in all layers)   (M/LT)
P_{av} = average value of P_r during pumping = $(P_0 + P_r)/2$
n_1', n_2', and n_3' = effective porosity in layers 1, 2, and 3, respectively
m_1, m_3 = thickness of layers 1 and 3, respectively

14.6.2 *Flow Mechanics for Small Vacuum at the Well Head*

For small drawdowns, the functions ϕ can be simplified and written in terms of pressure drawdowns,

$$P'_1 = (P_0 - P_1)$$
$$P'_3 = (P_0 - P_3)$$
$$P'_r = (P_0 - P_r)$$

by noting for example that

$$\phi_r = (P_0)^2 - (P_0 - P'_r)^2 \approx 2P'_r P_0$$

$$\phi_1 \qquad\qquad \approx 2P'_1 P_0$$

$$\phi_3 \qquad\qquad \approx 2P'_3 P_0 \tag{14.24}$$

In terms of drawdown P', the flow equation is

$$\frac{1}{r}\frac{\partial P'_r}{\partial r} + \frac{\partial^2 P'_r}{\partial r^2} + \frac{K_1}{T_r}\frac{\partial P'_1}{\partial z} - \frac{K_3}{T_r}\frac{\partial P'_3}{\partial z} = \frac{S'_g}{T_r}\frac{\partial P'}{\partial t} \tag{14.25}$$

14.6.3 Solutions for Transient Conditions

Following Hantush (1960), Eqs. 14.20 and 14.25 are already solved; the solutions are summarized in Section 14.6.6. The solutions may require three type curves, depending on the boundary conditions. For the case of no leakage ($K_1 = K_3 = 0$), or for a leaky case with an impervious boundary at the surface, the function $W(u)$ is plotted in Figure 14.5 (Fetter, 1988). For the leaky case at large time, it is assumed that there is no change in storage in the leaky layer; the function $W(u\delta_3, r/B)$ is plotted in Figure 14.6. Figure 14.7 is a plot of the function $H(u, \beta)$, which is applicable for the very early stages of pumping. The

Figure 14.5
The nonequilibrium reverse type curve (Theis curve) for a fully confined aquifer

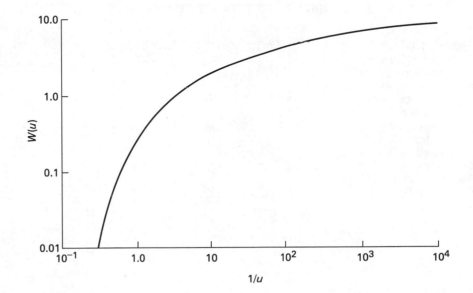

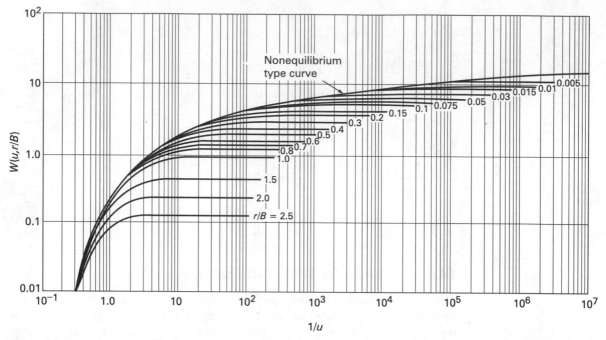

Figure 14.6 Type curves of leaky artesion aquifer in which no water is released from storage in the confining layer. *Applied Hydrogeology*, 2d ed., by C. W. Fetter, 1988. (Reprinted by permission of Prentice-Hall, Inc., Upper Saddle River, NJ.)

specified parameter for the small vacuum case is the flow rate Q (L^3/T). For the large vacuum case, the flow rate and the absolute pressure P_w at the well head must be known. An isothermal condition prevails and temperature therefore does not appear in the solution.

Figure 14.7
Type curves for a well in an aquifer confined by a leaky layer that releases water from storage. *Applied Hydrogeology*, 2d ed., by C. W. Fetter, 1988. (Reprinted by permission of Prentice-Hall, Inc., Upper Saddle River, NJ.)

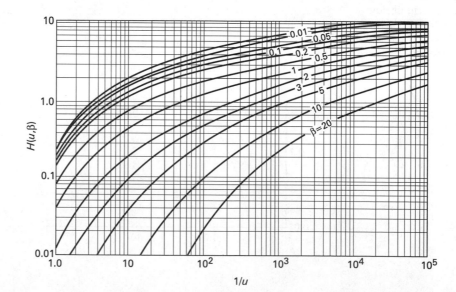

The solutions for small vacuum (large P_w) are in terms of Q'_g. The solutions for large vacuum (small P_w) are in terms of Q_g. Thus

$$Q'_g = \mu Q \text{ (small vacuum)} \tag{14.26}$$

and

$$Q_g = 2P_w \mu Q = 2P_w Q'_g \text{ (large vacuum)} \tag{14.27}$$

where P_w is the absolute pressure at the well head. Other parameters are defined as follows. For the small vacuum case,

$$u = \frac{r^2 S'_g}{4 T_r t} \tag{14.28}$$

$$\delta_2 = 1 + \frac{S'_1 + S'_3}{S'_g} \tag{14.28a}$$

$$\beta = \frac{1}{4} r \left(\sqrt{\frac{K_1/m_1}{T_r} \frac{S'_1}{S'_g}} + \sqrt{\frac{K_3/m_3}{T_r} \frac{S'_3}{S'_g}} \right) \tag{14.28b}$$

$$\delta_3 = 1 + \frac{S'_3 + \dfrac{S'_1}{3}}{S'_g} \tag{14.28c}$$

The parameter B in Figure 14.6 is defined as

$$\frac{1}{B} = \sqrt{\frac{K_1/m_1}{T_r}} \tag{14.29}$$

For large vacuum, the parameters in Eq. 14.28 are used with S_1, S_3, S_g replacing S'_1, S'_3, and S'_g, respectively.

$$\delta_2 = 1 + \frac{S'_1 + S'_3}{S'_g}$$

14.6.4 Solutions for the Steady-state Case with No Leakage

With prolonged pumping, the change of pressure with time is assumed to be zero. The solution for a large vacuum is

$$\phi_r = \frac{Q_g}{2\pi T_r} \ln \frac{R_i}{r} \tag{14.30}$$

and for a small vacuum the solution is

$$P_r' = \frac{Q_g'}{2\pi T_r} \ln \frac{R_i}{r}$$

(14.31)

Under a steady-state condition where there is no change in storage, the gas transmissivity can be calculated from either Eq. 14.30 or Eq. 14.31. The drawdowns (ϕ or P') are plotted on a drawdown-log r paper. The slope of the line is the change of drawdown over a log cycle of r. If we denote this slope as m, then for large vacuum, m is the slope of the ϕ-log r plot:

$$m = \frac{\Delta\phi_r}{\Delta \log r} = 2.3 \frac{Q_g}{2\pi T_r} = 2.3 \frac{P_w \mu Q}{\pi T_r}$$

(14.32)

and for a small vacuum, m is the slope of the P'-log r plot:

$$m = \frac{\Delta P'}{\Delta \log r} = 2.3 \frac{\mu Q_g}{2\pi T_r}$$

(14.33)

From the transmissivity, the gas permeability is calculated by $K_2 = T_r/m_2$. From Eqs. 14.30 and 14.31, it can be shown that the solution for a small vacuum is expressed by Eq. 14.34 and that the solution for a large vacuum is expressed by Eq. 14.35.

$$\frac{Q}{m_2} = \frac{2\pi \dfrac{K_2}{\mu} (P_0 - P_w)}{\ln \dfrac{R_i}{r_w}}$$

(14.34)

$$\frac{Q}{m_2} = \frac{\pi K_2 P_w}{\mu} \frac{\left(\dfrac{P_0^2}{P_w^2} - 1 \right)}{\ln \dfrac{R_i}{r_w}}$$

(14.35)

14.6.5 Solution for the Steady-state Case with Leakage

The equation to be solved in this case is

$$\frac{1}{r} \frac{d}{dr} \left(r \frac{d\phi}{dr} \right) - \frac{\phi_r}{B^2} = 0$$

(14.36)

and the solution (Strack, 1989) is

$$\phi_r = \frac{-\dfrac{Q_g}{T_r}}{2\pi r_w} \cdot B \, \frac{K_0(r/B)}{K_1(r/B)} \tag{14.37}$$

where $K_0(r/B)$ is the modified Bessel function of order 0 and second kind. $K_1(r/B)$ is the derivative of $K_0(r/B)$ and is the modified Bessel function of order 1. For $r/B < 0.2$, the drawdown can be approximated by

$$\phi_r = \frac{Q_g/T_r}{2\pi r} \, [\ln{(r)} - \ln{(1.123)B}] \tag{14.38}$$

From Eq. 14.38, it is clear that leakage reduces the drawdown, as expected. For small pressures, Eqs. 14.36 and 14.37 may be written in terms of P' (instead of ϕ) and Q'_g (instead of Q_g).

14.6.6 Use of Type Curves

The following steps are implemented in using the type curves for interpreting the results of the pumping test under transient conditions (Fetter, 1988).

The type curve in Figure 14.5 is applicable to the case of pumping from layer 2 and no leakage from either layers 1 or 3.

1. For a given observation well, plot the drawdown versus time t or t/r^2, where r is the distance from the pumped well. The scale used in plotting must be identical to that of the type curve (i.e., the same length of the logarithmic cycle).

2. Lay the graph paper with the data curve on a light table over the type curve of the same scale.

3. Adjust the position of the field data, keeping the axes of both sheets parallel, until the data points fall on the type curve.

4. Select an arbitrary match point (not necessarily on the type curve). $W = 1$, $1/u = 1$ may be selected for convenience.

5. For the match point, scale W, $1/u$, drawdown (P' or ϕ) and t (or t/r^2).

6. Knowing Q, W, and t (or Q', W, and t), calculate the transmissivity T_r and permeability $K = T_r/m_2$ (from Eq. 14.39 for small vacuum and 14.40 for large vacuum).

$$P'_r = \frac{Q'_g}{4\pi T_r} \, W(u) \tag{14.39}$$

$$\phi_r = \frac{2P_w Q'_g}{4\pi T_r} \, W(u) \tag{14.40}$$

Knowing u and t calculate (from Eq. 14.28) the storage term S'_g (for small vacuum) or S_g (for large vacuum). The type curve in Figure 14.5 can be used for a leaky upper layer (layer 1) but with an impervious boundary on the ground surface. In this case $u\delta_2$ is used instead of u and $W(u\delta_2)$ is used instead of $W(u)$. The parameter δ_2 is defined in Eq. 14.28.

The type curve in Figure 14.6 is applicable for a leaky layer 1 and/or 3 after a long pumping time. The parameter u in this case is treated as $u\delta_3$ and W as $W(u\delta_3, r/B)$. δ_3 is defined in Eq. 14.28.

Steps 1 and 2 are the same as described above.

3. Adjust the position of the field data, keeping the axes of both sheets parallel. The data curve should match one of the r/B lines or it may have to be matched to a line interpolated between two r/B lines.

4. Select any arbitrary match point ($W = 1$, $u\delta_3 = 1$).

5. For the match point, scale W, $u\delta_3$, drawdown, and t.

6. Knowing Q_g (or Q'_g), drawdown, and W, calculate T_r and K.

$$P'_r = \frac{Q'_g}{2\pi T_r} W\left(u\delta_3, \frac{r}{B}\right) \tag{14.41}$$

Knowing u and t, calculate S_g (or S'_g).

Knowing r/B or B, calculate K_1/K_2.

In a case of large vacuum pumping, ϕ_r is used instead of p'_r and $2P_w Q'_g$ is used instead of Q'_g in Eq. 14.41.

The type curve in Figure 14.7 is applicable to leaky layers 1 and/or 2 but after a short time. In this case,

$$P'_r = \frac{Q'_g}{2\pi T_r} H(u, \beta) \tag{14.42}$$

For the large vacuum case, $2P_w Q'_g$ is used instead of Q'_g and ϕ'_r is used instead of P'_g.

Steps 1 and 2 are the same as above.

3. Adjust the position of the field data, keeping the axes of both sheets parallel. The data curve should match one of the curves.

4. Select an arbitrary match point.

5. For the match point selected, read H, $1/u$, drawdown, and t.

6. Knowing Q_g (or Q'_g), drawdown, and m_2, calculate T_r and K.

Knowing u and t, calculate S_g (or S'_g).

Knowing β and if $K_3 = 0$, calculate $K_1 S'_1$.

As an example, consider Table 14.3. The first two columns (Massmann, 1989) show time and drawdown data from a pumping test. Column 2 shows the measured drawdown (vacuum in units of cm at an observation well). Column 3 is the absolute pressure (in units of cm of water) based on atmospheric pressure of 1000 cm of water. Column 4 is the function ϕ (in units of (pressure)^2). The drawdown in cm is converted to pressure (L(M/L^3) (L/T^2) or (cm · g/cm^3)

Table 14.3
Gas pump test data

Time (min)	Drawdown (cm water)	Absolute pressure (cm water)	$\phi = P_0^2 - P^2$ $((g)/cm/s^2)^2$
1	0.05	999.95	9.6037×10^7
2	0.08	999.92	15.3666×10^7
3	0.05	999.95	9.6037×10^7
5	0.03	999.97	5.762×10^7
6	0.18	999.82	34.571×10^7
7	0.38	999.62	7.2977×10^8
9	0.66	999.34	1.2673×10^9
11	1.02	998.98	1.9582×10^9
14	1.6	998.4	3.0708×10^9
18	2.29	997.71	4.394×10^9
21	2.82	997.18	5.41×10^9
26	3.43	996.57	6.577×10^9
31	4.06	995.94	7.782×10^9
41	5.05	994.95	9.675×10^9
46	5.44	994.56	10.421×10^9
61	6.53	993.47	12.502×10^9
101	7.95	992.05	15.21×10^9
166	9.37	990.63	17.914×10^9
306	10.11	989.89	19.321×10^9

$$P\left(\frac{g}{cm/s^2}\right) = P(cm)(980)$$

Source: Massmann, 1989

(acceleration of gravity $= 980$ cm/s^2). The vacuum at the well head is 100 cm of water. The viscosity of the gas mixture is 1.45×10^{-4} gm/cm/s. The observation well where the drawdowns were measured was 38 m (r) away. Layer 1 is glacial till with a thickness of $m = 6$ m. Layer 2 is sand and gravel with a thickness of $m_2 = 24$ m. The well screen is in layer 2.

For a small vacuum with no leakage, the drawdown data P' is plotted in Figure 14.8 and superimposed on the $W(u)$ versus $1/u$ type curve (Figure 14.5). From curve matching, the parameters T_r and S'_g are calculated as shown in Figure 14.8. The same data for a small vacuum with leakage (no release from storage) is plotted for curve matching with the function in Figure 14.6 assuming that the layer 1 recharges layer 2 (see Figure 14.9). It is clear that this type curve fits the data better and that leakage through the till is likely.

Finally, for a small vacuum with a leaky confining layer and a short time release from storage in the confining layer, the curve matching is illustrated in Figure 14.10 using the type curve in Figure 14.7. It is clear that Figure 14.9 still provides the best match.

The cases in Figures 14.8 and 14.9 are repeated in Figures 14.11 and 14.12 but for the high vacuum cases. The drawdowns in terms of ϕ are taken from the last column of Table 14.3. If we assume that the final drawdown of about 10 cm is a steady-state value and that the well radius is 5 cm, then from Eq. 14.33,

$$m = (100 - 10)(980)/(\log 3800 - \log 5) = 3.06 \times 10^4$$

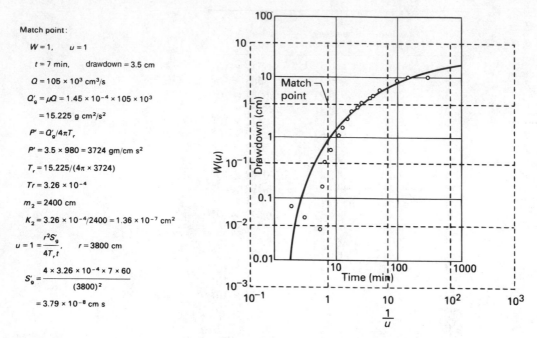

Match point:

$W = 1$, $u = 1$

$t = 7$ min,   drawdown = 3.5 cm

$Q = 105 \times 10^3$ cm^3/s

$Q'_g = \mu Q = 1.45 \times 10^{-4} \times 105 \times 10^3$

   $= 15.225$ g cm^2/s^2

$P' = Q'_g/4\pi T_r$

$P' = 3.5 \times 980 = 3724$ gm/cm s^2

$T_r = 15.225/(4\pi \times 3724)$

$Tr = 3.26 \times 10^{-4}$

$m_2 = 2400$ cm

$K_2 = 3.26 \times 10^{-4}/2400 = 1.36 \times 10^{-7}$ cm^2

$u = 1 = \dfrac{r^2 S'_g}{4T_r t}$, $r = 3800$ cm

$S'_g = \dfrac{4 \times 3.26 \times 10^{-4} \times 7 \times 60}{(3800)^2}$

   $= 3.79 \times 10^{-8}$ cm s

Figure 14.8   Small vacuum assumption field data overlain on type curve (Theis, fully confined)

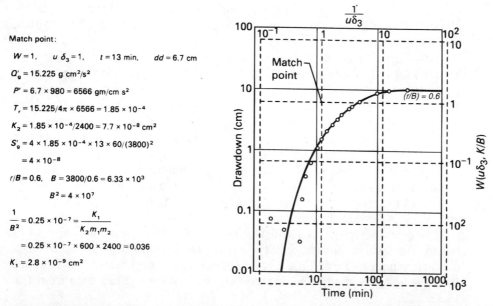

Match point:

$W = 1$, $u\,\delta_3 = 1$, $t = 13$ min, $dd = 6.7$ cm

$Q'_g = 15.225$ g cm^2/s^2

$P' = 6.7 \times 980 = 6566$ gm/cm s^2

$T_r = 15.225/4\pi \times 6566 = 1.85 \times 10^{-4}$

$K_2 = 1.85 \times 10^{-4}/2400 = 7.7 \times 10^{-8}$ cm^2

$S'_g = 4 \times 1.85 \times 10^{-4} \times 13 \times 60/(3800)^2$

   $= 4 \times 10^{-8}$

$r/B = 0.6$, $B = 3800/0.6 = 6.33 \times 10^3$

   $B^2 = 4 \times 10^7$

$\dfrac{1}{B^2} = 0.25 \times 10^{-7} = \dfrac{K_1}{K_2 m_1 m_2}$

   $= 0.25 \times 10^{-7} \times 600 \times 2400 = 0.036$

$K_1 = 2.8 \times 10^{-9}$ cm^2

Figure 14.9   Small vacuum assumption field data overlain on type curve (leaky confining layer — no release from storage)

Match point:

$H = 1$, $u = 1$, $t = 8$ min., $dd = 5$ cm

$Q'_g = 15.225$ g cm²/s²

$P' = 5 \times 980 = 4900$ gm/cm s²

$T_r = 15.225/4\pi \times 4900 = 2.47 \times 10^{-4}$

$S'_g = \dfrac{4 \times 2.47 \times 10^{-4} \times 8 \times 60}{(3800)^2} = 3.28 \times 10^{-8}$

$K_2 = 2.47 \times 10^{-4}/2400 = 1 \times 10^{-7}$ cm²

Assuming $K_3 = 0$,

$\beta = 0.11 = \dfrac{1}{4} r \sqrt{\dfrac{K_1/m_1}{T_r} \dfrac{S_1}{S'_g}}$

$\beta^2 = \dfrac{1}{16} r^2 \dfrac{K_1/m_1}{T_r} \dfrac{S_1}{S'_g}$

$m_1 = 600$ cm

$S'_1 = \dfrac{600 \times (0.11)^2 (2.47 \times 10^{-4})(3.28 \times 10^{-8}) \times 16}{(3800)^2}$

$= 6.5 \times 10^{-17}$

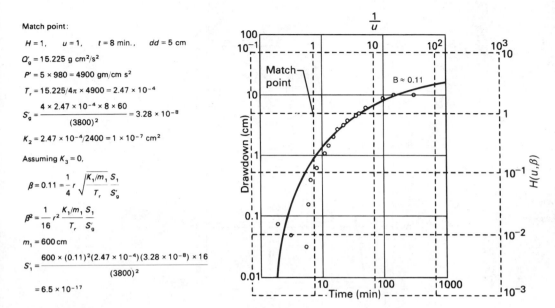

Figure 14.10 Small vacuum assumption field data overlain on type curve (Hantush, leaky confining layer with release from storage)

Match point:

$W - 1$, $u - 1$, $t = 6.8$ min

$\phi = 6.6 \times 10^9$ (g/cm s²)²

$P_w = 900$ cm $\times 980$

$= 8.82 \times 10^5$ g/cm s²

$Q_g = 2P_w Q'_g = 2 \times 8.82 \times 10^5 \times 15.225$

$= 2.69 \times 10^7$ gm² cm/s⁴

$T_r = \dfrac{2.69 \times 10^7}{4(\pi)(6.6 \times 10^9)} = 3.2 \times 10^{-4}$

$S_g = \dfrac{4 \times 3.2 \times 10^{-4} \times 6.8 \times 60}{(3800)^2}$

$= 3.6 \times 10^{-8}$

$K_2 = 3.2 \times 10^{-4}/2400 = 1.33 \times 10^{-7}$ cm²

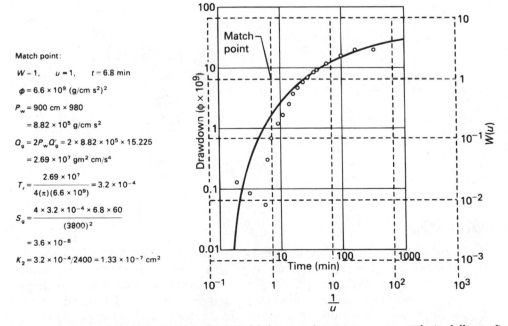

Figure 14.11 High vacuum assumption field data overlain on type curve (Theis, fully confined)

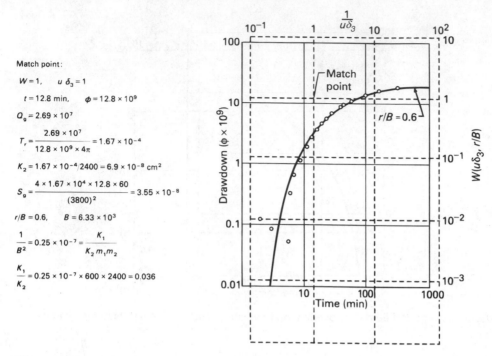

Match point:

$W = 1, \quad u\,\delta_3 = 1$

$t = 12.8 \text{ min}, \qquad \phi = 12.8 \times 10^9$

$Q_g = 2.69 \times 10^7$

$T_r = \dfrac{2.69 \times 10^7}{12.8 \times 10^9 \times 4\pi} = 1.67 \times 10^{-4}$

$K_2 = 1.67 \times 10^{-4}/2400 = 6.9 \times 10^{-8} \text{ cm}^2$

$S_g = \dfrac{4 \times 1.67 \times 10^4 \times 12.8 \times 60}{(3800)^2} = 3.55 \times 10^{-8}$

$r/B = 0.6, \qquad B = 6.33 \times 10^3$

$\dfrac{1}{B^2} = 0.25 \times 10^{-7} = \dfrac{K_1}{K_2\,m_1 m_2}$

$\dfrac{K_1}{K_2} = 0.25 \times 10^{-7} \times 600 \times 2400 = 0.036$

Figure 14.12   High vacuum assumption field data overlain on type curve (leaky confining layer—no release from storage)

From Eq. 14.26

$$Q_g{}' = 1.45 \times 10^{-4} \times 105 \times 10^3 = 15.225 \text{ cm}^2/\text{g/s}^2$$

And from Eq. 14.33,

$$T_r = 15.225(2.3)/3.06 \times 10^4(6.28) = 1.82 \times 10^{-4} \text{ cm}^3$$

$$K_2 = 1.82 \times 10^{-4}/2400 = 7.58 \times 10^{-8} \text{ cm}^2$$

14.7   GAS CONTROL

In broad terms, control of landfill gas is usually accomplished by either a passive or an active venting system. A passive system is one with no mechanical components (e.g., pumps, fans, etc.). An active system (forced venting) employs mechanical means to induce a sink (vacuum or suction) to where the gas flows, which gives positive control and greater flexibility but perhaps at a greater cost. The choice of the method depends on site conditions (refuse thickness, depth to water or leachate or low permeability soil, and ease of excavation). Other considerations include the adjacent site use. The presence of homes in close proximity may require active venting whereas open space may allow simple passive venting. Other factors include cost of installation and maintenance as well as the market for extracted gas as an energy source.

14.7.1 *Passive Venting*

In many landfills, the local geology allows lateral migration of gas to be controlled by a perimeter trench. Vertical migration through the cap is passively controlled as well. Figure 14.13 shows some examples of passive venting. For

Figure 14.13
Typical application of passive venting

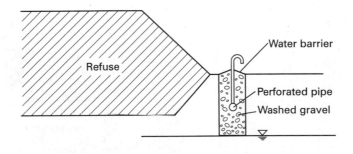
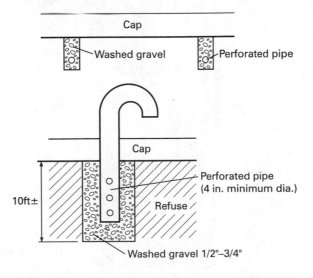

the passive system to work, it should act as a sink; this will require the gas pressure in the trench to be less than the pressure in the landfill. A high permeability material (such as washed gravel) in a passive system will cause a pressure sink that will attract gas just as a toe drain in a dam will collect water seeping through the dam.

Figure 14.13 shows a typical installation around the perimeter of a landfill where the situation is favorable for installation. The refuse below grade is shallow or the water table or low permeability soil is shallow. Thus the trench excavations can be performed by conventional equipment (backhoe). The system in Figure 14.13 could be adapted to future active venting (as shown in Figure 14.14). A second example of a passive system, cap venting, is shown in Figure 14.15. Some states require the gas to be flared, as illustrated in Figure 14.15, but others allow simple vents. A third example is a passive venting system beneath the floor of a structure, as illustrated in Figure 14.16. Fresh air

Figure 14.14
Hybrid system combination vent/ barrier trench

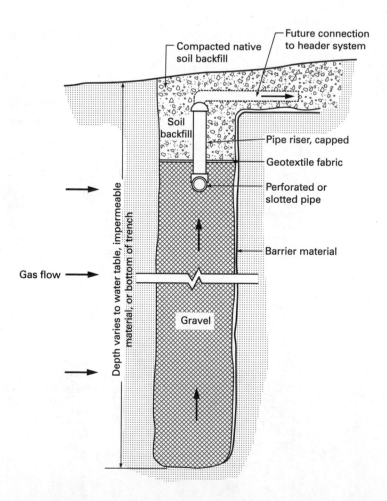

Figure 14.15
Gas flare detail with
horizontal collection
trench

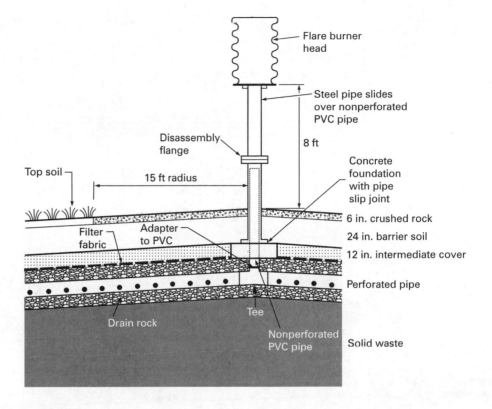

is introduced through suction pipes while gases are vented to the atmosphere above the roof. The air turbines are operated naturally by wind. (An active system would use powered turbines.) If future extraction of gases is contemplated, it may be desirable to install deep vents in the landfill that could be converted to extraction wells at a later date, as illustrated in Figure 14.17.

The spacing of vents and trenches is empirical although some estimates may be made based on the gas generation rates explained in this chapter. Underneath structures, the spacing should be less than 50 ft.

14.7.2 Active Venting

Where passive venting is considered insufficient for control of landfill gas, active, or forced, venting, which can provide full control of landfill gas migration, may be appropriate. A vacuum pump or blower may be connected to the discharge end of a passive system pipe to develop an active venting system or extraction wells may be used.

Where extraction wells are used (Figure 14.17) the vacuum induced at the well head provides the required sink for gas collection. The suction is induced away from the wells and is measured in a gas probe, as illustrated in Figure 14.18. The gas flow rate to the well initially will be derived mostly from

Figure 14.16
Passive gas venting below the floor slab

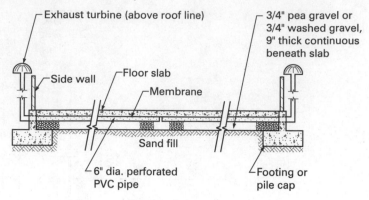

Section A–A

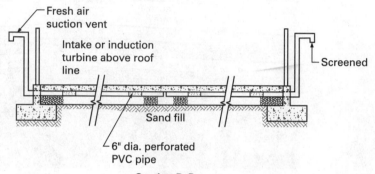

Section B–B

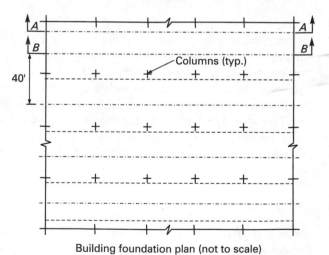

Building foundation plan (not to scale)

Notes

1. With slab-on-grade construction, use flexible corrugated pipe.

2. With pile supported slab, use rigid pipe hung from floor slab

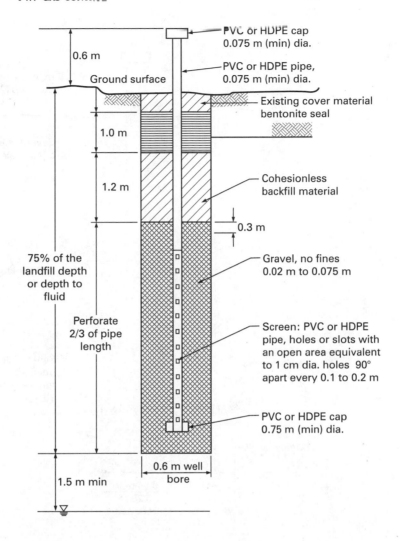

Figure 14.17
Gas extraction well
(after USEPA, 1996)

storage but in the long run, it would not exceed the gas generation rate in the landfill. The spacing of the gas extraction wells may be determined based on actual pumping tests. The rule of thumb is that the spacing is three times the refuse thickness (i.e., the radius of influence of 1.5 times the refuse thickness). Performance tests on some landfills indicated a radius of influence ranging from 100 ft to 500 ft (McBean et al., 1995).

USEPA (1996) mandates a method for the measurement of landfill gas (LFG) flow rates for use in calculating the emission of nonmethane organic compounds (NMOC). The method requires drilling a test well as shown in Figure 14.17. Shallow pressure probes are used to assess the infiltration of air into the landfill during the pumping test. Deep pressure probes are used to measure pressures. Deep probes (Figure 14.18) are specified at 120° apart at distances of 3 m, 15 m, 30 m, and 45 m from the pumped well (i.e., nine deep probes are required for each pumped well). Deep probes extend to the level of

Figure 14.18
Pressure probe
(after USEPA, 1996)

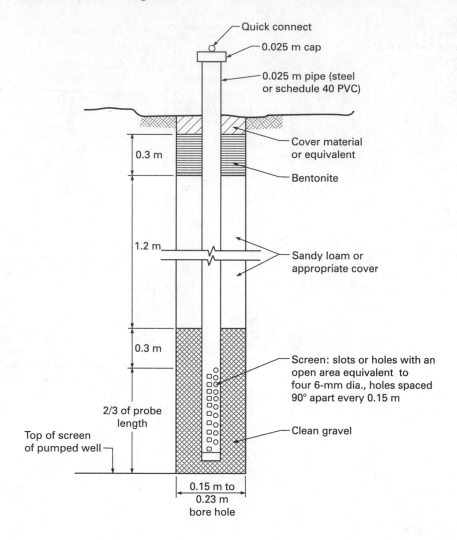

the top of the screened portion of the well. Three shallow probes are specified for each test well to extend to half the deep probes. The testing procedure requires a short-term testing (24 hours at each vacuum) to determine the maximum vacuum that can be applied to the well without infiltration of air to the landfill. Infiltration is characterized by LFG N_2 concentration greater than 20% or negative gauge pressure in the shallow probes. After the maximum vacuum is determined, the test at maximum vacuum is continued for 24 hours or longer. Prior to conducting the pumping test, the refuse thickness and age over the entire landfill need to be assessed, which determines the location and number of tests required. Considerable engineering judgment and experience is needed for this type of investigation.

Extracted gas is typically saturated with water vapor; thus condensates usually develop in the collection pipes. Traps for condensate collection are placed at all low points and at spacings compatible with the slopes of the header pipes. If the header pipes are designed with adjustable support, the settlement of the landfill can be accommodated so that condensate and gas flow in the same direction. Otherwise, steep slopes and/or more frequent traps may be needed. Methods for estimating condensate quantities are given by McBean et al., 1995.

In some old landfills designed without a leachate collection system, a leachate mound usually develops in the landfill. As gas is pumped, an upconing effect takes place because of the reduction in gas pressure above the leachate. Therefore, the bottom of the well is often limited to several feet above the highest expected leachate level. An alternative is to implement a combination gas-and-leachate extraction well.

SUMMARY

This chapter has presented the mechanics of gas flow and procedures for gas control. Gas is generated in sanitary landfills as a result of refuse decomposition and moisture. Methane and carbon dioxide are the major components of landfill gas; other components are the nonmethane organic compounds (NMOC), which have recently been regulated. Methane is explosive and needs to be controlled, although it does have the side benefit as a source of fuel. Landfill gas is measured in various ways that may include conduction gas-pumping tests and is vented by trenches, vent pipes, and extraction wells. Extraction wells can serve the dual purpose of methane and NMOC control and LFG utilization.

NOTATIONS

ζ decay constant (T^{-1})

ρ gas density at pressure (M/L^3)

ρ_0 gas density at standard pressure and temperature (M/L^3)

$(D)_{T_1}$ diffusion coefficient (L^2/T) at temperature T_1 (Kelvin)

$(D)_{T_2}$ diffusion coefficient (L^2/T) at temperature T_2 (Kelvin)

μ gas viscosity (M/LT)

a volumetric air content of the soil

C molar density (mol/m^3)

C_1 amount of methane-producing substrata at time t_1 (yr)

C_2 amount of methane-producing substrata at time t_2 (yr)

C_g gas concentration by volume

D diffusion coefficient that depends on the gas and flow medium (L^2/T)

D_1 diffusion coefficient for gas 1

D_2 diffusion coefficient for gas 2

E Young's modulus F/L^2

h_1 thickness of the trench (L)

h_2 thickness of the refuse (L)

J quantity of gas diffusing per unit time through a unit area perpendicular to the concentration gradient

K intrinsic permeability (L^2)

M_1 molecular weight for gas 1

M_2 molecular weight for gas 2

n porosity

P_1 absolute gas pressure at the trench (F/L^2)

P_2 absolute gas pressure

P_w absolute pressure at the well head (F/L^2)

q specific discharge (Darcy velocity) (L/T)

r average pore radius (L)

R universal gas constant

T absolute temperature (Kelvin)

PROBLEMS

14.1 A landfill receives municipal waste at the rate of 2000 tn/day. Assuming a maximum generation rate of 0.1 ft^3/yr/lb at standard temperature and pressure (S.T.P.) and a decay constant of 0.06 $year^{-1}$, determine the amount of methane generated in year 5 assuming a landfill operation of 260 days/yr.

14.2 A landfill containing municipal refuse is about 30 years old in an area with 50 in. of annual rainfall. Estimate the average gas generation rate. If the leachate is recirculated into the landfill, what would be the expected increase in the gas generation rate? Without leachate recirculation, estimate the gas generation rate after 30 years using a decay constant of 0.04.

14.3 Consider a 50-ft-thick landfill with a reasonably effective cap. Under the cap are gas collection trenches. Develop design charts for the spacing of the trench versus the depth of the trench for a gas generation rate of 0.04 ft^3/yr/lb of refuse. Vary the depth of the trench from 5 to 25 ft and the suction caused by the trench from 5 to 20 cm of water. Assume a unit weight of refuse of 45 pcf. Consider the pressure in the landfill to be atmospheric (1000 cm water).

14.4 A landfill clay cap is 3 ft thick with a porosity of 0.6. If the pressure in the landfill is 1.05 atm and methane concentration in the landfill is 40%, determine the total discharge of methane to the atmosphere. The permeability of the cap is 0.1 darcies.

14.5 A pumping test setup is shown in Figure 14.19. The measured vacuums at the monitoring well are as follows.

Time (minutes)	P (cm)
0.5	1.42
1.0	5.0
2	10.99
3	14.32
4	16.84
5	18.7
9.75	23.6
17	27.89
27.17	28.17
57.25	30.66

Assume that the average methane concentration during the test is 17.7% and carbon dioxide is 3%. The pumping rate is 37.83 l/s. The vacuum at the well head is 76.2 cm. Determine the gas transmissivity using curve matching techniques for the following conditions. Perform the curve matching for both small and large vacuums.

a. Fully confined flow

b. Leaking confined flow with no release from storage.

Figure 14.19

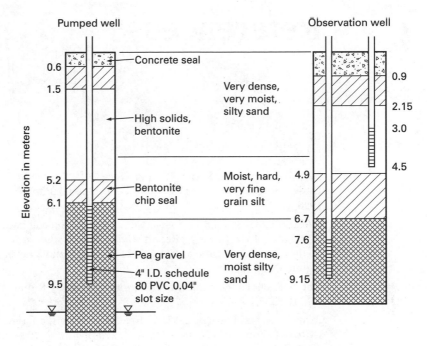

References

Aas, G., Lacasse, S., Lunne, T., and Hoeg, K., "Use of In Situ Tests for Foundation Design on Clay," *Proc. In Situ '86*, Geot. Special Publ. No. 6, S. P. Clemence, Ed., 1986, pp. 1–30.

AASHTO-AGA-ARTBA Joint Committee, "Subcommittee on New Highway Materials," Task Force 25 Report, *Guide Specifications and Test Procedures for Geotextiles*, American Association of State Highway and Transportation Officials, Washington, D. C., 1990.

Abt, S. R. et al., "Development of Riprap Criteria by Riprap Testing in Flumes: Phase 1," NUREG/CR-4651, 1987.

Acar, Y. B., and Olivieri, I., "Pore Fluid Effects on the Fabric and Hydraulic Conductivity of Laboratory-Compacted Clay," *Transportation Research Record 1219, Geotechnical Engineering 1989*, National Research Council, Washington, D. C., 1989, pp. 144–59.

Acar, Y. B., and Haider, L., "Transport of Low Concentration Contaminants in Saturated Earthen Barriers," *J. of Geot. Eng. Div.*, **116**(7), pp. 1031–52, 1990.

Acar, Y. B., Hamidon, A., Field, S., and Scott, L., "The Effect of Organic Fluids on Hydraulic Conductivity of Compacted Kaolinite," in *Hydraulic Barriers in Soil and Rock*, ASTM STP 874, A. I. Johnson, R. K. Frobel, N. J. Cavalli, and C. B. Pettersson, Eds., ASTM, Philadelphia, 1985, pp. 171–87.

Acar, Y. B., Seals, R. K., and Puppla, J., "Engineering and Compaction Characteristics of Boiler Slag," in *Geotechnics of Waste Fills — Theory and Practice*, ASTM STP 1070, A. Landva and G. D. Knowles, Eds., ASTM, Philadelphia, 1990, pp. 123–41.

Ajaz, A., and Parry, R. H. G., "Stress Strain Behavior of Two Compacted Clays in Tension and Compression," *Geotechnique*, **25**(3), 1975.

Albertson, M. I., and Simons, D. B., "Fluid Mechanics," Sec. 7 in *Handbook of Applied Hydrology*, V. T. Choe, Ed., McGraw-Hill, New York, 1964.

Alther, G. R., "The Methylene Blue Test for Bentonite Liner Quality Control," *Geot. Testing J.*, **6**(3), 1983.

Alther, G., Evans, J. C., Fang, H., and Witmer, K., "Influence of Inorganic Permeants upon the Permeability of Bentonite," in *Hydraulic Barriers in Soil and Rock*, ASTM STP 874, A. I. Johnson, R. K. Frobel, N. J. Cavalli, and C. B. Pettersson, Eds., Philadelphia, 1985, pp. 64–74.

Al-Yousfi, A. B., "Modelling of Leachate and Gas Production and Composition at Sanitary Landfills," Ph.D. Thesis, University of Pittsburgh School of Engineering, 1992.

American Concrete Pipe Association, *Concrete Pipe Design Manual*, Vienna, Va, 1980.

Andersland, O. B., and Mathew, P. J., "Consolidation of High Ash Papermill Sludge," *J. of the Soil Mechanics and Foundation Div.*, *ASCE*, **99**(SM5), May 1973, pp. 365–74.

Anderson, D. C., "Does Landfill Leachate Make Clay Liners More Permeable?" *Civil Eng., ASCE*, Sept. 1982, pp. 66–68.

Anderson, M. P., "Using Models to Simulate the Movement of Contaminants Through Groundwater Flow Systems," *CRC Critical Reviews in Environmental Control*, **9**, 1979, pp. 97–156.

Anderson, D. C., Crawley, W., and Zabcik, J. D., "Effects of Various Liquids on Clay Soil: Bentonite Slurry Mixtures," in *Hydraulic Barriers in Soil and Rock*, ASTM STP 874, A. I. Johnson, R. K. Frobel, N. J. Cavalli, and C. B. Pettersson, Eds., 1985, pp. 93–101.

API (American Petroleum Institute), *Detection of Hydrocarbons in Groundwater by Analysis of Shallow Soil Gas/Vapor*, API, Washington, D. C., 1985.

Applied Technology Council (ATC), "Improved Seismic Design Criteria for California Bridges," Report No. ATC-32, ATC, Redwood City, Calif., 1996.

ASCE and WPCF, "Gravity Sanitary Sewer Design and Construction," *ASCE Manuals and Reports on Engineering Practice*, No. 60, ASCE, New York, 1982.

ASCE, *Subsurface Investigation for Design of Foundations of Buildings*, ASCE Manual and Report on Eng. Pract., No. 56, 1976.

Aziz, M. A., "Surface Disposal of Refuse: Geotechnical Considerations, Technologies and Environmental Impacts," *Int. Symp. on Env. Geot.*, V. 1, H. Y. Fang, Ed., 1986, pp. 81–90.

Bagulein, F. J., Bustsmantem, M. G., and Frank, R., "The Pressuremeter for Foundations: French Experience," *Proc. In Situ '86*, Geot. Special Publ. No. 6., S. P. Clemence, Ed., 1986, pp. 321–46.

Baguelin, F. J., Jezequel, J. F., and Shields, D. H., *The Pressuremeter and Foundation Engineering*, Trans. Tech. Publications, 1978.

Bailey, S. W., "Summary of Recommendations of AIPEA Nomenclature Committee," *Clays and Clay Mineral*, **28**(1), 1980, pp. 73–78.

Baker, W. H., "Highway 71–Springdale, Arkansas, Dynamic Deep Compaction–Sanitary Landfill," Hayward Baker Company Report to Arkansas State Highway and Transportation Department, April 1982.

Baligh, M. M., and Levadoux, J. N., "Consolidation after Undrained Piezocone Penetration. II: Interpretation," *J. of Geot. Eng. Div., ASCE*, **112**(7), 1980.

Barlaz, M. A., and Ham, R. K., "Leachate and Gas Generation," Chapter 6 in *Geotechnical Practice for Waste Disposal*, D. Daniel, Ed., Chapman and Hall, New York, 1993.

Barlaz, M. A., Ham, R. K., and Schaefer, D. M., "Methane Production from Municipal Refuse: A Review of Enhancement Techniques and Microbial Dynamics," *Critical Review in Envir. Control*, **19**(6), 1990.

Barone, F. S., Rowe, R. K., and Quigley, R. M., "A Laboratory Estimation of Diffusion and Adsorption Coefficients for Several Volatile Organics in a Natural Clayey Soil," *J. of Contaminant Hydrogeology*, **10**, 1992, pp. 225–50.

Behnke, J. J., "Clogging in Surface Spreading Operations for Artificial Groundwater Recharge," *Water Resources Research*, **5**(4), 1969.

Belfiore, F., Manassero, M., and Viola, C., "Geotechnical Analysis of Some Industrial Sludges," *Geotechnics of Waste Fills — Theory and Practice*, ASTM STP 1070, A. Landva and G. D. Knowles, Eds., ASTM, Philadelphia, 1990, pp. 317–30.

Benson, C. H., Zhai, H., and Wang, X., "Estimating Hydraulic Conductivity of Clay Liners," *J. of Geot. Eng. Div., ASCE*, **120**(2), 1994.

Berend, J. E., "An Analytical Approach to the Clogging Effect of Suspended Matter," *Bull. Int. Assoc. of Scientific Hydrology*, **12**(2), 1967.

Bergstrom, W. R., Sweatman, M. B., and Dodt, M. E., "Slurry Trench Construction—Collier Road Landfill," *Geot. Practice for Waste Disposal '87*, Geot. Special Publ. No. 13, R. D. Woods, Ed., ASCE, Ann Arbor, June 1987, pp. 260–74.

Bjerrum, L., "Embankments on Soft Grounds," *Proc. of the Specialty Conf., Performance of Earth and Earth-Supported Structures*, Vol. 2, Purdue University, June 1972, pp. 1–45.

Bonaparte, R., Giroud, J. P., and Gross, B. A., "Rates of Leakage through Landfill Liners," *Proc. Geosynthetics '89*, Vol. 1, San Diego, 1989, pp. 18–29.

Boutwell, G. P., "The STEI Two-Stage Borehole Field Permeability Test," presented to "Containment Liner Technology and Subtitle D," Seminar Sponsored by Geotechnical Committee, ASCE, Houston Branch, March 1992.

Bowders Jr., J. J., "Discussion: Termination Criteria for Clay Permeability Testing," *J. of the Geot. Eng. Div.*, ASCE, **114**(8), Aug. 1988, pp. 947–49.

Bowders Jr., J. J., and Daniel, D. E., "Hydraulic Conductivity of Compacted Clay to Dilute Organic Chemicals," *J. of Geot. Eng. Div.*, ASCE, **113**(12), Dec. 1987, pp. 1432–48.

Bowders, J. J., Usmen, M. A., and Gidley, R. A., "Stabilized Fly Ash for Use as Low-Permeability Barriers," *Geot. Practice for Waste Disposal '87*, Geot. Special Publ. No. 13, R. D. Woods, Ed., ASCE, Ann Arbor, June 1987, pp. 320–33.

Boyd, M. B., Saucier, J. W., Keeley, R. O., Montgomery, R. D., Brown, R. D., Mathis, D. B., and Guice, C. J., "Disposal of Dredge Spoil," *Dredge Material Research Program Technical Report* H-72-8, U.S. Army Engineer Waterways Experiment Station, Vicksburg, Miss., 1972.

Bredario, A. W., Martin, J. P., and Cheng, S. C., "Flexural Cracking of Compacted Clay in Landfill Covers," *Geoenvironment 2000*, Vol. 2, New Orleans, Geot. Special Publ. No. 46, 1995.

Briaud, J. L., *Pressuremeter Testing*, Balkema, Rotterdam, 1992.

Broderick, G. P., and Daniel, D. E., "Stabilizing Compacted Clay against Chemical Attack," *J. of Geot. Eng. Div.*, ASCE, **116**(GT10), Oct. 1990, pp. 1549–60.

Bromwell, L. G., "Properties, Behavior and Treatment of Waste Fills," ASCE, Met. Section, Seminar—Improving Poor Soil Conditions, Oct. 1978, New York.

Bromwell, L. G., and Oxford, T. P., "Waste Clay Dewatering and Disposal," *Proc. Geot. Pract. for Disp. of Solid Waste Mat.*, ASCE, Ann Arbor, 1977, pp. 541–48.

Brooker, E. W., and Ireland, H. O., "Earth Pressures at Rest Related to Stress History," *Canadian Geot. J.*, **2**(1), 1965, pp. 1–15.

Brooks, R. H., and Corey, A. T., "Hydraulic Properties of Porous Media," Hydrology Paper No. 3, Colorado State University, 1964.

Brown, K. W., and Anderson, D. C., "Effect of Organic Solvents on the Permeability of Clay Soils," EPA/600/S2-83/061, Cincinnati, April 1983.

Brown, K. W., Thomas, J. C., Lytton, R. L., Jayawickrama, P., and Bahrt, S. C., "Quantification of Leak Rates through Holes in Landfill Liners," EPA/600/2-87/062, Cincinnati, 1987.

Byrne, R. J., Kendall, J., and Brown, S., "Cause and Mechanism of Failure: Kettleman

Hills Landfill B-19, Unit IA," *Proc. ASCE Spec. Conf. on Perf. and Stability of Slopes and Embankments — 11*, Vol. 2, ASCE, New York, 1992, pp. 1188–215.

Cancelli, A., and Cazzuffi, D., "Permittivity of Geotextiles in Presence of Water and Pollutant Fluids," *Geosynthetic '87 Conference*, Vol. 2, New Orleans, 1987.

Carra, J. S., and Cossu, R., "Introduction," in *International Perspectives on Municipal Solid Wastes and Sanitary Landfilling*, J. S. Carra and R. Cossu, Eds., Academic Press, New York, 1990, pp. 1–14.

Carrier III, W. D., and Beckman, J. F., "Correlations between Index Test and the Properties of Remoulded Clays," *Geotech.* **34**(2), 1984, pp. 211–28.

Chae, Y. S., and Gurdziel, T. J., "New Jersey Fly Ash as Structural Fill," in *New Horizons in Construction Materials*, Vol. 1, H. Y. Fang, Ed., Envo Publishing, Bethlehem, Penn., 1976, pp. 1–13.

Charles, J. A., Burford, D., and Watts, K. S., "Field Studies of the Effectiveness of Dynamic Consolidation," *Proc. 10th ICSMFE*, Vol. 3, Stockholm, 1981, pp. 617–22.

Charlie, W. A., "Pulp and Papermill Solid Waste Disposal — A Review," *Proc. Geot. Pract. for Disp. of Solid Waste Mat.*, ASCE, Ann Arbor, June 1977, pp. 71–86.

Chen, C. Y., Elnaggar, H. A., and Bullen, A. G. R., "Degradation and the Relationship between Shear Strength and Various Index Properties of Coal Refuse," in *New Horizons in Construction Materials*, Vol. 1, Envo Publishing, Bethlehem, Penn., 1976, pp. 41–52.

Chen, H. W., and Yamamoto, L. O., "Discussion: Field Permeability Test for Earthen Liners," *J. of Geot. Eng. Div., ASCE*, **115**(10), 1989.

Cheng, S. C., Larralde, J. L., and Martin, J. P., "Hydraulic Conductivity of Compacted Clay Soils under Distortion or Elongation Conditions," *ASTM STP 1142*, ASTM, Philadelphia, 1993.

Chow, T. V., *Handbook of Applied Hydrology*, McGraw-Hill, New York, 1964.

Clapp, T. L., Kosson, D. S., and Ahlert, R. C., "Leaching Characteristics of Residual Ashes from Incineration of Municipal Solid Waste," *Proc. 2nd Int. Conf. on New Frontiers for Hazardous Waste Management*, EPA/600/9-87/018F, 1987, pp. 19–27.

Clayton, C. R. I., Simons, N. E., and Matthews, M. C., *Site Investigation*, Halsted Press, New York, 1982.

COE (U.S. Corps of Engineers), *Laboratory Soils Testing*, EM 1110-2-1906, Washington, D. C., 1970.

COE (U.S. Corps of Engineers), "Plastic Filter — Cloth," Civil Works Construction Guide Presentation No. CW-02215, Chief of Engineers, Washington, D. C., 1977.

COE (U.S. Corps of Engineers), *Geophysical Exploration*, EM 1110-1-1802, Washington, D. C., 1979.

COE (U.S. Corps of Engineers), *Hydraulic Design of Flood Control Channels*, EM 1110-2-1601, Washington, D. C., 1991.

Collins, R. J., "Highway Construction of Incinerator Residue," *Proc. Geot. Pract. for Disp. of Solid Waste Mat.*, ASCE, Ann Arbor, June 1977, pp. 246–66.

Collins, R. J., and Ciesielski, S. K., "Highway Construction Use of Waste and By-Products," in *Utilization of Waste Materials in Civil Engineering Construction*, H. I. Inyang and K. L. Bergeson, Eds., ASCE, New York, 1992, pp. 140–52.

Cooper and Clark Consulting Engineers, "Stability Evaluation of Sunnyvale Sanitary Landfill," Report to the City of Sunnyvale, Calif., April 1982.

Cossu, R., "Sanitary Landfilling in Japan," in *International Perspectives on Municipal Solid Wastes and Sanitary Landfilling*, J. S. Carra and R. Cossu, Eds., Academic Press, New York, 1990, pp. 110–38.

Cunnigham, J. A., Lukas, R. G., and Anderson, T. C., "Impoundment of Fly Ash and Slag—A Case Study," *Proc. Geot. Pract. for Disp. of Solid Waste Mat.*, ASCE, Ann Arbor, June 1977, pp. 227–45.

Daniel, D. E., "In-Situ Hydraulic Conductivity Test for Compacted Clay," *J. of Geot. Eng. Div.*, *ASCE*, **115**(9), 1989.

Daniel, D. E., "Geosynthetic Clay Liners (GCLs) in Landfill Covers," *Thirty-First Annual Solid Waste Exposition of the Solid Waste Association of North America*, San Jose, Calif., 1993.

Daniel, D. E., "State-of-the-Art: Laboratory Hydraulic Conductivity Tests for Saturated Soils," in *Hydraulic Conductivity and Waste Contaminant Transport in Soil*, ASTM STP 1142, D. E. Daniel and S. J. Trautwein, Eds., ASTM, Philadelphia, 1994, pp. 30–54.

Daniel, D. E., *Geotechnical Practice for Waste Disposal*, Chapman and Hall, New York, 1993.

Daniel, D. E., and Liljestrand, H. M., "Effects of Landfill Leachates on Natural Liner Systems," Report GR83-6, Geotechnical Engineering Center, University of Texas, Austin, Tex., 1984.

Daniel, D. E., and Wu, Y. K., "Compacted Clay Liners and Covers for Arid Sites," *J. of Geot. Eng. Div.*, *ASCE*, **119**(2), 1993.

D'Appolonia, D. J., "Soil–Bentonite Slurry Trench Cutoffs," *J. of Geot. Eng. Div.*, *ASCE*, **106**(GT4), April 1980, pp. 399–417.

Darilek, G., Menzel, R., and Johnson, A., "Minimizing Geomembrane Liner Damage While Emplacing Protective Soil," *Geosynthetic '95*, Vol. 2, IFA I, Nashville, Tenn., 1995.

Davidson, R. R., Levallois, J., and Graybeal, K., "Seepage Cutoff for Mud Mountain Dam," STP 1129, ASTM, Philadelphia, 1992.

Day, S. R., "The Compatibility of Slurry Cutoff Wall Materials with Contaminated Groundwater," in *Hydraulic Conductivity and Waste Contaminate Transport in Soil*, ASTM STP 1142, D. E. Daniel and S. J. Trautwein, Eds., ASTM, Philadelphia, 1993.

Day, S. R., and Daniel, D. E., "Hydraulic Conductivity of Two Prototype Clay Liners," *J. of Geot. Eng. Div.*, *ASCE*, **111**(8), Aug. 1985, pp. 957–70.

Day, S. R., and Ryan, R. C., "Containment, Stabilization and Treatment Using Insitu Soil Mixing," in *Geotechnical Construction Applications for Site Remediation*, ASCE, North Jersey Section, June 4, 1994.

Del Greco, O., and Oggeri, C., "Shear Resistance Tests on Solid Municipal Wastes," *1st Int. Congress on Eng. Geot.*, Edmonton, Alberta, Canada, 1994, pp. 643–51.

DiGioia, A. M., Brendel, G. F., Glogowski, P. E., and Golden, D. M., "East Street Valley Expressway—A Structural Fill," *Proc. 9th Int. Ash Use Symp.*, Vol. 3, GS-7162, American Coal Ash Association, Orlando, 1991, Paper No. 63.

DiGioia, A. M., Meyers, J. F., and Niece, J. E., "Design and Construction of Bituminous Fly Ash Disposal Sites," *Proc. Geot. Pract. for Disp. of Solid Waste Mat.*, ASCE, Ann Arbor, June 1977, pp. 267–84.

Dobry, R., Oweis, I. S., and Urzua, A., "Simplified Procedures for Estimating the Fundamental Period of a Soil Profile," *Bull. Seis. Soc. Am.*, **66**, 1976.

Dodt, M. E., Sweatman, M. B., and Bergstrom, W. R., "Field Measurement of Landfill Surface Settlement," *Geot. Practice for Waste Disp. '87*, Geot. Special Publ. No. 13, R. D. Woods, Ed., ASCE, Ann Arbor, June 1987, pp. 406–17.

Driscoll, D. G., *Groundwater and Wells*, Johnson Division, St. Paul, Minn., 1986.

Duncan, J. M., and Buchignani, A. L., *An Engineering Manual for Slope Stability Studies*, Dept. of Civil Eng., Univ. of Calif., Berkeley, March 1975.

Dunn, R. J., and Mitchell, J. K., "Fluid Conductivity Testing of Fine-Grained Soils," *J. of Geot. Eng. Div.*, ASCE, **110**(10), Nov. 1984, pp. 1648–65.

Edil, T. B., Berthous, P. M., and Vesperman, K. D., "Fly Ash as a Potential Waste Liner," *Geot. Practice for Waste Disp. '87*, Geot. Special Publ. No. 13, R. D. Woods, Ed., ASCE, Ann Arbor, June 1987, pp. 447–61.

Edil, T. B., and Erickson, A. E., "Procedure and Equipment Factors Affecting Permeability Testing of a Bentonite-Sand Liner Material," in *Hydraulic Barriers in Soil and Rock*, ASTM STP 874, A. I. Johnson, R. K. Frobel, N. J. Cavalli, and C. B. Pettersson, Eds., ASTM, Philadelphia, 1985, pp. 155–70.

Edil, T. B., Park, J. K., and Heim, D. P., "Large-Size Test for Transport of Organics through Clay Liners," in *Hydraulic Conductivity and Waste Contaminant Transport in Soil*, ASTM STP 1142, D. E. Daniel and S. J. Trautwein, Eds., ASTM, Philadelphia, 1994, pp. 353–74.

Eliassen, R., "Why You Should Avoid Housing Construction on Refuse Landfills," *Engineering News Record*, May 1, 1947, pp. 90–94.

Engemon, W. O., and Hensley, P. J., "ECPT Investigation of Slurry Trench Cutoff Wall," *Proc. In-Situ '86*, Geot. Special Publ. No. 6, S. P. Clemence, Ed., 1986, pp. 514–28.

Ericksen, L. G., "Clogging Potential of Geotextiles Due to Municipal Landfill Leachate," Unpublished Masters Thesis, The Cooper Union School of Engineering, New York, 1991.

Ericson, W. A., and Carrier III, W. D., "Discussion" in *Limitations of Conventional Analysis of Consolidation Settlement*, by Duncan, J. M., *J. of Geot. Eng. Div.*, ASCE, **121**(6), June 1995, pp. 513–14.

Fang, H. Y., "Physical Properties of Compacted Disposal Materials," Unpublished report, 1983.

Fang, H. Y., and Slutter, R. G., "Stress-Strain Characteristic of Compacted Waste Disposal Materials," in *New Horizons in Construction Materials*, Vol. 1, H. Y. Fang, Ed., Envo Publishing, Bethlehem, Penn., 1976, pp. 127–37.

Farmer, W., Yang, M., Letey, J., and Spencer, W. F., "Hexachlorobenzene: Its Vapor Pressure and Vapour Diffusion in Soil," *Soil Science Society J. of Amer.*, **44**, 1980.

Fenn, D. G., Hanley, K. J., and DeGeare, T. U., *Use of the Water Balance for Predicting Leachate Concentration from Solid Waste Disposal Sites*, EPA-530/SW-168, U.S. Environmental Protection Agency, Cincinnati, 1975.

Fernandez, F., and Quigley, R. M., "Hydraulic Conductivity of Natural Clays Permeated with Simple Liquid Hydrocarbons," *Canadian Geot. J.*, **22**(2), May 1985, pp. 205–14.

Fernandez, F., and Quigley, R. M., "Viscosity and Dielectric Constant Controls on the Hydraulic Conductivity of Clayey Soils Permeated with Water-Soluble Organics," *Canadian Geot. J.*, **25**, 1988, pp. 582–89.

Fernandez, F., and Quigley, R. M., "Controlling the Destructive Effects of Clay-Organic Liquids Interaction by Application of Effective Stresses," *Canadian Geot. J.*, **28**(2), 1991, pp. 388–98.

Fetter, C. W., *Applied Hydrogeology*, 2nd Ed., Merril, Columbus, Oh., 1988.

Foster, C. R., Chapter 12, in *Field Problems: Compaction, Foundation Engineering*, G. A. Leonards, Ed. 1962, McGraw-Hill, New York, 1962.

Franklin Associates, Ltd., "Characterization of Products Containing Lead and Cadmium in Municipal Solid Waste in the United States, 1970–2000," EPA/530-SW-89-015A, NTIS PB89-151039, U.S. Environmental Protection Agency, 1989.

Franklin, A. G., and Chang, F. K., "Permanent Displacements of Earth Embankment by Newmark Sliding Block Analysis," *Earthquake Resistance of Earth and Rockfill Dams*, Rpt. 5, Misc. Paper No. S-71-17, U.S. Army Engr. Waterways Experiment Station, Vicksburg, Miss., 1977.

Freeze, R. A., and Cherry, J. A., *Groundwater*, Prentice-Hall, Englewood Cliffs, N. J., 1979.

Fungaroli, A. A., and Steiner, R. L., "Investigations of Sanitary Landfill Behavior," *Final Report*, EPA 600/2-79-053, V. 1, 1979, p. 314.

Gas Generation Institute, *Landfill Methane Recovery*, Part 1, 1981.

Gatti, G., and Tripiciano, L., "Mechanical Behavior of Coal Fly Ashes," *Proc. of the 10th ICSMFE*, V. 2, Stockholm, 1981, pp. 317–22.

Gauffreau, P. E., "A Review of Pile Protection Methods in a Corrosive Environment," *Int. Symp. on Env. Geotech.*, Vol. 2, Envo Publishing, Bethlehem, Penn., 1987, p. 372.

GCA Corp, "Procedures for Modelling Flow through Clay Liners to Determine Required Liner Thickness," EPA/530-SW-84-001, 1984.

Geological Society of London, "Tropical Residual Soils," *Quarterly J. Engineering Geology, Engineering Working Group Report*, **23**(1), 1990.

Gera, F., Mancini, O., Mecchia, M., Sarrocco, S., and Schneider, A., "Utilization of Ash and Gypsum by Coal Burning Power Plants," in *Waste Materials in Construction, Proc. Int. Conf. on Env. Implications of Const. with Waste Materials*, Elsevier Science Pub., Netherlands, 1991, pp. 433–40.

Gibbs, H. J., and Holz, W. G., "Research in Determining the Density of Sands by Spoon Penetration Testing," *Proc. 4th Int. Conf. Soil Mech. and Found. Eng.*, Vol. 1, London, 1957, pp. 35–39.

Gibson, R. E., England, G. L., and Hussey, M. L., "The Theory of One-Dimensional Consolidation of Saturated Clays. I. Finite Nonlinear Consolidation of Thin Homogeneous Layers," *Geotechnique*, **17**(3), Sept. 1967, pp. 261–73.

Gibson, R. E., Schiffman, R. L., and Cargill, K. W., "The Theory of One-Dimensional Consolidation of Saturated Clays. II. Finite Nonlinear Consolidation of Thick Homogeneous Layers," *Canadian Geot. J.*, **18**(2), May 1981, pp. 280–93.

Gilbert, R. B., Fernandez, F., and Horsfield, D. W., "Shear Strength of Geosynthetic Clay Liner," *GE, ASCE*, **122**(4), 1996.

Giroud, J. P., "Filter Criteria for Geotextiles," *Proc. 2nd Int. Conf. Geotextiles*, Vol. 1, Las Vegas, 1982.

Giroud, J. P., Badu-Tweneboah, K., and Bonaparte, R., "Rate of Leakage through a Composite Liner Due to Geomembrane Defect," Preprint, accepted for publication in *Geotextiles and Geomembranes*, 1990.

Gleason, M. H., Daniel, D. E., and Eykholt, G. R., "Calcium and Sodium Bentonite for Hydraulic Confinement Applications," *J. of Geot. Eng. Div., ASCE,* **123**(5), 1997.

Golden, D., and DiGioia, A., "Fly Ash for Highway Construction and Site Development," *Proc.; Shanghai 1991 Ash Utilization Conf.,* EPRI GS-7388, V. 3, Sept. 1991, Paper No. 95.

Goldman, L. J., "Design, Construction, and Evaluation of Clay Liners for Waste Management Facilities," EPA/530/SW-86/007F, 1988.

Gray, D. H., "Containment Strategies for Landfilled Waste," *Geoenvironment 2000,* Geot. Special Publ. No. 46, Y. B. Acar and D. E. Daniel, Eds., ASCE, New Orleans, Feb. 1995.

Green, W. H., and Ampt, G. A., (Quoted in EPA, 1988) "Studies in Soil Physics I: The Flow of Air and Water Through Soils," *J. Ag. Sci.,* **4,** 1911, pp. 1–24.

Griffith, M., Wilson, J., and Whittaker, A., *Earthquake Engineering Short Course '94,* University of Melbourne, 1994.

Grim, R. E., *Clay Mineralogy,* 2nd Ed., McGraw-Hill, New York, 1968.

Hadge, W., and Barvenik, M. J., "Upgrading Soil Bentonite Cutoff Wall Technology for Containment of Hazardous Waste," Presented at Geotechnical Aspects of Waste Management, Met. Sec., ASCE, Dec. 1985.

Hagerty, D. J., Ullrich, C. R., and Thacker, B. K., "Engineering Properties of FGD Sludges," *Geot. Pract. for Disp. of Solid Waste Mat.,* ASCE, Ann Arbor, June 1977, pp. 23–40.

Haliburton, T. A., "Development of Alternatives for Dewatering Dredged Materials," *Proc. Geot. Pract. for Disp. of Solid Waste Mat.,* ASCE, Ann Arbor, June 1977, pp. 615–31.

Haliburton, T. A., and Wood, P. D., "Evaluation of U.S. Army Corps of Engineers Gradient Ratio for Geotextile Performance," *Proc. 2nd Intl. Conf. on Geotextiles,* Las Vegas, 1982.

Ham, K. R. "Decomposition and Leachate and Landfill Gas Generation," Solid Waste Association of North America, Preprint, SWANA, Silver Spring, Md, 1993.

Ham, K. R., and Bookter, T. J., "Decomposition of Solid Waste in Test Lysimeters," *J. Env. Eng. Div. ASCE,* **108**(EE6), 1982.

Hansbo, S., "Consolidation of Clay by Band-shaped Prefabricated Drains," *Ground Engineering,* **12**(5), July 1979, pp. 16–25.

Hantush, M. S., "Modification of the Theory of Leaky Aquifers," *J. of Geophys. Research,* **65**(11), 1960, pp. 3713–25.

Harborth, P., and Hannert, H. H., "Bacterial Plugging Mechanism in Drain Lines and Filter Structures," Institute for Microbiology, Technical University, Braunschweig, West Germany, 1987.

Harrop-Williams, K., "Acceptance Sampling for Clay Liner Design," *Geot. Pract. for Waste Disp. '87,* Geot. Special Publ. No. 13, R. D. Woods, Ed., ASCE, Ann Arbor, June 1987, pp. 515–21.

Hart, H., and Schuetz, R. D., *Organic Chemistry: A Short Course,* 4th Ed., Houghton Mifflin, Boston, 1972.

Haxo, Jr., H. E., "Determining the Transport through Geomembranes of Various Permeants in Different Applications," ASTM, STP 1081, ASTM, Philadelphia, 1990.

Haxo, Jr., H. E., and Waller, M. J., "Laboratory Testing of Geosynthetics and Plastic Pipe for Double-Liner Systems," *Geosynthetics '87 Conference*, New Orleans, La., 1987.

Hazen, A., "Physical Properties of Sands and Gravels with Reference to Their Use in Filteration," Rept. Mass. State Board of Health, p. 539.

Heiland, C. A., *Geophysical Exploration*, Hafner, New York, 1963.

Hermanns, R., Meseck, H., and Reuter, E., "Sind Dichtwandmassen Beständig Gegenüber den Sickerwässern aus Altlasten?" Mitteilung des Instituts für Grundbau und Bodenmechanik, TU Braunschweig, Heft Nr. 23, Meseck H., Ed., Dichtwände und Dichtsohlen, Braunschweig, Federal Republic Germany, June 1987, pp. 113–54.

Hilf, J. W., "Compacted Fill," in *Foundation Engineering Handbook*, H. F. Winterkorn and H. Y. Fang, Eds. Van Nostrand Reinhold, New York, 1975.

Hillenbrand, S., "One Billion People Can't be Wrong," *Waste Age*, Oct. 1986.

Hinkle, R. D., "Landfill Site Reclamation for Commercial Use as Container Storage Facility," in *Geotechnics of Waste Fills — Theory and Practice*, ASTM STP 1070, A. Landva and G. D. Knowles, Eds., ASTM, Philadelphia, 1990, pp. 331–44.

Holtz, R. D., and Wagner, O., "Preloading by Vacuum-Current Prospects," *Transportation Research Record 548*, 1975, pp. 26–29.

Holubec, I., "Geotechnical Aspects of Coal Waste Embankments," *Canadian Geot. J.*, **13**(1), Feb. 1976, pp. 27–39.

Howard, A. K., "The USBR Equation for Predicting Flexible Pipe Deflections," Proc. of the Int. Conf. on Underground Plastic Pipe, ASCE, March 1981.

Hsuan, Y. G., Koerner, R. M., and Lord, A. E., "Stress-Cracking Resistance of High-Density Polyethylene Geomembranes," *J. of Geot. Eng. Div.*, ASCE, **119**(11), 1993.

Huang, W-H., and Lowell, C. W., "Bottom Ash as Embankment Material," in *Geotechnics of Waste Fills — Theory and Practice*, ASTM STP 1070, A. Landva and G. D. Knowles, Eds., ASTM, Philadelphia, 1990, pp. 71–85.

Hvorslev, M. J., *Subsurface Exploration and Sampling of Soils for Civil Engineering Purposes*, ASCE, Engineering Foundation, 1962.

Indraratna, B., "Problems Related to Disposal of Fly Ash and Its Utilization as a Structural Fill," in *Utilization of Waste Materials in Civil Engineering Construction*, H. I. Inyang and K. L. Bergeson, Eds., ASCE, New York, Sept. 1992, pp. 274–85.

Indraratna, B., Nutalaya, P., Koo, K. S., and Kuganenthira, N., "Engineering Behavior of a Low Carbon, Pozzolanic Fly Ash and Its Potential as a Construction Fill," *Canadian Geot. J.*, **28**, 1991, pp. 542–56.

Jacob, C. E., "On the Flow of Water in an Elastic Aquifer," *Trans. Amer. Geophys. Union*, **21**, 1940, pp. 574–686.

Jaros, D. C., "Closing Hazardous Waste Landfills," *Civil Engineering*, American Society of Civil Engineers, New York, April 1991.

Jedele, L. P., "Evaluation of Compacted Inert Paper Solids as a Cover Material," *Geot. Practice for Waste Disposal '87*, Geot. Special Publ. No. 13, R. D. Woods, Ed., ASCE, Ann Arbor, June 1987, pp. 562–95.

Jessberger, H. L., "Geotechnical Aspects of Landfill Design and Construction, Part 2: Material Parameters and Tests," *Proc. Instn. Viv. Engrs. Geotech. Eng.*, **109**, Apr. 1994, pp. 105–13.

Jessberger, H. L., "Waste Containment with Compacted Clay Liners," *Geoenvironment 2000*, Geot. Special Publ. No. 46, Y. B. Acar and D. E. Daniel, Eds., ASCE, New York, Feb. 1995, pp. 463–83.

Kavazanjian Jr., N., Matasovit, R., Bonaparte, G. R., and Schmertmazin, E., "Evaluation of MSW Properties for Seismic Analysis," *Geoenvironment 2000*, Geot. Special Publ. No. 46, Y. B. Acar and D. E. Daniel, Eds., ASCE, New Orleans, Feb. 1995, pp. 1126–41.

Keene, P., "Sanitary Landfill Treatment, Interstate Highway 84," *Proc. Geot. Pract. for Disp. of Solid Waste Mat.*, ASCE, Ann Arbor, June 1977, pp. 632–44.

Kelly, J. M., DiGioia Jr., A. M., and Glogowski, P. E., "Fly Ash Field Compaction Test Program," *Proc. 9th Int. Ash Use Symp.*, Vol. 3, GS-7162, American Coal Ash Association, Orlando, 1991, Paper No. 64.

Khera, R. P., "Calcium Bentonite, Cement, Slag, and Fly Ash as Slurry Wall Materials," *Geoenvironment 2000*, Geot. Special Publ. No. 46, Y. B. Acar and D. E. Daniel, Eds., ASCE, New Orleans, Feb. 1995, pp. 1237–49.

Khera, R. P., Kasturi, R. M. R., Oweis, I. S., and Alam, M. K., "Depth and Width Effect on Pullout Resistance of Woven Geotextiles in Sand," *Geosynthetics '97, IFAI*, **2**, 1997, pp. 851–62.

Khera, R. P., Oweis, I. S., and Kasturi, R. M. R., "Pullout Resistance of Geotextiles," *7th Int. Conf. on Solid Waste Management and Secondary Materials*, Philadelphia, December 1991, 4C.

Khera, R. P., and Thilliyar, M., "Slurry Wall Backfill Integrity and Desiccation," in *Physico-Chemical Aspects of Soil and Related Materials*, ASTM STP 1095, K. B. Hoddinott and R. O. Lamb, Eds., ASTM, Philadelphia, 1990, pp. 81–90.

Khera, R. P., and Tirumala, R. K., "Materials for Slurry Walls in Waste Chemicals," in *Slurry Walls: Design, Construction, and Quality Control*, ASTM STP 1129, D. B. Paul, R. R. Davidson, and N. J. Cavalli, Eds., ASTM, Philadelphia, 1992, pp. 172–80.

Kim, W. H., and Daniel, D. E., "Effects of Freezing on Hydraulic Conductivity of Compacted Clay," *J. of Geot. Eng. Div., ASCE*, **118**(7), 1992.

Klee, A. J., and Carruth, D., "Sample Weights in Solid Waste Composition," *J. of the Sanitary Eng. Div., ASCE*, Aug. 1970, **96**(SA4), pp. 945–53.

Kleppe, J. H., and Olson, R. E., "Desiccation Cracking of Soil Barriers," in *Hydraulic Barriers in Soil and Rock*, ASTM STP 874, A. I. Johnson, R. K. Frobel, N. J. Cavalli, and C. B. Pettersson, Eds., ASTM, Philadelphia, 1985.

Knox, D. P., and Najjar, R. A., "Solidification and stabilization of Lagoon Sludge and Dredged Spoils," *Proc. of the 1st Int. Congress on Env. Geot.*, W. D. Carrier III, Ed., Edmonton, Canada, July 1994, pp. 41–48.

Koerner, G. R., and Koerner, R. M., "Constructing a Sound Leachate Collection and Removal System." *Solid Waste Technologies*, **VIII**(6), 1994.

Koerner, G. R., Koerner, R. M., and Martin, J. P., "Field Performance of Leachate Collection Systems and Design Implications," Preprint, Solid Waste Association of North America, SWANA, Silver Spring, Md, 1993.

Koerner, R. M., *Designing with Geosynthetics*, Prentice-Hall, Englewood, N. J., 1990.

Koerner, R. M., "Collection and Removal Systems," in *Geotechnical Practice for Waste Disposal*, D. Daniel, Ed., Chapman and Hall, New York, 1993.

Koerner, R. M., Koerner, G., and Hwu, B. L., "Three-Dimensional, Axi-Symmetric Geomembrane Tension Test," ASTM STP 1081, ASTM, Philadelphia, 1990.

Koerner, R. M., and Wilson-Fahmy, R., *Geosynthetics for Geoenvironmental Applications*, Vol. 1, ASCE, New York, 1995.

Kosson, D. S., van der Sloot, H., Holmes, T., and Wiles, C., "Leaching Properties of Untreated and Treated Residues Tested in the USEPA Program for Evaluation of Treatment and Utilization Technologies for Municipal Waste Combustor Residues," in *Waste Materials in Construction*, J. J. J. R. Goumans, H. A. van der Sloot, and Th. G. Aalbers, Eds., Elsevier, Netherlands, 1991, pp. 119–34.

Koutsoftas, D. C., and Kiefer, M. L., "Improvement of Mine Spoil in Southern Illinois," in *Geotechnics of Waste Fills — Theory and Practice*, ASTM STP 1070, A. Landva and G. D. Knowles, Eds., ASTM, Philadelphia, 1990, pp. 153–67.

Krizek, R. J., and Salem, A. M., "Time-Dependent Development of Strength in Dredging," *J. of the Geot. Eng. Div.*, ASCE, **103**(GT3), March 1977, pp. 169–84.

Krizek, R. J., and Salem, A. M., "Field Performance of a Dredge Disposal Area," *Proc. Geot. Pract. for Disp. of Solid Waste Mat.*, ASCE, Ann Arbor, June 1977, pp. 358–83.

Krizek, R. J., Chu, S. C., and Atmatzidis, D. K., "Geotechnical Properties and Landfill Disposal of FGD Sludge," *Geot. Practice for Waste Disposal '87*, Geot. Special Publ. No. 13, R. D. Woods, Ed., ASCE, Ann Arbor, June 1987, pp. 625–39.

Krizek, R. J., Giger, M. W., and Legatski, L. K., "Engineering Properties of Sulfur Dioxide Scrubber Sludge with Fly Ash," in *New Horizons in Construction Materials*, V. 1, H. Y. Fang., Ed., Envo Publishing, Bethlehem, Penn., 1976, pp. 67–81.

Kutter, B. L., Abghari, A., and Cheney, J. A., "Strength Parameters for Bearing Capacity of Sand," *J. of Geot. Eng. Div.*, ASCE, **114**(4), Apr. 1988, pp. 491–98.

Lacasse, S. M., Lambe, T. W., Maar, W. A., and Neff, T. L., "Void Ratio of Dredged Material," *Proc. Geot. Pract. for Disp. of Solid Waste Mat.*, ASCE, Ann Arbor, June 1977, pp. 153–68.

LaGatta, D. M., Boardman, B. T., Cooley, B. H., and Daniel, D. E., "Geosynthetic Clay Liners Subjected to Differential Settlement," *J. of Geot. and Geoenviron. Eng.*, ASCE, **123**(5), 1997.

Lambe, T. W., and Whitman, R. V., *Soil Mechanics*, Wiley, New York, 1969.

Landreth, R., "Landfill Containment and Cover Systems," EPA/600/A-92/200, U.S. Environmental Protection Agency, Cincinnati, 1992.

Landva, A., and Clark, J. I., "Geotechnics of Waste Fills," in *Geotechnics of Waste Fills — Theory and Practice*, ASTM STP 1070, A. Landva and G. D. Knowles, Eds., ASTM, Philadelphia, 1990, pp. 86–106.

Lane, E. W., "Design of Stable Channels," *Transactions*, ASCE, **120,** ASCE, New York, 1955.

Lee, I. K., White, W., and Ingles, O. G., *Geotechnical Engineering*, Pitman, Boston, 1983.

Legget, R. F., *Cities and Geology*, McGraw-Hill, New York, 1973.

Lentz, R. W., Horst, W. D., and Uppot, J. O., "The Permeability of Clay to Acidic and Caustic Permeants," in *Hydraulic Barriers in Soil and Rock*, ASTM, STP 874, A. I. Johnson, R. K. Frobel, N. J. Cavalli, and C. B. Pettersson, Eds., ASTM, Philadelphia, 1984, pp. 127–39.

Leonards, G. A., *Foundation Engineering*, McGraw-Hill, New York, 1962.

Leonards, G. A., and Bailey, B., "Pulverized Coal Ash as Structural Fill," *J. of the Geot. Eng. Div.*, ASCE, **108**(GT4), April 1982, pp. 517–32.

Leonards, G. A., Cutter, W. A., and Holtz, R. D., "Dynamic Compaction of Granular Soils," *J. of the Geot. Eng. Div.*, ASCE, **106**(GT1), Jan. 1980, pp. 35–44.

Leonards, G. A., and Narain, J., "Flexibility of Clay and Cracking of Earth Dams," *J. of Soil Mechanics and Foundation Div.*, **89**(SM2), 1963.

Leroueil, S., Le Bihan, J. P., and Bouchard, R., "Remarks on the Design of Clay Liners Used in Lagoons as Hydraulic Barriers," *Canadian Geot. J.*, 1992.

Long, J. H., Daly, J. J., and Gilbert, R. B., "Structural Integrity of Geosynthetic Liner and Cover Systems for Solid Waste Landfills," Rep. prepared for Illinois Office of Solid Waste Research, University of Illinois, Urbana, 1993, p. 111.

Long, R. P., and Demars, K. R., "Shallow Ocean Disposal of Contaminated Dredge Material," *Geot. Practice for Waste Disposal '87*, Geot. Special Publ. No. 13, R. D. Woods, Ed., ASCE, Ann Arbor, June 1987, pp. 655–67.

Long, J. H., Gilbert, R. B., and Daly, J. J., "Geosynthetic Loads in Landfill Slopes: Displacement Compatibility," *J. of Geot. Eng. Div. ASCE*, **120**(11), 1994.

Lu, James C. S., Eichenenberger, B., and Stearns, R. J., *Leachate from Municipal Landfills*, Noyes Publications, Park Ridge, N. J., 1985.

Lukas, R. G., "Densification of Loose Deposits by Pounding," *J. of the Geot. Eng. Div., ASCE*, **106**(GT4), April 1980, pp. 435–46.

Lukas, R. G., "Dynamic Compaction of Sanitary Landfills," *Geotechnical News*, **10**(3), 1992.

Lukas, R. G., and Seiler, N. H., "Settlement of Dynamically Compacted Deposits," *Settlement '94, Vert. and Horiz. Settlement of Found. and Embankments*, A. T. Yeung and G. Y. Felio, Eds., ASCE, College Station, June 1994, pp. 1590–601.

Lum, K. M., and Tay, J-K., "A Study on the Utilization of Incinerator Residue for Asphalt Concrete," in *Utilization of Waste Materials in Civil Engineering Construction*, H. I. Inyang and K. L. Bergeson, Eds., ASCE, New York, 1992, pp. 217–29.

Lumb, P., "Application of Statistics in Soil Mechanics," in *Soil Mechanics — New Horizons*, I. K. Lee, Ed., Newnes-Butterworths, London, 1974, pp. 44–111.

Lutton, R. J., Regan, G. L., and Jones, L. W., "Design and Construction of Covers for Solid Waste Landfills," EPA-600/2-79-165, U.S. Environmental Protection Agency, Cincinnati, 1979.

Mabes, D. L., Hardcastle, J. H., and Williams, R. E., "Physical Properties of Pb-Zn Mine-Process Waste," *Proc. Geot. Pract. for Disp. of Solid Waste Mat.*, ASCE, Ann Arbor, June 1977, pp. 103–17.

Madsen, F. T., and Mitchell, J. K., "Chemical Effects on Clay Hydraulic Conductivity and Their Determination," Open File Report, Environmental Institute for Waste Management Studies, University of Alabama, Tuscaloosa, 1987.

Makdisi, F. I., and Seed, H. B., "Simplified Procedure for Estimating Dam and Embankment Earthquake-Induced Deformations," *J. of Geot. Eng. Div., ASCE* **104**(7), 1978.

Manassero, M., "Hydraulic Conductivity Assessment of Slurry Wall Using Piezocone Test," *J. of Geot. Eng. Div., ASCE*, **120**(10), 1994.

Manassero, M., Fratalocchi, E., Pasqualini, E., Spanna, C., and Verga, F., "Containment with Vertical Cutoff Wall," *Proc. Geoenvironment 2000*, Vol. 2, ASCE, New Orleans, 1995.

Mantell, C. L., *Solid Wastes: Origin, Collection, Processing, and Disposal*, Wiley, New York, 1975.

Manz, O. E., and Manz, B. D., "Utilization of Fly Ash in Road Bed Stabilization: Some Examples of Western U.S. Experience," *Fly Ash and Coal Conversion Byproducts: Characterization, Utilization and Disposal I. Materials Research Soc.*, Vol. 43, 1985, pp. 129–44.

Massmann, J. W., "Application of Groundwater Models in Vapor Extraction System Design," *J. Env. Eng. Div.*, ASCE, **115**(1), 1989.

Matrecon Inc., *Lining of Waste Impoundment and Disposal Facilities*, EPA 530/SW-870C, 1980.

Maynard, T. R., "Incinerator Residue Disposal in Chicago," *Proc. Geot. Pract. for Disp. of Solid Waste Mat.*, ASCE, Ann Arbor, June 1977, pp. 773–92.

Mayne, P. W., Jones Jr., J. S., and Dumes, J. C., "Ground Response to Dynamic Compaction," *J. of the Geot. Eng. Div.*, ASCE, **110**(6), June 1984, pp. 757–74.

McBean, E. A., Rovers, F. A., and Farquher, G. J., *Solid Waste Landfill Engineering and Design*, Prentice-Hall, Englewood Cliffs, N. J., 1995.

McEnroe, B. M., "Drainage of Landfill Covers and Bottom Liners: Unsteady Case," *J. of Geot. Eng. Div.*, ASCE, **115**(6), 1989.

McEnroe, B. M., "Steady Drainage of Landfill Covers and Bottom Liners," *J. of Env. Eng. Div.*, ASCE, **115**(6), Dec. 1989, pp. 1114–22.

McLaren, R. J., and DiGioia, A. M., "The Typical Engineering Properties of Fly Ash," *Geot. Practice for Waste Disposal '87*, Geot. Special Publ. No. 13, R. D. Woods, Ed., ASCE, Ann Arbor, June 1987, pp. 683–97.

McNabb, A., "Mathematical Treatment of One-Dimensional Soil Consolidation," *Quarterly of Applied Mathematics*, **17**(4), 1960, pp. 337–47.

McVay, M., Townsend, F., and Bloomquist, D., "Quiescent Consolidation of Phosphatic Waste Clays," *J. of Geot. Eng. Div.*, ASCE, **112**(11), Nov. 1986, pp. 1033–52.

Meigh, A. C., "Cone Penetration Testing: Methods and Interpretation, Ground Engineering Report: In-Situ Testing," Construction Industry Research and Information Assoc. (CIRIA), Butterworths, London, 1987.

Ménard, L., (1984), company literature.

Ménard, L., and Broise, Y., "Theoretical and Practical Aspects of Dynamic Consolidation," *Geotechnique*, **15**(1), 1975, pp. 3–18.

Michalowski, R. L., "Limit Analysis of Weak Layers under Embankments," *Soils and Foundation*, **33**(1), 1993.

Millet, R. A., Perez, J., and Davidson, R. R., "USA Practice Slurry Wall Specification 10 Years Later," STP 1129, ASTM, Philadelphia, 1992.

Mitchell, J. K., *Fundamentals of Soil Behavior*, Wiley, New York, 1976.

Mitchell, J. K., *Fundamentals of Soil Behavior*, 2nd Ed., Wiley, New York, 1993.

Mitchell, J. K., Hooper, D. N., and Campanella, R. G., "Permeability of Compacted Clay," *J. of Soil Mechanics and Foundations Div.*, ASCE, **91**(S4), Part-1, July 1965, pp. 41–66.

Mitchell, J. K., Seed, R. B., and Seed, H. B., "Kettleman Hills Waste Landfill Slope Failure. I: Liner-System Properties," *J. of Geot. Ent.*, **116**(4), April 1990, pp. 647–68.

Mittal, H., and Morgenstern, N., "Parameters for the Design for Tailings Dams," *Canadian Geot. J.*, **13**, 1975, pp. 277–93.

Mlynarek, J., and Rollin, A. L., "Bacterial Clogging of Geotextiles — Overcoming Engineering Concerns," *Geosynthetics '95*, IFIA, 1995.

Mlynarek, J., Vemeersch, O. G., and DeBerardino, S., "Evaluation of Filtration Design Criterion for Nonwoven Heat-Bonded Geotextiles," *Geosynthetics '95*, IFIA, 1995.

Monahan, E. J., *Construction of Fills*, Wiley, New York, 1994.

Moore, C. A., "Landfill and Surface Impoundment Performance Evaluation," USEPA SW869, Cincinnati, 1983.

Moore, P. J., and Pedler, I. V., "Some Measurements of Compressibility of Sanitary Landfill Materials," *Speciality Session, Proc. 9th ICSME*, Tokyo, 1977, pp. 319–30.

Morgenstern, N. R., and Scott, J. D., "Geotechnics of Fine Tailings Management," *Geoenvironment 2000*, Geot. Special Tech. Publ. No. 46, Y. B. Acar and D. E. Daniel, Eds., ASCE, New Orleans, Feb. 1995, pp. 1663–83.

Morris, D. V., and Woods, C. E., "Settlement and Engineering Considerations in Landfill and Final Cover Design," *Geotechnics of Waste Fills — Theory and Practice*, ASTM STP 1070, A. Landva and G. D. Knowles, Eds., ASTM, Philadelphia, 1990.

Moulton, L. K., Rao, S. K., and Seals, R. K., "The Use of Coal-Associated Wastes in the Construction and Stabilization of Refuse Landfills," in *New Horizons in Construction Materials*, Vol. 1, H. Y. Fang, Ed., Envo Publishing, Bethlehem, Penn., 1976, pp. 53–65.

Murdock, A., and Zeman, A. J., "Physico Chemical Properties of Dredge Spoil," *J. of the Waterways, Harbors and Coastal Eng. Div., ASCE*, **101**(WW2), 1975, pp. 201–14.

Nakayama, F. S., and Bucks, D. A., "Water Quality in Drip/Trickle Irrigation: A Review," *Irrigation Science*, **12,** 1991.

NAVFAC DM 7.1, "Soil Mechanics," Chapter 6, Naval Facilities Engineering Command, Alexandria, Va., 1982.

NAVFAC DM 7.2, "Foundations and Earth Structures," Naval Facilities Engineering Command, Alexandria, Va., 1982.

NAVFAC DM 7.3, "Soil Dynamics, Deep Stabilization and Special Construction," Naval Facilities Engineering Command, Alexandria, Va., 1982.

Nelson, J. D., Abt, S. R., Volpe, R. L., Van Zyl, D., Hinkle, N. E., Staub, W. P., "Methodology for Evaluating Long-Term Stabilization of Uranium Mill Tailings Impoundments," NUREG/CR-4620, ORNL/TM-10067, U.S. Nuclear Regulatory Commission, Washington, D. C., 1986.

Nemark, N. M., "Effects of Earthquakes on Dams and Embankments," *Geotechnique*, **15**(2), 1965.

New York State Department of Environmental Conservation, GNYCR Port 360, NYDEC, Albany, N. Y., 1993.

Newman, F. B., McGee, J., and Burns, D., "Embankment over Fly Ash Pond at Portsmouth Power Station, "*Geot. Practice for Waste Disposal '87*, Geot. Special Publ. No. 13, R. D. Woods, Ed., ASCE, Ann Arbor, June 1987, pp. 713–27.

Noble, G., *Sanitary Landfill Design Handbook*, Technomic, Westport, Conn., 1976.

Noble, G., *Siting Landfills and Other LULUs*, Technomic Publishing, Lancaster, Penn., 1992.

Nußbaumer, M., "Beispiele für die Herstellung von Dichtwänden im Schlitzwandverfahren," *Mitteilung des Instituts für Grundbau und Bodenmechanik*, TU Braunschweig, Heft Nr. 23, H. Meseck, Ed., Dichtwände und Dichtsohlen, Braunschweig, Federal Republic Germany, June 1987, pp. 21–34.

Ogata, A., "Theory of Dispersivity in a Granular Medium," *U.S. Geological Survey*, Professions Paper, 1970.

Olsen, H. W., Nicols, R. W., and Rice, T. L., "Low Gradient Permeability Measurements in Triaxial Systems," *Geotechnique*, **35**(2), 1985, pp. 145–57.

Olsen, H. W., Wilden, A. T., Kiusalaas, N. J., Nelson, K. R., and Poeter, E. P., "Volume Controlled Hydrologic Property Measurements in Triaxial Systems," in *Hydraulic Conductivity and Waste Contaminant Transport in Soil*, ASTM STP 1142, D. E. Daniel and S. J. Trautwein, Eds., ASTM, Philadelphia, 1994, pp. 482–504.

Olsen, R. E., and Daniel, D. E., "Measurement of the Hydraulic Conductivity of Fine-Grained Soils," *Permeability and Groundwater Contaminant Transport*, ASTM STP 746, 1981, pp. 18–64.

Olsen, R. S., and Farr, J. V., "Site Characterization Using the Cone Penetrometer Test," *Proc. In Situ '86*, Geot. Special Pub. No. 6, S. P. Clemence, Ed., June 1986, pp. 854–68.

Orman, M. E., "Interface Shear-Strength Properties of Roughened HDPE," *J. of Geot. Eng. Div., ACSE*, **120**(4), 1994, pp. 758–63.

OTA Congress of the United States, "Managing Industrial Solid Waste from Manufacturing, Mining, Oil and Gas Production, and Utility Coal Combustion," 1992.

Othman, M. A., and Benson, C. H., "Effect of Freeze–Thaw on the Hydraulic Conductivity of Three Compacted Clays from Wisconsin," Transportation Research Board, No. 1369, 1992, pp. 118–29.

Othman, M. A., Benson, C. H., Chamberlain, E. J., and Zimmie, T. F., "Laboratory Testing to Evaluate Changes in Hydraulic Conductivity of Compacted Clays Caused by Freeze–Thaw: State-of-the-Art," in *Hydraulic Conductivity and Waste Contaminant Transport in Soil*, ASTM STP 1142, D. E. Daniel and S. J. Trautwein, Eds., Philadelphia, 1994, pp. 227–54.

Oweis, I. S., "The Relevancy of One-Dimensional Shear Models in Predicting Surface Accelerations," *Proc. of the 2nd Int. Conf. on Microzonation*, Vol. II, San Francisco, 1978.

Oweis, I. S., "Sanitary Landfill Caps: Do They Inhibit Leachate Generation?" *J. of Resource Management and Technology*, **17**(3).

Oweis, I. S., "Stability of Landfills," in *Geotechnical Practice for Waste Disposal*, D. Daniel, Ed., Chapman and Hall, New York, 1993.

Oweis, I. S., and Biswas, G., "Leachate Mound Changes in Landfills Due to Change in Percolation by a Cap," *J. of Assoc. of Groundwater Scientists and Engineers*, 1993.

Oweis, I. S., Dakes, G., Marturano, T., and Weierer, R., "Geotechnical Observations Saves $ Millions," *Civil Engineering*, October 1994.

Oweis, I. S., Dakes, G., Marturano, T., and Weierer, R., "Soil-Cover Success," *Civil Engineering*, 1994.

Oweis, I. S., and Khera, R., "Criteria for Geotechnical Construction on Sanitary Landfills," *Int. Symp. on Env. Geot.*, Vol. 1, H. Y. Fang, Ed., 1986, pp. 205–22.

Oweis, I. S., and Marturano, T. R., "Ingradient Design of a Sanitary Landfill," *Proc. of the 6th Int. Conf. on Solid Waste Management and Secondary Materials*, Philadelphia, Dec. 1990.

Oweis, I. S., Mills, W. T., Leung, A., and Scarino, J., "Stability of Sanitary Landfills," *Geot. Aspects of Waste Management, Seminar*, Met. Section, ASCE, Dec. 1985.

Oweis, I. S., Smith, D. A., Ellwood, R. B., and Greene, D. S., "Hydraulic Characteristics of Municipal Refuse," *J. of Geot. Eng. Div.*, *ASCE*, **116**(4), April 1990, pp. 539–53.

Oweis, I. S., Zamiskie, E. M., and Kabir, G. M., "Evaluation of Slurry Wall by Piezocone," *Proc. 1st Int. Conf. on Env. Geot.*, Int. SMFE and Canadian Geotechnical Society, Edmonton, Alberta, Canada, July 1994.

Parker, D. G., Thornton, S. I., and Cheng, C. W., "Permeability of Fly Ash Stabilized Soils," *Proc. Geot. Pract. for Disp. of Solid Waste Mat.*, ASCE, Ann Arbor, June 1977, pp. 63–70.

Parry, R. H. G., "Estimating Bearing Capacity of Sand from SPT Values," *J. of Geot. Eng. Div.*, *ASCE*, **103**(9), 1977.

Peck, R. B., Hanson, W. E., and Thornburn, T. H., *Foundation Engineering*, 2nd Ed., Wiley, New York, 1974.

Peggs, I. D., "We Need to Reassess HDPE Seam Specifications," *GFR*, **15**(5), June–July 1997.

Peirce, J. J., and Witter, K. A., "Termination Criteria for Clay Permeability Testing," *J. of Geot. Eng. Div.*, *ASCE*, **112**(9), Sep. 1986, pp. 841–54.

Peyton, R. L., and Schroeder, P. R., "Evaluation of Landfill Liner Design," *J. of Env. Eng. Div.*, *ASCE*, **116**(3), May/June, 1990.

Philip, J. R., "The Theory of Infiltration," *Soil Science*, **83,** 1957, pp. 345–57.

Poran, C. J., and Ahtchi-Ali, F., "Properties of Solid Waste Incinerator Fly Ash," *J. of Geot. Eng. Div.*, *ASCE*, **115**(8), Aug. 1989, pp. 1118–33.

Puig, J., Gouy, J. L., and Labrou, L., "Ferric Clogging of Drains," *Proc. of 3rd Int. Conf. on Geotextiles*, Vol. IV, Vienna, 1986.

Raghava, D., and Atwater, J., "A Review of Clogging Procession Porous Media as Related to Wastewater Collection and Drainage," *Proc. of 1st Int. Conf. on Environmental Geotechnics*, Intl. Soc. SMFE and Canadian Geotechnical Society, Edmonton, Alberta, Canada, July 1994.

Rainbow, A. K. M., and Nutting, M., "Geotechnical Properties of British Minestone Considered Suitable for Landfill Projects," *Int. Symp. on Env. Geot.*, Vol. 1, H. Y. Fang, Ed., 1986, pp. 531–39.

Reades, D. W., Poland, R. J., Kelly, G., and King, S., "Discussion: Hydraulic Conductivity of Two Prototype Clay Liners," *J. of Geot. Eng. Div.*, *ASCE*, **113**(7), July 1987, pp. 809–13.

Reimbold, M. W., (Quoted in EPA, 1993) "An Evaluation Model for Predicting Infiltration Rates in Unsaturated Compacted Soils," M.S. Thesis, The University of Texas at Austin, Austin, Tex., 1988.

Reinhart, D. R., and Carson, D., "Experiences with Full-Scale Application of Landfill Bioreactor Technology," Solid Waste Association of North America, Preprint, SWANA, Silver Spring, Md, 1993.

Rhee, V. C., and Bezuigen, A., "Influence of Seepage on Sandy Slope," *J. GEN*, *ASCE*, **118**(8), 1992.

Rizkallah, V., "Geotechnical Properties of Polluted Dredged Material," *Geot. Practice for Waste Disposal '87*, Geot. Special Publ. No. 13, R. D. Woods, Ed., June 1987, pp. 759–71.

Robertson, P. K., and Campanella, R. G., "Interpretation of Cone Penetration Tests. Part I: Sand," *Canadian Geot. J.*, **20**(4), Nov. 1983a, pp. 718–33.

Robertson, P. K., and Campanella, R. G., "Interpretation of Cone Penetration Tests. Part II: Clay," *Canadian Geot. J.*, **20**(4), Nov. 1983b, pp. 734–45.

Robinson, W. D., Ed., *The Solid Waste Handbook*, Wiley, New York, 1986.

Rollin, A. L., and Denis, R., "Geosynthetic Filtration in Landfill Design," *Proc. Geosynthetics '87*, Vol. 2, Industrial Fabrics Association Int., St. Paul, Minn., 1987.

Rollings, M. P., "Geotechnical Considerations in Dredged Material Management," *Proc. of the 1st Int. Congress on Environmental Geotechnics*, W. D. Carrier III, Ed., Edmonton, Canada, July 1994, pp. 21–32.

Rowe, R. K., Quiegley, R. M., and Booker, J. R., *Clay Barrier Systems for Waste Disposal Facilities*, E&FN SPON, London, 1995.

Ryan, C. R., "Vertical Barriers in Soil for Pollution Containment," *Geot. Practice for Waste Disposal '87*, Geot. Special Publ. No. 13, R. D. Woods, Ed., June 1987, pp. 182–204.

Salem, A. M., and Krizek, R. J., "Stress-Deformation-Time Behavior of Dredgings," *J. of Geot. Eng. Div.*, ASCE, **102**(GT2), Feb. 1976, pp. 139–58.

Sargunan, A., Mallikarjun, N., and Ranapratap, K., "Geotechnical Properties of Refuse Fills of Madras, India," *Int. Symp. on Env. Geot.*, Vol. 1, H. Y. Fang, Ed., 1986, pp. 197–204.

Saxena, S. K., Lourie, D. E., and Rao, J. S., "Compaction Criteria for Eastern Coal Waste Embankments," *J. of Geot. Eng. Div.*, ASCE, **110**(2), Feb. 1984, pp. 262–84.

Schmertmann, J. H., Hartman, J. P., and Brown, P. R., "Improved Strain Influence Factor Diagram," Tech. Note, *J. of the Geot. Eng. Div.*, ASCE, **104**(GT8), Aug. 1978, pp. 1131–35.

Schroeder, P. R., Lloyd, C. M., Zappi, P. A., and Aziz, N. A., "The Hydrologic Evaluation of Landfill Performance (HELP) Model," Version 3, U.S. Environmental Protection Agency, EPA/600/R-94/168a, Cincinnati, 1994.

Schroeder, P. R., and Palermo, M. R., "The Automated Dredging and Disposal Alternatives Management Systems," EEPD Technical Note 06-12, USAEWES, Vicksburg, Miss., 1990.

Schroeder, P. R., and Peyton, R. L., "Evaluation of Landfill-Liner Designs," *J. of Env. Eng. Div.*, ASCE, **116**(3), 1990.

Schumcker, B. O., and Buffalini, J. R., "Pulverized Glass and Landfill Liner System," *Waste Age*, April 1995.

Scott, R. F., *Principles of Soil Mechanics*, Addison-Wesley, Reading, Mass., 1963.

Scully, R. W., Schiffman, F. L., Olsen, H. W., and Ko, H-K., "Validation of Consolidation Properties of Phosphatic Clay at Very High Void Ratio," in *Sedimentation/Consolidation Models*, R. N. Yong and F. C. Townsend, Eds., ASCE, New York, 1984, pp. 1–29.

Seals, R. K., Moulton, L. K., and Kinder, D. L., "In Situ Testing of a Compacted Fly Ash Fill," *Proc. Geot. Pract. for Disp. of Solid Waste Mat.*, ASCE, Ann Arbor, June 1977, pp. 493–516.

Seed, H. B., Idriss, I. M., and Arango, I., "Evaluation of Liquefaction Potential Using Field Performance Data," *J. of Geot. Eng. Div.*, ASCE, **109**(3), 1983.

Seed, H. B., Tokimatsu, K., Harder, L. F., and Chang, R. M., "Influence on SPT Procedures in Soil Liquefaction Resistance Evaluations," *J. of Geot. Eng. Div.*, ASCE, **111**(12), 1985.

Shackelford, C. D., "Contaminant Transport," in *Geotechnical Practice for Waste Disposal*, D. Daniel, Ed., Chapman & Hall, New York, 1993.

Shackelford, C. D., "Analytical Model for Cumulative Mass Column Testing," *Geoenvironment 2000*, Geot. Special Tech. Publ. No. 46, Y. B. Acar and D. E. Daniel, Eds., ASCE, New Orleans, Feb. 1995, pp. 355–72.

Shackelford, C. D., "Waste-Soil Interactions That Alter Hydraulic Conductivity," *Hydraulic Conductivity and Waste Contaminant Transport in Soil*, ASTM STP 1142, D. E. Daniel and S. J. Trautwein, Eds., ASTM, Philadelphia, 1994, pp. 111–68.

Shackelford, C. D., Chang, C. K., Chiu, T. F., "The Capillary Barrier Effect in Unsaturated Flow through Soil Barriers," *Proc. of 1st Intl. Congress on Environmental Geotechnics*, Edmonton, Alberta, Canada, 1994.

Sharma, H. D., Dukes, M. T., and Olsen, D. M., "Field Measurements of Dynamic Moduli and Poisson's Ratio of Refuse and Underlying Soils at a Landfill Site," in *Geotechnics of Waste Fills—Theory and Practice*, ASTM STP 1070, A. Landva and G. D. Knowles, Eds., ASTM, Philadelphia, 1990, pp. 57–70.

Sherard, J. L., Dunnigan, L. P., and Talbot, J. R., "Basic Properties of Sand and Gravel Filters," *J. of Geot. Eng. Div., ASCE*, **110**(6), 1984.

Sheurs, R. E., and Khera, R. P., "Stabilization of a Sanitary Landfill to Support a Highway," National Academy of Science, TRR 754, 1980, pp. 46–53.

Shoemaker, N. B., "Construction Techniques for Sanitary Landfill," *Waste Age*, March/April 1972.

Siegel, R. A., Robertson, R. J., and Anderson, D. G., "Slope Stability Investigations at a Landfill in Southern California," in *Geotechnics of Waste Fills—Theory and Practice*, ASTM STP 1070, A. Landva and G. D. Knowles, Eds., ASTM, Philadelphia, 1990, pp. 259–84.

Silka, R. L., and Jordan, D. L., "Vapor Analysis/Extraction," Chapter 16 in *Geotechnical Practice for Waste Disposal*, D. Daniel, Ed., Chapman and Hall, New York, 1993.

Simons, Li & Associates, "Minimizing Embankment Damage During Overtopping Flow," Prepared for USDOT, 1988.

Singh, S., and Murphy, B., "Evaluation of the Stability of Sanitary Landfills," in *Geotechnics of Waste Fills—Theory and Practice*, ASTM STP 1070, A. Landva and G. D. Knowles, Eds., ASTM, Philadelphia, 1990, pp. 240–58.

Skempton, A. W., "The Pore Pressure Coefficients A and B," *Geotechnique*, **4,** 1954, pp. 143–47.

Smith, C. L., "FDG Waste Engineering Properties Are Controlled by Disposal Choice," in *Utilization of Waste Materials in Civil Engineering Construction*, H. I. Inyang and K. L. Bergeson, Eds., ASCE, Philadelphia, Sept. 1992, pp. 44–59.

Smith, A. D., and Kraemer, S. R., "Creep by Geocomposite Drains," *Proc. Geosynthetics '87*, Industrial Fabrics Association Int., Vol. 2, St. Paul, Minn., 1987.

Smith, M. E., and Criley, K., "Interface Friction Is Not for the Uninitiated," *Geotextile Fabrics Report*, **13**(3), April 1995, pp. 28–31.

Soil Conservation Service, "Guidelines for Determining the Gradation of Sand and Gravel Filters," Water Resources Publication, Littleton, Col., 1986.

Soil Conservation Service, "Urban Hydrology for Small Watersheds," Technical Release Number 55, 1986.

Soliman, N. N., "Laboratory Testing of Lime Fixed Fly Ash and FDG Sludge," in

Geotechnics of Waste Fills — Theory and Practice, ASTM STP 1070, A. Landva and G. D. Knowles, Eds., ASTM, Philadelphia, 1990, pp. 168–84.

Soliman, N., Houlik Jr., C. W., and Schneider, M. H., "Geotechnical Properties of Lime Fixed Fly Ash and FGD Sludge," *Int. Symp. on Env. Geot.*, Vol. 1, 1986, pp. 549–63.

Somogyi, F., "Large Strain Consolidation of Fine Grained Slurries," Presented at the Canadian Society for Civil Engineers, Winnipeg, Manitoba, May 1980.

Somogyi, F., and Gray, D. H., "Engineering Properties Affecting Disposal of Red Muds," *Proc. Geot. Pract. for Disp. of Solid Waste Mat.*, ASCE, Ann Arbor, June 1977, pp. 1–22.

Sowers, G. F., "Settlement of Waste Disposal Fills," *8th Int. CSMFE*, Moscow, Russia, 1973, pp. 207–10.

Spigolon, S. J., and Kelly, M. F., "Geotechnical Quality Assurance of Construction of Disposal Facilities," EPA-600/2-84-040, 1984, p. 49.

Srinivasan, V., Beckwith, G. H., and Burke, H. H., "Geotechnical Investigations of Power Plant Wastes," *Proc. Geot. Pract. for Disposal of Solid Waste Mat.*, ASCE, Ann Arbor, June 1977, pp. 169–87.

Stark, T. D., and Poeppel, A. R., "Landfill Liner Interface Strength from Torsional-Ring-Shear Tests," *J. of Geot. Eng. Div.*, ASCE, **120**(3), March 1994, pp. 597–615.

"State of the Art on Current Soil Sampling," The Subcommittee on Soil Sampling, International Society of Soil Mechanics and Foundation Engineering, 1979.

Stone, K. J. L., Randolph, M. F., Toh, S., and Sales, A. A., "Evaluation of Consolidation Behavior of Mine Tailings," *J. of Geot. Eng. Div.*, ASCE, **120**(3), March 1994, pp. 473–90.

Strack, D. L., *Groundwater Mechanics*, Prentice-Hall, Englewood Cliffs, N. J., 1989.

Stulgis, R. P., Soydemir, C., and Telgener, R. J., "Predicting Landfill Settlement," *Geoenvironment 2000*, Geot. Special Tech. Publ. No. 46, Y. B. Acar and D. E. Daniel, Eds., ASCE, New Orleans, Feb. 1995, pp. 980–94.

Surprenant, G., and Lemke, J., "Landfill Compaction: Setting a Density Standard," *Waste Age*, Aug. 1994.

Taha, R., and Saylak, D., "The Use of Flue Gas Desulfurization Gypsum in Civil Engineering Construction," in *Utilization of Waste Materials in Civil Engineering Construction*, H. I. Inyang and K. L. Bergeson, Eds., ASCE, Philadelphia, Sept. 1992, pp. 264–73.

Taha, R., and Seals, R., "Engineering Properties and Potential Uses of Byproduct Phosphogypsum," in *Utilization of Waste Materials in Civil Engineering Construction*, H. I. Inyang and K. L. Bergeson, Eds., ASCE, Philadelphia, Sept. 1992, pp. 250–63.

Tchobanoglous, G., Theisen, H., and Eliassen, R., *Solid Wastes*, McGraw-Hill, New York, 1977.

Tchobanoglous, G., Theisen, H., and Vigil, S., *Integrated Waste Management*, McGraw-Hill, New York, 1993.

Tempe, D. M. et al., "Stability Design of Grass-lined Open Channels," *Agricultural Handbook Number 667*, U. S. Department of Agriculture, 1987.

Terzaghi, K., *Theoretical Soil Mechanics*, Wiley, New York, 1943.

Terzaghi, K., and Peck, R. B., *Soil Mechanics in Engineering Practice*, 2nd Ed., Wiley, New York, 1967.

Thornthwaite, C. W., and Mather, J. R., "Instructions and Tables for Computing Potential Evapotranspiration and the Water Balance," Publications in Climatology, Lab. of Climatology, Drexel Institute of Technology, Centerton, N. J., 1975.

Timoshenko, S., and MacCullough, G. H., "Elements of Strength of Materials," Van Nostrand, 1949, p. 197.

Torii, K., and Kawamura, M., "Effective Utilization of Coal Ashes in Road Construction," in *Waste Materials in Construction*, J. J. J. M. Goumans, H. A. van der Sloot, and Th. G. Albers, Eds., Elsevier Science Publishers, Netherlands, 1991, pp. 561–68.

Toth, P. S., Chan, H. T., and Cragg, C. B., "Coal Ash as Structural Fill, with Special Reference to Ontario Experience," *Canadian Geot. J.*, **25**, 1988, pp. 694–705.

Townsend, F. C., and McVay, M. C., "SOA: Large Strain Consolidation Predictions," *J. of Geot. Eng. Div., ASCE*, **116**(2), Feb. 1990, pp. 222–43.

Townsend, T. G., Miller, W. L., and Earle, J. F. K., "Leachate–Recycle Infiltration Ponds," *J. of Env. Eng. Div., ASCE*, **121**(6), 1995.

Trautwein, S. J., and Boutwell, G. P., "In Situ Hydraulic Conductivity Tests for Compacted Soil Liners and Caps," ASTM STP 1142, ASTM, Philadelphia, 1994.

Ullrich, C. R., and Hagerty, D. J., "Stabilization of FGC Wastes," *Geot. Practice for Waste Disposal '87*, Geot. Special Publ. No. 13, R. D. Woods, Ed., ASCE, Ann Arbor, June 1987, pp. 797–811.

Uriksen, P. F., *Application of Impulse Radar to Civil Engineering*, Geophysical Survey Systems, Inc., Hudson, N. H., 1982.

U.S. Bureau of Reclamation, *Earth Manual*, U.S. Government Printing Office, Washington, D. C., 1974.

U.S. Congress, *Technologies and Management Strategies for Hazardous Waste Control*, Office of Technology Assessment, Washington, D. C., 1986.

U.S. Department of Energy, "Technical Approach Document, Uranium Mill Tailings Remedial Action Project," UMTRA-DOE/AL 050425, Washington, D. C., 1989.

U.S. Department of Transportation, "Hydraulic Design of Energy Dissipators for Culverts and Channels," Hydraulic Engineering Circular No. 14, Washington, D. C., 1983.

U.S. Nuclear Regulatory Commission, *Final Staff Technical Position, Design of Erosion Protection Covers for Stabilization of Uranium Mill Tailings Sites*, 1990.

USEPA, 1975, *An Evaluation of Landfill Gas Migration and a Prototype Gas Migration Barrier*, EPA/530/SW-79d.

USEPA, 1980, *Classifying Solid Waste Disposal Facilities*, March, SW-828, Cincinnati.

USEPA, 1983, *Lining of Waste Impoundment and Disposal Facilities*, SW-870, Cincinnati.

USEPA, 1985, *Covers for Hazardous Waste Sites*, EPA/540/2-85/002, Washington, D. C.

USEPA, 1986, *Design, Construction, and Evaluation of Clay Liners for Waste Management Facilities*, EPA/530-SW-86/007, March, p. 663.

USEPA, 1986, Subtitle D Study, Phase Report, U.S. Environmental Protection Agency EPA/50-SW-86/054.

USEPA, 1987, *A Compendium of Superfund Field Operation Methods*, EPA/540/P-87/001, Washington, D. C.

USEPA, 1988, *Design, Construction, and Evaluation of Clay Liners for Waste Management Facilities*, EPA/530/SW-86/007F, Washington, D. C.

USEPA, 1989, *Technical Guidance Document: The Fabrication of Polyethylene FML Field Seams*, EPA/530/SW-89/069, Washington, D. C.

USEPA, 1989, *Stabilization/Solidification of CERCLA and RECRA Wastes*, EPA/625/6-89/022, May.

USEPA, 1992, *Action Leakage Rate for Leak Detection Systems*, EPA/530/R-92/004, Washington, D. C.

USEPA, 1992, *RCRA Ground-Water Monitoring, Draft Technical Guidance*, EPA/530/R-93/001, Washington, D. C.

USEPA, 1992, *Characterization of Municipal Solid Waste in U.S.*, Update Franklin Associates, Washington, D. C.

USEPA, 1993, *Anthropogenic Methane Emissions in the United States: Estimates for 1990*, EPA/430/R-93/003, Washington, D. C.

USEPA, 1993, *Solid Waste Disposal Facility Criteria*, EPA/530/R-93/017, Washington, D. C.

USEPA, 1993, *Technical Guidance Document: Quality Assurance and Quality Control for Waste Containment Facilities*, D. E. Daniel and R. M. Koerner, Authors, EPA/600/R-93/182, Cincinnati.

USEPA, 1996, Federal Register, March 12, Vol. 61, No. 49, "Standards of Performance for New Stationary Sources, and Guidelines for Control of Existing Sources," Municipal Solid Waste Landfills, 40 CFR Parts 51, 52, and 60.

Usmen, M. A., "Properties, Disposal and Stabilization of Combined Coal and Refuse," *Int. Symp. on Env. Geot.*, Vol. 1, 1986, pp. 515–26.

Vallee, R. P., and Andersland, O. B., "Field Consolidation of High Ash Papermill Sludge," *J. of the Geot. Eng. Div.*, ASCE, **100**(GT3), March 1974, pp. 309–28.

Van Genuchten, M. Th., and Alves, W. J., *Analytical Solutions of the One-Dimensional Convection-Dispersive Solute Transport Equation*, Technical Bulletin 1661, U.S. Department of Agriculture, Washington, D. C., 1982.

Van Zanten, R. V., *Geotextiles and Geomembranes in Civil Engineering*, Wiley, New York, 1986.

Vick, S. G., *Planning, Design, and Analysis of Tailing Dams*, Wiley, New York, 1988.

Wait, J. R., *Geo-Electromagnetism*, Academic Press, New York, 1982.

Wall, K. D., and Zeiss, C., "Municipal Landfill Biodegradation and Settlement," *J. of Env. Eng. Div.*, ASCE, **121**(3), 1995.

Walsh, K. D., Houston, W. N., and Houston, S. L., "Evaluation of In-Place Wetting Using Soil Suction Measurements," *J. of Geot. Eng. Div.*, ASCE, **119**(5), 1993.

Wardwell, R. E., and Nelson, J. D., "Settlement of Sludge Landfills with Fiber Decomposition," *Proc. of the 10th ICSMFE*, Vol. 2, Stockholm, 1981, pp. 397–401.

Washington State Department of Ecology, "Solid Waste Landfill Design Manual."

Waste Age, "An Introduction," Jan. 1986, p. 124.

Way, S. G., *Terrain Analysis: A Guide to Site Selection Using Aerial Photographic Interpretation*, Dowden, Hutchinson, and Russ, Inc., Stroudsburg, Penn., 1973.

Wayne, M. H., Susavidge, M. A., Hulling, D. E., Martin, J. P., and Cheng, S. C., "An Assessment of the Geotechnical Strength Characteristics of Fly Ash," *Proc. of the 9th*

Int. Ash Use Symp., Vol. 3, GS-7162, American Coal Ash Association, Orlando, 1991, Paper No. 65.

Weast, R. C., and Melvin, J., *Handbook of Chemistry and Physics,* 61st Ed., CRC Press, Boca Raton, Fla., 1980.

Weiss, W., Siegmund, M., and Alexiew, D., "Field Performance of Geosynthetic Clay Liner Landfill Capping System under Simulated Waste Subsidence," *Geosynthetic '95,* Vol. 2, Nashville, 1995.

Welsh, J. P., "Dynamic Compaction of Sanitary Landfill to Support Superhighway," *Proc. of the 8th European Conf. on Soil Mech. and Found. Eng.,* Helsinki, 1983.

Wiles, C. C., Kosson, D. S., and Holmes, T., "The US EPA Program for Evaluation of Treatment and Utilization Technologies for Municipal Waste Combustion Residues," in *Waste Materials in Construction,* J. J. J. M. Goumans, H. A. van der Sloot, and Th. G. Albers, Eds., Elsevier Science Publishers, Netherlands, 1991, pp. 57–69.

Williams, E. A., Massa, A. K., Blau, D. H., and Schaal, H. R., *Siting of Major Facilities,* McGraw-Hill, 1992.

Winterkorn, H. F., and Fang, H. Y., *Foundation Engineering Handbook,* Van Nostrand Reinhold, New York, 1975.

Wiss, J. F., "Construction Vibrations: State-of-the-Art," *J. of Geot. Eng. Div., ASCE,* **107**(GT2), Feb. 1981, pp. 167–82.

Woodward-Clyde Consultants, *Geotechnical Investigation Global Sanitary Landfill,* Report obtained from the New Jersey Department of Environmental Protection, 1984.

WPCF Manual of Practice No. FD-5, Water Pollution Control Federation, Washington, D. C., 1982.

Wu, J. Y., and Khera, Raj P., "Properties of a Treated-Bentonite/Sand Mix in a Contaminant Environment," in *Physico-Chemical Aspects of Soil and Related Materials,* ASTM STP 1095, K. B. Hoddinott and R. O. Lamb, Eds., ASTM, Philadelphia, 1990, pp. 47–59.

Xanthakos, P. P., *Slurry Walls,* McGraw Hill, New York, 1979.

Yegian, M. K., Marciano, E. A., and Ghahraman, V. G., "Earthquake-Induced Permanent Deformations: Probabilistic Approach," *J. of Geot. Eng. Div., ASCE,* **117**(1), 1991.

Yen, B. C., and Scanlon, B., "Sanitary Landfill Settlement Rates," *J. of Geot. Eng. Div., ASCE,* **101**(GT5), May 1975, pp. 475–90.

Yong, R. N., "Selected Leaching Effects on Some Mechanical Properties of Sensitive Clay," *Int. Symp. on Env. Geot.,* Vol. 1, 1986, pp. 349–62.

Yong, R. N., and Ludwig, C. A., "Large-Strain Consolidation Modeling of Land Subsidence," *Symposium on Geotechnical Aspects of Mass and Materials Transport,* Asian Institute of Technology, Bangkok, Thailand, 1984.

York, D., Lesser, N., Bellatty, T., Israi, E., and Patel, A, "Terminal Development on a Refuse Fill Site," *Geot. Pract. for Disp. of Solid Waste Mat.,* ASCE, Ann Arbor, 1977, pp. 810–30.

Zamiskie, E. M., Kabir, M. G., and Haddad, A., "Settlement Evaluation for Cap Closure Performance," in *Vertical and Horizontal Deformations of Foundations and Embankments,* Vol. 1, Geot. Special Publ. No. 40, ASCE, New York, 1994.

Zohdy, A. A., "Automatic Interpretation of Schlumberger Sounding Curves, Using Modified Dar Zarrouk Functions," Bulletin 1313-E, U.S. Geologic Survey, Reston, Va., 1975.

Index

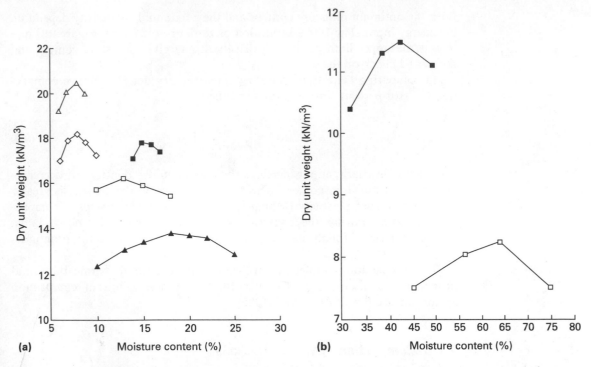

Figure 9.2 Compaction tests on waste: (a) ■, fly ash, western U.S.; □, fly ash, New Jersey; ◇, coal waste, eastern U.S.; △, coal waste, western U.S.; ▲, fly ash, Pennsylvania; (b) ■, FGD sludge; □, paper solid waste

For moisture content 6.4% to 13.4% below the laboratory optimum moisture content (OMC) value, none of the rollers significantly improved percent compaction after four passes. With moisture content from 0.4% to 6.4% below the OMC, density continued to increase from 1 through 12 passes, with an asphalt roller giving the highest value (98.5% of the maximum dry unit weight). After 7 and 12 passes, the asphalt roller performed much better at 1850 vpm than at

Table 9.1
Compaction of fly ash
(modified compaction)

Source	Optimum moisture (%)	Max. dry[a] unit weight (kN/m³)	Specific gravity of solids
Arizona (low sulfur)	22.9	14.2[a]	2.35
Michigan	32–20	11.7–14.6	2.36–2.61
New Jersey	13.6	16.2	2.54
Pennsylvania	31–19	12.0–14.0	2.29–2.59
West Virginia	33–14	11.2–17.5	2.20–2.50
Shelby tune samples	—	14.9–16.0	2.26–2.37
Wisconsin	28–19	12.2–14.9	2.33–2.44
Wyoming (low sulfur)	9	18.5	2.75

[a] Standard compaction.
Source: Chae and Gurdziel, 1976; Moulton et al., 1976; Parker et al., 1977; Srinivasan et al., 1977